T0337845

Damage Prognosis

Damage Prognosis

For Aerospace, Civil and Mechanical Systems

EDITED BY

Daniel J. Inman
Virginia Polytechnic Institute and State University, USA

Charles R. Farrar
Los Alamos National Laboratory, USA

Vicente Lopes Junior
Universidade Estadual de São Paulo, SP, Brazil

Valder Steffen Junior
Federal University of Uberlândia, Brazil

John Wiley & Sons, Ltd

This material is based upon work supported by the National Science Foundation under Grant No. 0221222.
Any opinions, findings, and conclusions or recommendations expressed in this material are those of the
authors and do not necessarily reflect the views of the National Science Foundation.

Other Wiley Editorial Offices

John Wiley & Sons Inc., 111 River Street, Hoboken, NJ 07030, USA

Jossey-Bass, 989 Market Street, San Francisco, CA 94103-1741, USA

Wiley-VCH Verlag GmbH, Boschstr. 12, D-69469 Weinheim, Germany

John Wiley & Sons Australia Ltd, 33 Park Road, Milton, Queensland 4064, Australia

John Wiley & Sons (Asia) Pte Ltd, 2 Clementi Loop #02-01, Jin Xing Distripark, Singapore 129809

John Wiley & Sons Canada Ltd, 22 Worcester Road, Etobicoke, Ontario, Canada M9W 1L1

Wiley also publishes its books in a variety of electronic formats. Some content that appears in print may not be
available in electronic books.

Library of Congress Cataloging in Publication Data

Damage prognosis for aerospace, civil and mechanical systems / edited by
    Daniel J. Inman . . . [et al.].
      p.   cm.
    Includes index.
    ISBN 0-470-86907-0
    1. Structural analysis (Engineering)   2. Materials—Deterioration.   I. Inman, D.J.
    TA646.D27 2005
    624.1'71—dc22

                                                                2004026798

British Library Cataloguing in Publication Data

A catalogue record for this book is available from the British Library

ISBN 0-470-86907-0 (HB)

Typeset in 10/12pt Times by Integra Software Services Pvt. Ltd, Pondicherry, India
Printed and bound in Great Britain by Antony Rowe Ltd, Chippenham, Wiltshire
This book is printed on acid-free paper responsibly manufactured from sustainable forestry
in which at least two trees are planted for each one used for paper production.

Contents

List of Contributors

EDITORS

Daniel J. Inman
Center for Intelligent Material Systems
  and Structures
310 Durham Hall, Mail Code 0261
Virginia Polytechnic Institute
Blacksburg, VA 24061
USA

Charles R. Farrar
Los Alamos National Laboratory
Engineering Sciences and Applications, ESA-WR
Mail Stop T001
Los Alamos
New Mexico 87545
USA

Vicente Lopes Jr
Av. Brasil, 56
15.385-000 Ilha Solteira, SP
Brazil

Valder Steffen Jr
Federal University of Uberlândia
School of Mechanical Engineering
Campus Santa Monica, Uberlândia
Minas Gerais, 38400-902
Brazil

AUTHORS

Douglas E. Adams
Purdue University
School of Mechanical Engineering – Ray W. Herrick Laboratories
140 S. Intramural Drive
West Lafayette, IN 47907-2031
USA

José Roberto de França Arruda
Departamento de Mecânica Computacional
Faculdade de Engenharia Mecânica
Universidade Estadual de Campinas
Caixa Postal 6122
13083-970 Campinas, SP
Brazil

José M. Balthazar
State University of São Paulo at Rio Claro
PO Box 178, 13500-230 Rio Claro, SP
Brazil

Matthew T. Bement
Los Alamos National Laboratory
Engineering Sciences and Applications, ESA-WR
Mail Stop T080, Los Alamos
New Mexico 87545
USA

Irene J. Beyerlein
Los Alamos National Laboratory
Theoretical Division, T-3
Mail Stop B216, Los Alamos
New Mexico 87545
USA

Eric Blaise
Department of Aeronautics and Astronautics
Stanford University
Stanford, CA 94305
USA

Geraldo C. Brito Jr
ITAIPU Binacional
Itaipu Power Plant – Laboratory of Maintenance
Foz do Iguaçu, PR 85867-090
Brazil

Carlos E.S. Cesnik
Department of Aerospace Engineering
The University of Michigan
1320 Beal Avenue—3024 FXB
Ann Arbor, MI 48109-2140
USA

Fu-Kuo Chang
Department of Aeronautics and Astronautics
Stanford University
Stanford, CA 94305
USA

Phillip J. Cornwell
Rose-Hulman Institute of Technology
5500 Wabash Ave.
Terre Haute
IN 47803
USA

R.M. Doi
Unesp – Ilha Solteira
Department of Mechanical Engineering
Av. Brasil 56
15385-000 Ilha Solteira – SP
Brazil

Charles R. Farrar
Los Alamos National Laboratory
Engineering Sciences and Applications, ESA-WR
Mail Stop T001, Los Alamos
New Mexico 87545
USA

François M. Hemez
Los Alamos National Laboratory
Engineering Sciences and Applications, ESA-WR
Mail Stop T001, Los Alamos
New Mexico 87545
USA

Norman F. Hunter
Los Alamos National Laboratory
Engineering Sciences and Applications, ESA-WR
Mail Stop P946, Los Alamos
New Mexico 87545
USA

Jeong-Beom Ihn
Department of Aeronautics and Astronautics
Stanford University
Stanford, CA 94305
USA

Daniel J. Inman
Center for Intelligent Material Systems
    and Structures
310 Durham Hall, Mail Code 0261
Virginia Tech
Blacksburg, VA 24061
USA

Gregory J. Kacpryznski
Impact Technologies
125 Tech Park Drive
Rochester
NY 14616
USA

Nick A.J. Lieven
Department of Aerospace Engineering
University of Bristol
Queens Building
University Walk
Bristol BS8 1TR
UK

Vicente Lopes Jr
Universidade Estadual Paulista
Av. Brasil, 56
15.385-000 Ilha Solteira, SP
Brazil

Bruce R. Marshall
Machinery Systems Information Branch Head
Naval Surface Warfare Center
5001 S. Broad Street, Code 952
Philadelphia, PA 19112-1403
USA

Rolf F. Orsagh
Impact Technologies
125 Tech Park Drive

Rochester
NY 14616
USA

Gyuhae Park
Los Alamos National Laboratory
Engineering Sciences and Applications, ESA-WR
Mail Stop T001, Los Alamos
New Mexico 87545
USA

J.A. Pereira
Unesp – Ilha Solteira
Department of Mechanical Engineering
Av. Brasil 56,
15385-000 Ilha Solteria – SP
Brazil

Juan E. Perez Ipiña
Universidad Nacional del Comahue
Grupo Mecánica de Fractura
Buenos Aires 1400
Neuquén
Argentina

Bento R. Pontes
Department of Mechanical Engineering
State University of São Paulo at
    Bauru (UNESP-Bauru)
PO Box 473, CEP 17030-360
Bauru, SP
Brazil

Domingos A. Rade
Federal University of Uberlândia
School of Mechanical Engineering
Campus Santa Monica, PO Box 593-38400-902
Uberlândia, MG
Brazil

Ajay Raghavan
Department of Aerospace Engineering
The University of Michigan
1320 Beal Avenue
Ann Arbor, MI 48109-2140
USA

Amy Robertson
HYTEC, Inc.
4735 Walnut, Suite W-100
Boulder, CO 80301
USA

Michael J. Roemer
Director of Engineering
Impact Technologies
125 Tech Park Drive
Rochester, NY 14616
USA

Samuel da Silva
Av. Brasil, 56
15.385-000 Ilha Soltiera, SP
Brazil

Hoon Sohn
Civil and Environmental Engineering
Carnegie Mellon University
119 Porter Hall
Pittsburgh, PA 15213-3890
USA

Paulo Sollero
Computational Mechanics Department
Faculty of Mechanical Engineering
State University of Campinas
CP6122, Campinas, SP, 13083-970
Brazil

Valder Steffen Jr
Federal University of Uberlândia
School of Mechanical Engineering
Campus Santa Monica, Uberlândia
Minas Gerais, 38400-902
Brazil

Michael D. Todd
University of California, San Diego
Dept. of Structural Engineering
9500 Gilman Drive Dept. 0085
La Jolla, CA 92093-0085
USA

Todd O. Williams
Los Alamos National Laboratory
Theoretical Division, T-3
Mail Stop B216, Los Alamos
New Mexico 87545
USA

Alejandro A. Yawny
Centro Atómico Bariloche
8400 San Carlos de Bariloche
Argentina

Preface

From Sunday 19 October 2003 to Thursday 30 October, 2003 a workshop on the topic of damage prognosis was held at the Praiatur Hotel in Florianopolis, Brazil, as part of the US National Science Foundation's Pan American Advanced Studies Institutes program. This book is an outgrowth of that workshop. Damage prognosis, defined in more detail in the first chapter, is an outgrowth of structural health monitoring, and is an attempt to estimate the remaining useful life of a damaged structure when subjected to various future loads.

Damage prognosis is the prediction in near real time of the remaining useful life of an engineered system given the measurement and assessment of its current damaged (or aged) state and accompanying predicted performance in anticipated future loading environments. As interpreted here, damage prognosis incorporates hardware, software, modeling and analysis in support of prediction. A key element in damage prognosis is obviously that of structural health monitoring (SHM). However, unlike SHM, damage prognosis requires a model of the system's failure modes such as fatigue, cracking, etc. In addition, damage prognosis is keenly tied to the hardware required to make current measurements of key parameters in these models. Because of the interdisciplinary nature of the damage prognosis solutions, participation in the institute was by invitation only and consisted of a mix of researchers from the Americas in each of the fundamental disciplines underlying solutions to damage prognosis problems. In addition, a group of lectures addressed applications. All participants expenses where paid for by either the National Science Foundation or by Los Alamos National Laboratories' Damage Prognosis Section.

Researchers from industry, government laboratories and universities were also invited to lecture, bringing the total attendance to around 50. The mornings featured lectures on the fundamental disciplines needed to address damage prognosis. The afternoons featured lectures on applications of component technologies related to damage prognosis and time for student groups to form. During the first week, each student participant was paired up into a group, assigned a mentor from the speakers, given an application (either an airplane wing subject to fatigue or a building structure subject to an earthquake load) and asked to apply the topic of an assigned fundamental lecture to the damage prognosis problem for that application. The students then summarized their findings during the last day and a half. Thus the 30 advanced students who participated (15 from the US, 13 from Brazil, 1 from Argentina and 1 from Chile) took an active role in the learning of and the development of the Damage Prognosis discipline. The student's presentations can be found in 'Slides of Lectures' at http://www.cimss.vt.edu/PASI along with the slides from each of the research lectures given at the workshop. The PASI site also contains the agenda and a group photograph.

Each speaker was asked to provide the students with a sample chapter before the workshop. Following the workshop these chapters were reviewed and revised, making up the contents of this book. Hence, this book follows the workshop's format of trying to integrate the various disciplines required to solve damage prognosis problems. As such, the first chapter provides a general overview of the topic. Following this the book is divided into four main parts: Part I, Damage Models; Part II, Monitoring Algorithms; Part III, Hardware, and Part IV, Applications. It is our collective opinion that these topics must be integrated in order to solve damage prognosis problems, and we hope that this book will be of assistance in achieving this.

D.J. Inman, C.R. Farrar, V. Lopes, Jr, and V. Steffen, Jr

1

An Introduction to Damage Prognosis

Charles R. Farrar,[1] Nick A.J. Lieven[2] and Matthew T. Bement[1]

[1] *Los Alamos National Laboratory, Los Alamos, New Mexico, USA*
[2] *Department of Aerospace Engineering, University of Bristol, UK*

1.1 INTRODUCTION

This book is intended to provide an overview of the emerging technology referred to as *damage prognosis*. To this end, this introduction will summarize, in very general terms, the damage prognosis process and provide motivation for developing damage-prognosis technology. Also, the 'Grand Challenge' nature of this problem will be discussed. The development of a damage-prognosis capability requires a multidisciplinary approach and these requisite technologies are also summarized in this introduction. To begin this discussion, several terms related to this technology are defined.

Damage in a structural and mechanical system will be defined as intentional or unintentional changes to the material and/or geometric properties of the system, including changes to the boundary conditions and system connectivity, which adversely affect the current or future performance of that system. As an example, a crack that forms in a mechanical part produces a change in geometry that alters the stiffness characteristics of the part. Depending on the size and location of the crack and the loads applied to the system, the adverse effects of this damage can either be immediate or may take some time to alter the system's performance. In terms of length scales, all damage begins at the material level and then, under appropriate loading conditions, progresses to component- and system-level damage at various rates. In terms of time scales, damage can accumulate incrementally over long periods of time, such as damage associated with fatigue or corrosion. Damage can also occur on much shorter time scales as a result

Damage Prognosis – For Aerospace, Civil and Mechanical Systems   Edited by D.J. Inman, C.R. Farrar, V. Lopes Junior and V. Steffen Junior   © 2005 John Wiley & Sons, Ltd

of scheduled discrete events, such as aircraft landings, and from unscheduled discrete events, such as enemy fire on a military vehicle. Implicit in this definition of damage is the concept that damage is not meaningful without a comparison between two different system states.

Usage monitoring is the process of measuring responses of, and in some cases the inputs to, a structure while it is in operation.

Structural health monitoring (SHM) is the process of damage detection for aerospace, civil, and mechanical engineering infrastructure. SHM involves the observation of a system over time using periodically sampled dynamic response measurements from an array of sensors, the extraction of damage-sensitive features from these measurements, and the statistical analysis of these features to determine the current state of the system. For long-term SHM, the output of this process is periodically updated information regarding the ability of the structure to perform its intended function in light of the inevitable aging and degradation resulting from operational environments. After extreme events, such as earthquakes or blast loading, SHM is used for rapid condition screening and aims to provide, in near real time, reliable information regarding the integrity of the structure. This process is also referred to as *condition monitoring*, particularly when it is applied to rotating machinery, or simply *diagnosis*.

Damage prognosis is the **estimate of a system's remaining useful life**. This estimate is based on the output of predictive models, which develop such estimates by coupling information from usage monitoring, structural health monitoring, past, current and anticipated future environmental and operational conditions, the original design assumptions regarding loading and operational environments, and previous component and system level testing. Also, 'softer' information such as user 'feel' for how the system is responding should be used to the greatest extent possible when developing damage-prognosis solutions. Stated another way, damage prognosis attempts to forecast system performance by measuring the current state of the system, estimating the future loading environments for that system, and predicting through simulation and past experience the remaining useful life of the system. Figure 1.1 depicts the relationship between usage monitoring, structural health monitoring, and damage prognosis.

Figure 1.2, which shows damage to the USS *Denver*, emphasizes the challenge presented by damage prognosis. Clearly, it is not difficult to identify and locate such extreme damage. However, estimating the impact of this damage on various ship systems and coupling this information with estimates of future loading in order to predict the remaining useful life of the ship, so that authorities can determine its ability to return safely to port, is a very difficult problem. This difficulty suggests that developing damage prognosis solutions is an appropriate grand-challenge problem for engineers involved with aerospace, civil, and mechanical infrastructure. Some features of a grand-challenge problem include the following:

1. The problem must be difficult and will not be readily solved in the next few years.
2. Solutions to the problem must be multidisciplinary in nature.
3. Solutions to the problem must require developments in experimental, analytical and computational methods.
4. There must be quantifiable measures indicating progress toward the problem solution.
5. The problem must be of interest to many industries.
6. Solving the problem will have significant economical and social impact.

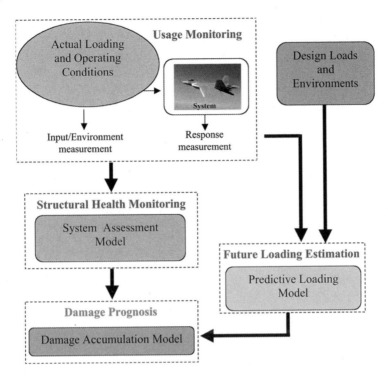

Figure 1.1   The relation between usage monitoring, structural health monitoring, and damage prognosis

Figure 1.2   Damage to the USS *Denver*

The authors believe that the development of damage-prognosis solutions meets the list of criteria for a grand-challenge problem and that quantifiable measures can be developed to indicate that a solution has been developed.

1.2   THE DAMAGE-PROGNOSIS SOLUTION PROCESS

The actual implementation of a damage-prognosis-solution strategy will be application specific. However, there are major components of such a strategy that are generic to many applications; these components are outlined in Figure 1.3. In this figure, shaded areas indicate system-specific information that will define how the three main technology components: instrumentation and data-processing hardware, data interrogation, and predictive modeling (shown as unshaded areas) are implemented in a damage-prognosis-solution strategy.

An important first step in defining damage-prognosis solutions is the classification of the damage-prognosis problem. While it is unlikely that all, or even most, damage-prognosis applications will fit nicely within a rigid, precise classification scheme, through the course of the workshop, general categories were seen to emerge. To understand the categories, one must first answer three general questions: (i) what is causing the damage of concern, (ii) what techniques should be used to assess and quantify the damage, and (iii) once the damage has been assessed, what is the goal of the prognosis? Figure 1.4 represents these questions graphically. As discussed below, the categories do not have sharp boundaries, and many applications will overlap the various categories.

For each potential failure mode, the source of the damage falls into three general categories. The first category is gradual wear, where damage accumulates slowly at the material or component level, often on the microscopic scale. Examples of this damage source include fatigue cracking and corrosion. The second category is predictable discrete events. While the damage typically still originates on the microscopic scale, it

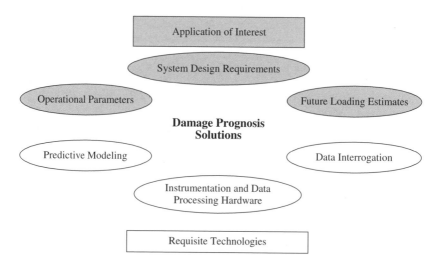

Figure 1.3   Major components of a damage-prognosis strategy

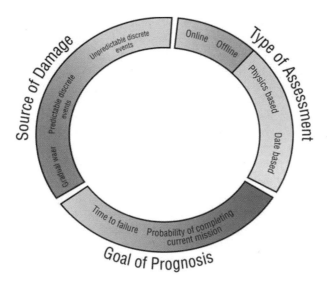

Figure 1.4 Classification of the damage-prognosis problem

accumulates at faster rates during sudden events that can be characterized *a priori*. Examples include aircraft landings and explosions in confinement vessels. Unpredictable discrete events make up the third category, in which unknown and usually severe damage is inflicted upon the system at essentially unpredictable times. Examples include foreign-object-induced fan blade-off in turbine engines, earthquake-induced damage in civilian infrastructure, or battle damage in military hardware. Of course, there do exist applications where one failure scenario will fall into all three of these categories at various times. An example is damage to jet engine turbine blades caused by gradual thermal creep, increased bending stresses during takeoffs and landings, and impact with foreign objects.

After identifying the type(s) and source(s) of damage, it is then important to determine which techniques should be used in the damage assessment. The first question that arises concerns whether the assessment should be done on-line, in near real time, or off-line at discrete intervals. This consideration will strongly influence the data-acquisition and data-processing requirements, as well as set limits on the computational requirements of potential assessment and prognosis techniques. While it is obvious that measurements should be taken on-line, the type of damage will influence whether the assessment needs to be done on-line. That is, for unpredictable discrete events, the assessment must be done on-line to be of any use, thus limiting the choice of assessment techniques. However, for gradual wear, assessments may not need to be performed in near real time, and virtually any assessment technique may be used.

Assessment techniques can generally be classified as either physics based or data based, though practically speaking, a combination of the two will usually be employed. Physics-based assessment techniques, as their name implies, use mathematical equations that theoretically predict the system behavior by simulating the actual physical processes that govern the system response. These assessments are especially useful for predicting system response to new loading conditions and/or new system configurations (damage states). However, physics-based assessment techniques are typically computationally intensive.

Data-based assessment techniques, on the other hand, rely on previous measurements from the system to assess the current damage state, typically by means of some sort of pattern recognition method such as neural networks. However, although data-based assessment techniques may be able to indicate a change in the presence of new loading conditions or system configurations, they will perform poorly when trying to classify the nature of the change. Thus, it is not uncommon to use the results from a physics-based model to 'train' a data-based assessment technique to recognize damage cases for which no experimental data exists. Typically the balance between physics-based models and data-based techniques will depend on the amount of relevant data available and the level of confidence in the physics-based models, as illustrated in Figure 1.5.

Once the current damage state has been assessed, the prognosis problem can begin to be addressed by determining the goal for the prognosis. Perhaps the most obvious and desirable type of prognosis is an estimate of how much time remains until maintenance is required, the system fails, or the system is no longer usable. While this estimate is of high value in systems where damage accumulates gradually and at predictable rates, it is of less value in more extreme conditions such as aircraft in combat, where the users of the system (the pilot and mission commander) really want to know the probability of completing the current mission given the current assessment of the damage state. Because predictive models typically have more uncertainty associated with them when the structure responds in a nonlinear manner, as is often the case when damage accumulates, *an alternate goal might be to estimate how long the system can continue to perform safely in its anticipated environments before one no longer has confidence in the predictive capabilities of the models that are being used to perform the prognosis.*

Having classified the damage-prognosis problem for the purpose of identifying appropriate measurement, assessment, and prognostic techniques, a general solution procedure is depicted in Figure 1.6, where the relationship between data-based and physics-based assessments and predictions are identified. The process begins by collecting as much initial system information as possible. This information is used to develop

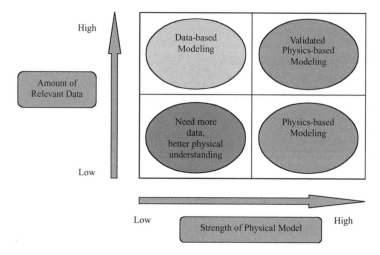

Figure 1.5 A comparison of regimes appropriate for physics-based modeling and regimes appropriate for data-based modeling

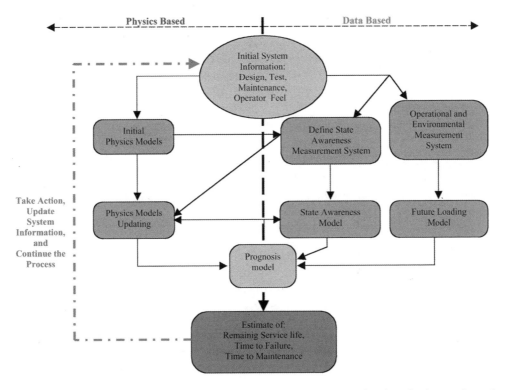

Figure 1.6 A general procedure for a damage-prognosis solution showing the interaction of data-based and physics-based assessments and predictions with the other issues

initial physics-based numerical models of the system as well as to define the sensing system for state-awareness assessments and whatever additional sensors are needed to monitor operational and environmental conditions. The physics-based models can also be used to define the necessary sensing system properties (e.g., sensor locations, band-width, sensitivity). As data become available from these sensing systems, they can be used to validate and update the physics-based models. These data, along with output from the physics-based models, can also be used to assess the current state of the structure (existence, location, type, and extent of damage). In addition, detailed infor-mation on system configurations and damage states may become available from time to time (e.g., via destructive testing, system overhauls, or system autopsies) that can be used to update the physics-based models. Data from the operational and environmental sensors can be used to develop data-based models that predict future system loading. The output of the future-loading model, state-awareness model, and the updated physics-based model can all be input into a reliability-based predictive tool that can be used to estimate the remaining system life. Note that 'remaining life' can take on a variety of meanings depending on the specific application. From Figure 1.6 it is clear that various models have to be employed in the prognosis process. Also, the data-based and physics-based portions of the process are not independent. As indicated, the solution process is iterative, relying on experience gained from past predictions to improve future predictions. It is again emphasized that results of past diagnoses and prognoses and their correlation with observed response can be used continually to enhance the system model.

Several additional points are brought out by this general solution procedure that should be mentioned:

(1) As much information as possible should be used to describe the initial system state, including 'soft' information such as operator 'feel' for the system response.
(2) Damage prognosis will rely on numerous models of different forms.
(3) The sensing system that is used to assess damage may be different from the one used to monitor environmental and operational loading, but, most likely, the two systems will overlap.
(4) Both the sensing and predictive modeling portions of the process will have to be updated as more information becomes available.
(5) Numerical simulations will need to be used when defining the sensing system in order to have confidence that this system has adequate properties (e.g., resolution, bandwidth) to detect the damage of interest.
(6) Numerical simulations will have to be combined with data-based models in order to estimate the extent of damage in the SHM portion of the problem.

1.3 MOTIVATION FOR DAMAGE-PROGNOSIS SOLUTIONS

The interest in damage prognosis is based on the tremendous potential for life safety and economic benefits that this technology can provide. The consequences of an unpredicted system failure were graphically demonstrated by the Aloha Airlines fuselage separation in 1988 (see Figure 1.7). Despite the catastrophic failure of the structure, all but one of the passengers and crew survived. In response to this event, the Federal Aviation Administration (FAA) established the Aging Aircraft Program to ensure that the integrity of aircraft is maintained throughout their useable lives. As part of this program, an Airworthiness Assurance Nondestructive Inspection (NDI) Validation

Figure 1.7 Damage to Aloha Airline fuselage resulting from fatigue cracks

Center was established.[1] The center was established to provide the developers, users, and regulators of aircraft with validated NDI, maintenance, and repair processes and with comprehensive, independent, and quantitative evaluations of new and enhanced inspection, maintenance, and repair techniques. Clearly, this catastrophic event has raised the importance of maintenance, reliability and safety with the public and industry alike.

Beyond the life-safety issues, the impetus from airframe and aircraft-engine manufacturers, as well as other manufacturers of high-capital-expenditure products such as large construction equipment, for effective damage-prognosis capabilities is business models whereby manufacturers charge for usage by some type of lease arrangement and meet the cost of maintenance themselves. Damage prognosis will allow these manufacturers to move from time-based maintenance to the more cost-effective approach of condition-based maintenance. Also, with effective damage-prognosis capabilities these companies can establish leasing arrangements that charge by the amount of system life used during the lease instead of charging simply by the time duration of the lease.

The escalating cost of aging aircraft is not simply restricted to the civil fleet. The US Air Force, through its Engine Rotor Life Extension program (ERLE) expects to commit 63 % of its capital budget to sustainment, and 16 % and 18 % respectively to development and acquisition [1]. On this basis alone there will be a major thrust in the next few years in the aerospace and defense industries to reduce costs of maintenance. The current processes applied to military engine-turbine rotors illustrate a need for new prognosis procedures. These rotors are safety-critical components of the aircraft, and failure through disc burst results in catastrophic loss of the engine at best, or the aircraft loss at worst. Guidelines outlined by Larson et al. [1] dictate that a disc is discarded on the probability of a one-in-one-thousand chance of failure after inspection. The cost of a single disk is in the range of $300 000 to $400 000. In this case, the probability is that 999 out of 1000 discs are being discarded before they have reached their full, safe, operating life. On this basis, the minimum component cost saving – excluding benefits derived by reduced maintenance and inventory achieved by even a 10 % increase in life resulting from improved prognosis methods – is approximately $30–40 million for 1000 discs.

Perhaps the most advanced and most scrutinized health monitoring systems are used in helicopters. The Health and Usage Monitoring Systems (HUMS) have already been operating successfully in transmission monitoring and engine applications. Their effectiveness and reliability has been endorsed by the Civil Aviation Authority (CAA) and FAA, and are now being considered for more general structural health monitoring. The most recent cost estimates for implementation of HUMS depend on the level of coverage required [2]:

1. Specific HUMS: partial implementation of a specific component e.g. engine or transmission $10 000–$50 000/unit
2. Mid-range HUMS: instrumentation of a range of structurally significant items (SSIs), but not full coverage, $75 000–$125 000/unit.
3. Complete HUMS: $150 000–$250 000/unit.

[1] See http://www.sandia.gov/aanc/AANC.htm

Although the cost of these units is falling because of adoption of commercial off-the-shelf (COTS) technology, Forsyth identifies data management as the largest single cost associated with these systems. This issue is significant because Forsyth acknowledges that 'considerably more data' would be required for robust verification of parameters needed for prognosis. Currently such systems are used for identifying damage. Work is ongoing in the field to use HUMS for prognosis through the Seeded Defect Program [3] whereby electrical discharge machining (EDM) is used to introduce tightly defined faults within gears. The outcome of this program has been an effective demonstration of damage diagnosis and life extension. The options for prognosis have not been fully explored though HUMS, but the achieved life extension from the current implementation of usage monitoring of 73 SSIs has led to cost savings on replacement parts of US$175 per hour of operation, excluding the savings associated with installation and removal of components [4]. With multichannel component usage monitoring, White reports an average increase in available structural fatigue life of 380 % over the original design life. On this basis there seems to be an overwhelming argument in favor of development of robust, low-cost, systems for health and usage monitoring.

In the civil engineering field, the driver for prognosis is largely governed by large-scale discrete events rather than more continuous degradation. Typical examples are aerodynamic gust loads on long span bridges and earthquake loading on all types of civil infrastructure. Although cyclic loads caused by traffic are also a consideration, the discrete events are the ones that require immediate prognosis for future use. Using the Kobe earthquake as an example, some buildings were subject to 2 years of scrutiny before a decision was made on their future use or demolition [5]. Clearly this delay had a significant commercial impact on the economic capacity of the city beyond the reconstruction costs. The most effective way to mitigate costs associated with building reoccupation is through timely prognosis of the system's load-bearing elements and connections. In this case, prognosis is required to ensure that the building can withstand the aftershock associated with a significant seismic event. This application requires a much denser array of sensors to identify local structural degradation than is typical for most current strong-motion instrumentation systems designed for seismic monitoring [6]. Current wired technology in the seismic field has a cost of $10 000 per node, including installation, which limits the sensor density for this application [7]. In California, the most densely instrumented structures have on the order of 10–30 sensors to measure seismic response. For damage prognosis, a 1 to 2 order of magnitude increase in sensor density is required. It is the authors' opinion that this increase can only be achieved economically by the use of new sensing technology such as wireless, self-assembling, embedded devices based on integrated circuit (IC) fabrication technology.

Damage identification may significantly mitigate the economic impact of seismic disruption to civil engineering infrastructure. The figures shown below are the results of Coburn and Spence's [8] work developing models of the fiscal impact caused by earthquake damage. The average annual worldwide repair and reconstruction costs associated with mechanical failures and earthquake damage is in the region of $60 billion. This figure does not include consequential losses (e.g., loss of revenues resulting from damage to a manufacturing facility), and costs caused by operator errors. Significantly, in-service mechanical failures contribute 20–40 % of all losses within a given engineering sector. Some of the major components of the $60 billion loss are (approximately): $1.5 billion resulting from commercial aircraft hull losses; $1.5 billion resulting from repair and reconstruction following petrochemical industry disasters, and $45 billion associated with earthquake damage. The trends of most concern in these statistics are the costs associated with the petrochemical industry, which have risen tenfold in real terms over the last 30 years, and with post-earthquake costs, which are rising up to 20 % per annum. During 1995, for example, the earthquakes in Northridge and Kobe were estimated to cost $220 billion. In 1999, the severe damage caused by earthquakes in Colombia, Turkey, Greece, and Taiwan showed that this is a problem of global proportions.

1.4 DISCIPLINES NEEDED TO ADDRESS DAMAGE PROGNOSIS

As discussed in detail in the body of this report, damage-prognosis solutions may be realized by the integration of a robust, densely populated sensing array and a system-specific adaptable modeling capability, both deployed on-board the system via advanced microelectronic hardware. This integration requires that many disciplines be brought to bear on the damage-prognosis problem. These disciplines include, but are not limited to, the following:

(1) **Engineering mechanics**: transient, nonlinear, computer simulation, system performance analysis, and damage evolution material models.
(2) **Reliability engineering**: probabilistic inference, probabilistic risk assessment, and reliability methods.
(3) **Electrical engineering**: micro-electro-mechanical systems (MEMS), wireless telemetry, power management, and embedded computing hardware.
(4) **Computer science**: networking, machine learning.
(5) **Information science**: data compression and communication, large-scale data management, signal processing.
(6) **Material science**: smart materials, material failure mechanisms, self-healing materials.
(7) **Statistics and mathematics**: statistical process control, model reduction, pattern recognition, and uncertainty propagation.

When so many disciplines are required effectively to tackle the damage-prognosis problem, technology integration becomes a major issue that must also be addressed.

1.5 SUMMARY

This introduction has defined damage prognosis and motivated the need to develop damage prognosis capabilities for a wide variety of civilian and defense infrastructure. The multidisciplinary nature of this field and the technologies that must be integrated to develop damage prognosis solutions has also been stressed. The rest of this book provides a summary of the current state of the art in damage prognosis. This summary includes more detailed discussions of the three main technology areas that form the damage-prognosis solution: instrumentation and data processing hardware, data interrogation, and predictive modeling. Also, various applications of damage prognosis are discussed. It is emphasized that the editors believe damage prognosis is, in most cases, emerging technology. Therefore, considerable technology development and integration is necessary before damage prognosis can make the transition from research to practice. As such, the editors hope this book will provide the necessary background that will allow researchers to start addressing the various technical issues associated with the development of damage-prognosis solutions in an efficient manner.

REFERENCES

[1] J. Larson, S. Russ, *et al.*, 'Engine rotor life extension (ERLE),' *Damage Prognosis Workshop*, Phoenix, AZ, March 2001.

[2] G.F. Forsyth and S.A. Sutton, 'Using econometric modeling to determine and demonstrate affordability,' *Tiltrotor/Runway Aircraft Technology and Applications Specialists' Meeting of the American Helicopter Society*, Arlington, VA, 2001.

[3] A.J. Hess and W. Hardman, 'SH-60 helicopter integrated design system (HIDS) program experience and results of seeded fault testing,' *DSTO Workshop on Helicopter Health and Usage Monitoring Systems*, Melbourne, Australia, 1999.

[4] D. White, 'Helicopter usage monitoring using the MaxLife system,' *DSTO Workshop on Helicopter Health and Usage Monitoring Systems*, Melbourne, Australia, 1999.

[5] C.A. Taylor (Ed.), 'Shaking table modeling of geotechnical problems,' *ECOEST/PREC8, Report No. 3*, p. 190, 1999.

[6] C.R. Farrar and H. Sohn, 'Condition/damage monitoring methodologies,' *Proceedings Invited Workshop on Strong-Motion Instrumentation of Buildings*, J.C. Stepp and R.L. Nigbor (Eds.), The Consortium of Organizations for Strong Motion Observation Systems (COSMOS) publication CP-2001/04, Emeryville, CA, November 2001.

[7] S.D. Glaser, 'Smart dust and structural health monitoring,' *Damage Prognosis Workshop*, Phoenix, AZ, March 2001.

[8] A. Coburn and R. Spence, *Earthquake Protection*, John Wiley & Sons, Inc., New York, 2002.

Part I
Damage Models

2

An Overview of Modeling Damage Evolution in Materials

Todd O. Williams and Irene J. Beyerlein

Los Alamos National Laboratory, Los Alamos, New Mexico, USA

2.1 INTRODUCTION

Mankind is always attempting to develop better structures. What constitutes a 'better' structure can be defined either by improvements in properties or in performance, i.e. better structures can be stronger, lighter, tougher, stiffer, more damage tolerant, more reliable, more durable, etc. In order to achieve these performance goals it is necessary to utilize materials capable of stepping up to meet the associated structural demands.

Material behavior has always represented a limiting factor in the performance of any structure. For example, the aerodynamic advantages of forward swept wings have been recognized since World War II (WWII). However, the materials available during WWII were unable to accept the stringent loading requirements imposed by this design concept, thus leaving the application unrealizable until the introduction of advanced composite materials. In this case, composite materials exhibited the requisite superior material properties necessary to meet performance requirements.

If materials maintained their perfect atomic structures during service, their behavior would be orders of magnitude better and more reliable and thus would hardly represent a limiting factor. However, actual material behavior deviates from the theoretical ideal. This deviation is due to the nucleation and evolution of material damage. Damage is responsible for those unacceptable and irreversible changes that occur in the response of a material or structure under loading. Signs of damage are typically reductions in some measurable form of strength, stiffness, or toughness.

Damage Prognosis – For Aerospace, Civil and Mechanical Systems   Edited by D.J. Inman, C.R. Farrar, V. Lopes Junior and
V. Steffen Junior   © 2005 John Wiley & Sons, Ltd

To control material behavior is to control its response to damage. Unfortunately, it is not possible to enhance simultaneously every aspect of material response. Usually, it is necessary to optimize or balance different aspects of material response to achieve the best solution for a given application. This is a nontrivial task as the relationships between *changes* in material response and *changes* in external factors, such as loading conditions or structural geometry, are generally complex. As it is prohibitive to rely solely on experimentation to determine these relations, it becomes clear that the structural designer and analyst must have access to material models in order to predict such relationships.

The selection and development of material models depends on five basic factors. First, the boundary conditions or service environments imposed on the structure or material need to be defined. Generally speaking, the more technologically advanced the application the harsher is the service environment and the more demanding the boundary conditions. Next, the type of material being modeled, traditional or nontraditional, single phase or multiphase, is an important factor in determining which model to use. Traditional materials are likely to have a well-established data base and associated material models, whereas nontraditional materials typically do not. Third, each type of material has a characteristic microstructure. Microstructure plays an important role in determining material behavior as the morphology of initial damage, and the nucleation and evolution of load-induced damage are controlled by material microstructure. The fourth basic factor is the identification of the dominant forms of damage and the associated length scales at which they operate. Fifth, the evolution of damage needs to be considered, i.e. how it transitions through length scales and how it interacts with other forms of damage.

Often claims are made for a particular model's suitability for many applications, but in fact there is no unified material theory capable of dealing with every situation. The reality is that there is a large number of theories that have been developed to model material behavior. There is, however, a relatively limited number of distinct classes of models, and within each class, theories share similar sets of assumptions and methodologies and often build upon one another.

The purpose of this chapter is to present many of the general considerations and issues associated with material damage modeling. This article considers the relative strengths and weaknesses of a variety of older, classical models, as well as more recently developed models. The particular models considered were chosen as they are representative of the different classes.

The outline for this chapter is as follows. First, the general considerations and issues associated with material modeling are discussed. Next, the identification of these issues from experimental data is considered. What constitutes an appropriate material model is then reviewed. This part of the chapter is fairly generic and accessible to undergraduate readers. Subsequently, various classes of approaches intended to model different types of nonlinear, history-dependent behavior ('damage') are examined beginning with the more classical approaches. A discussion of newer phenomenological continuum level as well as micromechanical modeling approaches follows. For these portions of the chapter it has been assumed that the reader has some basic familiarity with geometrically linear continuum mechanics (Malvern, 1969, or Eringen, 1989). However, in order to broaden the accessibility of the chapter, the major strengths and weaknesses of each class of models are outlined at the end of each section.

The discussions of various types of theories for modeling history-dependent behavior in materials given in Sections 5–7 are brief and attempt simply to emphasize the central concepts. The theories discussed in these sections are examined in greater detail in Williams and Beyerlein (2004) and the reader is referred to this work for the intricacies of these theories.

2.2   OVERVIEW OF GENERAL MODELING ISSUES

This section examines in more detail the previously mentioned five basic factors governing the selection and development of appropriate material models.

2.2.1   Service Environment

The service environment to which a material is subjected impacts material modeling in two ways. First, service conditions (or 'environments') are a significant factor in determining the type of materials that can be considered for a particular structural application. For example, high temperature environments severely limit candidate materials to metallic-based materials, carbon–carbon materials, or ceramic-based systems. The second avenue is through the boundary conditions imposed on the material. Material response is highly sensitive to the type and magnitude of the boundary conditions. Possible types of boundary conditions include simple quasistatic, uniform loading in one direction, broadly varying regimes of strain rates, strain gradients (nonlocal effects), changes in mechanical loading paths, extremes in temperature, varying thermal loading rates, corrosive environments, or a combination of these conditions. The duration over which these service conditions are applied is important as extremely short- or long-term applied loading states can change material response characteristics. Typically the more advanced the application, the more extreme the loading conditions and environments.

2.2.2   Considerations in Material Specification

The next major factor in the development of appropriate material models is the type of material to be modeled. In the following, materials are classified as either traditional or advanced.

Traditional materials, such as steel, aluminum, or plastics, see widespread use in a broad range of structural applications. These materials continue to be the mainstay for a number of reasons. First, they can be easily manufactured or processed and, thus, are, in general, relatively inexpensive materials. Second, they have a long history of testing and characterization; hence, a large data base on their behavior exists. Compared with less traditional materials they are relatively well understood. Consequently, they can be utilized with a high degree of confidence. Third, the fact that these materials can, as a first simplification, be reasonably considered as homogeneous, isotropic materials has meant that their behavior can be modeled by relatively simple theories.

As structural applications have become more demanding the capabilities of traditional materials are often exceeded. In order to meet more stringent design requirements, advanced materials are being developed. Currently, two different paths can be pursued in the development of advanced materials. The first method utilizes the concept of synergy, that is, to combine two or more materials in such a way that the overall behavior is more desirable than the sum of its constituents. The second approach is to engineer aspects of the material microstructure to achieve the desired response characteristics. Familiar examples of advanced materials are enhanced metallic alloys, fiber composites, and ultrafine-grained metals.

While advanced materials exhibit many superior qualities compared with more conventional materials, they are not in widespread use for a number of reasons. These types of material are frequently heterogeneous, meaning their local properties can vary substantially from point to point. Additionally, some of these materials are highly anisotropic, meaning their behavior is sensitive to the direction in which they are loaded. The degree of heterogeneity and anisotropy together usually result in highly complex behaviors and a multiplicity of interactive failure modes. These complex and interactive response characteristics often challenge conventional thinking and modeling. For example, reinforcement fibers containing more flaws (a higher variation in strength) make their parent composite more flaw tolerant. Third, advanced materials often require more sophisticated theories. The development of such advanced theories is often hindered due to the scarcity of large experimental databases and the general lack of practical experience with these materials in structural applications. Consequently, engineers cannot yet use advanced materials reliably in critical structural applications without huge safety factors. Finally, advanced materials must, typically, be fabricated using nontraditional manufacturing methods and thus, to date, are relatively expensive compared to conventional materials.

2.2.3 *Material Microstructural Considerations*

Every material has associated microstructural characteristics. At some level material models must appreciate material microstructure as it plays a role in damage initiation and evolution.

It is important to specify what constitutes a material microstructure. As illustrated in Figure 2.1 there is a wide variety of material microstructures. One way to characterize microstructures is to classify them first as either a single phase or a multiphase microstructure. As the name indicates, a single-phase microstructure is composed of only one material phase, e.g. pure metals, polymers or plastics, and ceramics. A multiphase microstructure incorporates two or more distinct phases, such as in fiber or particulate composites. In both types of microstructures, the interfaces that exist between the similar or distinct phases are an inherent and important aspect of the microstructure. These interfaces can be mechanical, chemical or atomic in nature.

Microstructure determines how external boundary conditions are related to the internal stress, strain or energy states in a material. The evolution of these internal states governs the initiation and evolution of damage within the microstructure and, ultimately, material failure. The greater the degree of heterogeneity the greater the spatial variation in these local fields. As an example of the above considerations,

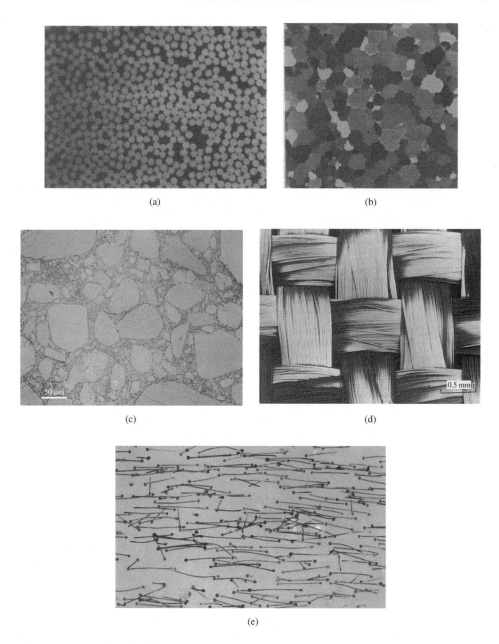

Figure 2.1 Examples of different types of material microstructures: (a) graphite/epoxy composite; (b) polycrystalline metal; (c) high explosive composite; (d) woven graphite/epoxy composite; (e) short fiber composite

processing effects are most often accentuated in heterogeneous materials where significant thermal residual stresses (internal stresses) can evolve due to the mismatch in thermal expansion coefficients and elastic properties. The magnitude of these types of stresses can have a strong impact on subsequent material performance.

2.2.3.1   Single-Phase Materials: Metals, Ceramics and Polymers

The microstructures of metals are composed of aggregates of single metallic crystals. Such an aggregate is called a 'polycrystal'. Microstructural features of polycrystalline metals include texture, grain size, interface characteristics, and crystalline lattice imperfections.

Texture is the crystallographic orientation distribution of the individual crystals within the polycrystal. Texture can be measured using X-ray diffraction, orientation imaging microscopy, and neutron diffraction. Texture, in conjunction with the properties of the individual crystals and interfaces, determines the anisotropic response of the bulk polycrystal.

The individual crystals or grains composing a polycrystal have an associated size and shape. A polycrystal can have substantially different properties as its average grain size changes. For example, large grains are more likely to twin or develop dislocation boundaries than smaller grains and fine-grained materials are stronger than coarse-grained materials.

Interfaces influence how individual crystals within the polycrystal interact, with the resulting interactions driving the bulk material behavior. There are many examples of interfaces in single phase metals, such as grain boundaries, twin boundaries, shear-band boundaries, deformation-induced dislocation boundaries within grains, and the list goes on. The properties of interfaces can vary widely throughout a microstructure and their variation can have a strong impact on localization processes and the bulk material behavior.

The basic lattice or crystal structure determines many aspects of metal behavior, such as the elastic properties of the crystal and the associated degree of anisotropy of these properties (i.e., the ratio between the greatest and smallest stiffness terms). Typically, imperfections, such as impurities, inclusions, or alloying atoms as well as dislocations, exist in the crystalline lattice. The local arrangements of these imperfections, for example dislocation structures and pileups, can have a significant impact on the local and bulk material response. The presence of such features changes the single crystal behavior as compared with that of a perfect crystal.

For more detailed discussions concerning the behavior of metallic systems and the associated microstructure the reader is referred to Dieter (1986) or Kocks *et al.* (1998).

Ceramics represent another type of material with microstructures that belong to the single-phase category. In general, they exhibit many of the same microstructural features as are present in metals and, thus, the above discussion is pertinent as a basis for considering ceramic microstructures.

The remaining general class of single phase materials is polymers and plastics. The basic microstructure is composed of an amorphous mixture of polymer chains. These polymer chains interact in two ways. The first type of interaction is due to mechanical effects, which arise from the interwoven arrangement of the chains and the associated sliding of these chains under loading. These effects, typically, have an overriding influence on the material's resistance to loading and damage. The second type is due to atomic effects and arise from secondary atomic bonding between different atoms in the different chains. Polymer chain structures can be considered, in general, to consist of hard (stiffer) and soft (more compliant) segments. The interactions of these hard and soft regions have an impact on the overall behavior of the material. The concept of interfaces becomes somewhat fuzzy within the context of polymeric materials but can be considered to exist between the hard and soft regions, at the regions between different chains, and at the mechanical contact points between the chains.

2.2.3.2 *Multiphase Materials*

The second broad category of microstructures are those that incorporate two or more materials (Figure 2.1). Materials of this type are typically called multiphase, heterogeneous, or composite materials. Generally most advanced materials fall within this category. There are two sub-categories within this class of materials.

The first sub-category is composed of materials in which the material microstructure consists of a binder or matrix material surrounding a reinforcement or inclusion material (Figure 2.1). The matrix is the continuous phase that permeates the microstructure, holds the entire composite together, and controls interactions between phases. There are many types of matrices, such as epoxies, metals, and ceramics. The inclusion phases are typically separated at the microstructural level. Often the inclusions represent the strength or stiffness enhancing phase. The inclusion phase can take many forms, such as continuous fibers or particulates of any shape and aspect ratio.

The second subcategory consists of materials in which the phases are equally intermingled and, as such, the distinct roles of binder and inclusion cannot be readily defined. Common examples are laminated composites, which are composed of layered sheets of different materials and welded joints between dissimilar metals where the different materials are equally comingled in the joint region. The so-called 'functionally graded materials' (FGMs) are another example. In these materials the microstructure consists of a gradual spatial transition from one phase to another phase.

There are a number of microstructural features in multiphase materials that can have an impact on the deformation response. First, the spatial arrangements of the different phases may be dictated or random. FGMs represent materials with dictated microstructures, while particulate composites are good examples of random microstructures. The second important consideration is the type of constituents that compose a microstructure. Typically, the component materials are specified based on imposed performance constraints for the bulk material behavior. For example, if a heterogeneous material is required to survive in high temperature environments, potential candidates would be metal matrix composites and carbon–carbon composites. For aerospace applications, where high performance is usually required (which often implies high specific stiffness and strength), composites composed of carbon or graphite fibers embedded in an epoxy matrix are common. Third, the relative properties of the components can have a strong impact on the material behavior. For instance, severe stress concentrations and residual stresses can form in multiphase materials whose component phases are strongly dissimilar. Fourth, the relative shapes and orientations of the different phases determine the degree of anisotropy of the bulk material. Continuous fiber composites, for example, have highly directional properties that can be useful in tailoring structural response. The aeroelastic tailoring of aircraft structures is a common application of this concept. Alternatively, chopped fiber composites, where the fibers have random orientations, typically exhibit relatively isotropic behavior. Fifth, as was the case in metallic microstructures, the interfaces between the different phases in a heterogeneous material become key players in governing the local and bulk material behavior by determining how the different phases interact. Finally, the component materials can exhibit their own microstructures. Graphite fibers, for example, are typically composed of concentric cylinders of layered materials with varying properties from layer to layer.

2.2.4 Forms of Damage

Damage comes in many forms even within the same material. The generic working definition used in this chapter for damage is any structural deviation from a perfect material state, such as material separations, imperfections, or displacement discontinuities, that affects performance. Damage occurs at virtually all length scales. Damage results in nonlinear, irreversible or dissipative deformation mechanisms manifesting as history-dependent behavior. The forms of damage that manifest depend on the service conditions, type of material, and material microstructure.

One common form of damage is material separation. Material separation can occur at any length scale. At the macroscopic length scale, these phenomena manifest in the form of macroscopic cracks. At smaller length scales, familiar examples are microcracks within or between phases, voids or pores, chain scissions in polymers, and delaminations between layers in laminated structures.

Other forms of damage that do not fall in this category are: dislocations and the resulting pileups in metals; atomic vacancies in a crystalline lattice, grain boundary steps and ledges in metals; chain sliding and disentanglement in polymers; and fiber buckling and kink banding in fiber composites.

Determining the dominant damage mechanisms is often part of the modeling process, as clearly, not all types of damage mechanisms need to be included in a failure model. Dominant failure mechanisms largely depend on material system, microstructure, and loading conditions. In continuous fiber composites, for instance, fiber buckling and kink banding occur only under compression and bending loadings, creep occurs under constant applied stress loading, interface debonding and sliding are predominant when the interface is weak, while fiber bridging is observed when the fiber strain to failure is higher than that of the matrix. Alternatively, in laminated structures, ply delamination can be initiated by residual stresses created by the difference in the directional properties between two adjacent plies or, alternatively, by impact loading and the subsequent stress wave propagation. Furthermore, as loading conditions change different damage mechanisms may become dominant within a given material system.

The above discussion presents various types of damage as individual events. Often various forms of damage exist in a material simultaneously. In general, when more than one type of damage mechanism is present, these failure modes interact in a highly complex fashion.

Finally, it is noted that damage can exist in a material prior to loading. Typically, this initial damage state is induced during material manufacturing. For example, in metal matrix composites it is common to observe radial cracking induced by thermal residual stresses.

2.2.5   Damage Evolution

There are several common issues that need to be considered in modeling damage evolution. Damage typically occurs over many length scales. Usually damage initiates at lower length scales, evolves and interacts at these scales, and then coalesces and manifests at higher length scales. If multiple forms of damage are present, then the interactions of the different forms of damage often have a synergistic effect resulting in greater changes to the material behavior than would be the case if each type of damage operated alone. The evolution of damage mechanisms is strongly dependent on the imposed loading conditions and the material microstructure.

Traditional thought has often considered a structure or a material to be useless once any macroscopic evidence of damage is detected. Under this practice, once a single damaged region or damage event achieved a critical size, the material or structure was considered to have failed. In most cases, this viewpoint is too simplistic. Increasingly, the concept of 'damage tolerant' structures is coming into greater use. Damage tolerant structures must, by definition, survive the imposed service environment in the presence of continually evolving and, ultimately, extensive damage without failure. The pursuit of damage tolerant or tough materials and structures introduces many of the previously discussed considerations into modeling, such as types of material, microstructure, and service environment. Also, it is frequently necessary to consider more than one type of damage in such systems, how the different types of damage interact, and how these interactions influence the continued evolution of damage. One way to achieve damage tolerant materials is to control or direct the evolution of the damage through changes in the microstructure or types of material phases employed. Regardless of how the material is manufactured, designing damage tolerant materials is a difficult problem that cannot be tackled without appropriate material models.

2.3   CHARACTERIZATION OF MATERIAL BEHAVIOR: DAMAGE INITIATION AND EVOLUTION

The previous section has given an overview of the general considerations that govern the identification and development of appropriate material models. These considerations alone are not sufficient for developing a complete material model. It is necessary to obtain specific information concerning a given material's behavior. This information is obtained from experimental data. The key outcomes of such experiments are the characterization of the stress–strain curves and/or the failure conditions of a material.

Simple mechanical tests, such as compression, tension, or shear tests, that generate uniform fields in the test section typically provide the starting point for developing a data base that characterizes a material's behavior. The simplest versions of these tests use a monotonically increasing load in one direction, often until extreme localized deformation or complete material failure is achieved.

As primary tools for characterizing material behavior, it is useful to consider a monotonic stress-strain curve. A generic stress–strain curve showing material behavior from initial loading up to material failure is given in Figure 2.2. The initial linear response regime represents the elastic (or recoverable) response of the material.

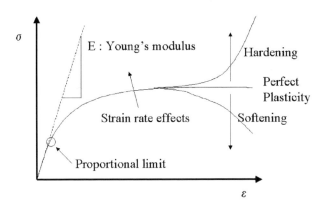

Figure 2.2   Generic monotonic stress–strain curve indicating various types of material responses

This part of the curve provides information about the elastic properties (such as the Young's and shear moduli, and the Poisson's ratio).

The next relevant region of material response begins once the material behavior departs from linearity and ends at the first peak stress. In particular, the stress at which the material begins to deviate from linear behavior is termed the proportional limit. In the case of metals this point is also called the yield stress. In this response regime the material begins to exhibit a gradual deviation from the initial slope in the form of increasing compliance or 'softening'. Under continued loading the reduction in stiffness continues, often at a greater rate. There are two possible explanations for the material response in this region. Which explanation is correct can be determined by unloading the material. If the unloading response is the same as the loading response then the material can be considered to be nonlinearly elastic. This aspect of material behavior will not be considered further in the chapter. Alternatively, if at zero stress, the material continues to exhibit unrecoverable ('inelastic') deformations, then this softening is due to the onset of damage. The deviation from linearity, in conjunction with the unrecovered nature of the deformations, indicates that behavior in this region is nonlinear and irreversible. Furthermore, the damage evolution responsible for the response in this region is usually a function of loading history. The implication is that damage is accumulating in the material at the microstructural length scale but has not yet reached a critical value. The behavior in this portion of the stress–strain curve can be used to characterize the initial damage evolution behavior of the material.

After the initial peak stress a number of potential types of material responses can occur. These types of potential responses are indicated in Figure 2.2. The first potential type of response is that the material fails abruptly at this peak stress. This type of behavior is termed 'brittle'. Examples of brittle materials are ceramics and unidirectional composites loaded in the fiber direction. It is noted that for many brittle materials the peak stress and the proportional limit stress coincide. Failure in brittle materials is due to the rapid propagation and coalescence of damage. Alternatively, the material may continue to deform without any change in stress, i.e. to flow inelastically, until failure. A generic term for this type of response is 'perfectly inelastic', although for metals this type of behavior is referred to as 'perfectly plastic'. This type of damage is often dictated by the continuous migration of damage through a material.

If, after the peak stress is reached, the stress carrying capability decreases as the strain increases the material is considered to exhibit 'strain softening'. Strain softening behavior is usually associated with the coalescence of damaging effects, such as microcracking or extreme localization of the plastic deformations (such as occurs in a shear band).

There are two additional possibilities for the continued response of the material after strain softening occurs. The first potential type of response is that the material continues to strain soften until failure. In this case, the material is failing gracefully. Typically, the peak stress and the ultimate material strength (the maximum loading carrying capability of the material) coincide. The second type of response that can occur after the initial strain softening response is that the stress again increases with continued straining. This type of behavior is called 'strain hardening'. Strain hardening behavior is associated with the development of increased restrictions to inelastic deformations in the microstructure. For example, in metals, this type of behavior can occur due to increasing dislocation densities developing with straining. Under these conditions the material continues to harden until failure. If greater than the initial peak stress, the final stress state represents the ultimate material strength; otherwise the initial peak stress is the ultimate material strength.

The area under the entire stress–strain curve is the work of fracture of the material, i.e. the total energy dissipated in failing the material. The higher the work of fracture, the tougher the material and the less catastrophic the failure. The work of fracture is not to be confused with the fracture toughness, which is determined by considering the failure processes at crack tips and has to do with the propensity for crack propagation.

The localized failure processes associated with the different key points on the stress–strain curve, such as initial deviation from linearity and peak value, are micromechanically complex and differ between metals, ceramics, polymers and composites. Despite material differences, in all cases, material responses are sensitive to the details of the failure processes of the constituent(s) and the associated defects and flaws, and microstructure. One of the objectives of experimental work is to assist in determining these sensitivities.

Composite materials are particularly notable in the above regard since they often exhibit extremes in material behavior. The highly heterogeneous and stochastic nature of these types of material and the associated flaw structures that simultaneously exist at multiple length scales within these materials result in the existence of many different potential failure modes and mechanisms. In most cases, composite strength is not deterministic. Thus, one should not be concerned with the average strength of the composite, but rather the strength of the composite associated with low probabilities of failure (or, alternatively, high reliability). Substantial variation in strength implies the need for larger data sets generated using identical specimens. Also, when a composite contains brittle phases it exhibits a size effect. The size effect means that the larger the volume exposed to a uniform stress, the higher the probability of finding a significant flaw, i.e. a decrease in bulk strength with an increase in volume. The size effect is important to characterize in order to relate test data obtained from comparatively smaller laboratory specimens, or to relate the data from tests on the same scale specimen to structural strengths. It is noted that, typically, analogous failure phenomena are not present in metals due to the relatively low variation in local material properties between the different grain orientations, as well as between the grains and interfaces, and the typically ductile nature of these materials.

The general features of a stress–strain curve have been discussed above within the context of an unchanging set of applied loading conditions. Ideally, multiple stress–strain responses are generated at many different applied loading states. Experimentally characterizing the material response at a number of different conditions is important as this information provides greater insight into the material response characteristics as a function of the boundary conditions. For example, as the applied strain rate is increased, the stress for a given strain typically increases as well. Alternatively, at higher temperatures the stress at a given strain will, in general, decrease. Often a given material may exhibit one response for one set of boundary conditions, for example, brittle behavior, and a completely different set of responses, such as inelastic flow, for another set of boundary conditions. Finally, it is desirable to conduct experimental tests that involve more complex loading states (such as cyclic loading or multiaxial loading) in order to achieve greater insight into a material's behavior.

2.4   MATERIAL MODELING: GENERAL CONSIDERATIONS AND PRELIMINARY CONCEPTS

The experimentally characterized behavior discussed in the previous section can be represented by constitutive models or failure criteria. Generically speaking, material models attempt to construct a tractable mathematical version of the actual material behavior that relates the applied loading state (applied strain/stress, temperature, strain/stress and temperature rate) to the bulk response of the material by considering the pertinent deformation and damage mechanisms over the appropriate length scales. Sometimes the dominant mechanisms or interactions between mechanisms governing material response are not known *a priori*, and determining these mechanisms is part of the modeling process. The resulting models may be quantitative or qualitative in nature. Examples of constitutive laws are the relations linking polycrystal behavior to that of the individual grains, the relationship between the bulk response of a ceramic and the evolution of microcracks, or the response of a particulate composite based on debonding at the inclusion-matrix interfaces.

Appropriate material models are needed for successfully predicting material behavior and structural response and failure amd developing appropriate damage prognosis solutions. Material and structural responses are inextricably coupled in the sense that the material behavior at every point in the structure helps to determine the structural response while the overall structural response connects the material behavior at every point in the structure.

Inherent in the development of material models is the assumption that a separation of length scales can be achieved. There are at least three characteristic length scales to be considered; the structural length scale, the material length scale, and a microlength scale associated with the length scales of individual features in the material microstructure. Material models provide the bridge between, at least, the material and structural length scales. A material model can provide the bridge between the structural and material states if (1) the size of the characteristic material volume of interest is much smaller than the characteristic structure length scale and (2) the bulk material fields associated with structural response do not significantly vary over the characteristic material length scale. Homogenization theories, which bridge even greater length scale differences by incorporating the influence of microlength scale effects on the material response, are good examples of material models that satisfy these criteria. When these assumptions are not appropriate the alternative is to carry out the structural and material analysis simultaneously. This last option can prove prohibitively expensive with regard to computation time for large structures.

A factor that is not always considered is the feasibility of implementing a model into a structural analysis. Some material models are intended for embedding in another model of a larger scale, which, in turn, is embedded in an even larger scale model. This multiscale or hierarchical modeling approach eventually makes its way to the structural level. In this case, the computational efficiency of such hierarchical implementation may be an issue. Alternatively, there are other models, not meant to be directly embedded in another level of model, that are intended to provide insight into the basic physics governing material behavior.

There are two different types of constitutive model: local and nonlocal theories. Local theories depend only on the state of the material at a point. Nonlocal theories incorporate the effects of spatial gradients of the stress or strain fields. This chapter will deal only with local constitutive theories.

The development of local material constitutive relations is based on conditions where the bulk (or applied or far field) stress, strain, and temperature fields are uniform. Under these types of loading conditions traditional bulk elastic properties (such as Young's modulus E or Poisson's ratio ν) can be defined.

It is useful to consider some of the potential uses and benefits of material models. First, accurate material modeling can save time and money otherwise dedicated to extensive testing in the laboratory by reducing experimental parameter spaces. Second, material modeling can provide a tool for considering material behavior outside the envelop of available experimental data. For example, oftentimes structures are expected to last for years or decades, timescales that can be prohibitively long for laboratory testing. Models are needed to forecast the behavior of a material/structure over these long periods of time. Third, if damage is introduced during service, the ability to predict the remaining service life of a structure based on how it was introduced or how extensive it is, is particularly critical in structural and associated damage prognosis applications. Fourth, models that can reliably predict damage evolution can have a huge economical impact, as uncertainties and safety factors can then be reduced. Fifth, materials models are especially crucial for materials exhibiting a size effect, such as brittle ceramics, concrete and fiber composites. It is most often the case that laboratory-size test specimens are much smaller than actual structures. Sixth, a useful application of materials models is to extend predictive capabilities obtained using simple loading states, experienced in the laboratory, to the multiaxial loading states or time sequences of different loading conditions experienced in service. Finally, some material models can be used to explore the possibility of engineering novel materials with enhanced performance characteristics.

Reliable extrapolation of the laboratory test data to the general behavior of the material within a structure operating in service is a difficult task and must be executed with caution. Oftentimes it is prohibitively expensive or impossible to obtain experimental data that completely characterizes a given material's bulk behavior under all expected service conditions. The characterization of a composite material provides a useful example. Often, due to size effects, full characterization demands large statistically significant sample sizes of identical specimens at many volumes. Also, the composite parameter space due to potential variations in the material properties (e.g., interface treatments, microstructures, and manufacturing effects) is enormous. In light of these considerations, successful application of a material must often rely more heavily on the development of a reliable material model than on extensive experimental data. Given this last consideration, models that rely less on macroscopic testing for free parameter characterization and that have a strong physical rather than empirical basis are more desirable.

A material model can be classified as being either a continuum or a micromechanical approach. Continuum theories represent the more classic approach to material modeling and have formed the foundation for understanding material behavior. Micromechanical theories are a newer approach. Both approaches consider damage or failure events occurring at length scales lower than that of the structure. There are a number of major characteristics associated with these approaches that deserve consideration when determining which type of model is most appropriate.

Continuum level approaches are based on an analysis of material behavior at the bulk or macroscopic length scale and, thus, involve damage in an 'average' sense. Hence, these approaches do not directly consider what is occurring at the microstructural length scale but rather incorporate microstructural effects empirically through the use of the experimental determination of associated model parameters. There are a number of implications of the above characteristics. It is often necessary to have a degree of *a priori* insight into the different aspects of the bulk material behavior (for example its inelastic deformation characteristics) in order to construct appropriate continuum models. The accuracy of these theories is dependent on these insights which, unfortunately, are often limited by the lack of extensive experimental data. Due to their inability to consider material microstructure, continuum approaches cannot be used to engineer material behavior. If some aspect of the material or its processing history changes, then the experimental data base used to characterize the material response (the model parameters) must be regenerated. One of the greatest strengths of continuum level theories is that these approaches are (typically) computationally efficient relative to the alternative, micromechanical theories. This characteristic has important implications for structural analysis and design and means that these theories can be particularly useful for examining trends in material and structural behavior under different loading conditions.

The alternative approaches to continuum level theories are micromechanical (or homogenization) theories. These theories directly analyze the effects of the microstructural geometry and the underlying microphysics, such as response characteristics and failure mechanisms, to predict the local behavior in a heterogeneous material. Subsequently, these theories use an appropriate averaging scheme to predict the bulk material responses. In a general sense, the accuracy of the predictions is determined by the microstructural idealizations and the sophistication of the microphysics used. As was the case with continuum level approaches, the above characteristics have implications concerning the applicability of micromechanical models for different types of analyses. These approaches directly model microlevel damage initiation and evolution. The ability to directly consider microstructural effects implies that micromechanical theories require insight into the mechanisms governing the local and global material behavior. In turn, the ability to directly consider both microstructural and microphyics effects allows such theories to be used to engineer the material microstructure and (potentially) the microphysics to control local and, consequently, the bulk material response characteristics. If the material microstructure is changed (such as the relative volume fractions of the constituents changes) or its processing history changes then these types can still predict the material behavior without the need for additional experimental data. One drawback to the theories in this category can be their greater computational requirements as compared to continuum level. In particular, highly accurate micromechanical theories involve detailed descriptions of the material microstructure, which results in large numbers (an infinite number in reality) of material states locally. The resulting large numbers of local material states implies a correspondingly large number of local history-dependent behaviors that must be tracked resulting in high computational demands.

2.5 CLASSICAL DAMAGE-MODELING APPROACHES

The discussions in this section center around more classical approaches to modeling damage. In all cases, these approaches are continuum level theories with the attendent characteristics (see Section 2.4).

2.5.1 Linear Elastic Fracture Mechanics (LEFM)

The presence of flaws and defects in materials is unavoidable and significantly impacts material strength and toughness. Classical fracture mechanics (CFM) is a set of techniques concerned with predicting material behavior in the presence of flaws and cracks. CFM theories are associated with the analyses of stress fields generated (most often) by a single crack or for a few idealized configurations of multiple cracks in a body. This field of research was first pioneered by Griffith (1920) who recognized the fact that materials fail at far lower applied stresses than their theoretical strength. He realized that the local stress field in the vicinity of cracks is significantly amplified over the applied far-field stress state, and this amplification can provide the necessary strain energy required to advance a crack.

To apply CFM or linear elastic fracture mechanics (LEFM) one needs to solve for the stress distributions in the body. Frequently, a single crack in a relatively larger body subjected to a uniform far field loading is analyzed, see Figure 2.3. The existence, location and size of the crack is assumed *a priori*; CFM is not concerned with crack nucleation phenomena. The material surrounding the crack is considered to be a homogeneous, linear elastic continuum with the exception of a small nonlinear zone at the crack tip which is much smaller than the crack dimensions. This nonlinear crack tip region is so small that it does not affect the elastic stress or strain field produced by the crack. These analyses also typically assume plane strain or plane stress conditions and these two cases are considered the bounding responses.

2.5.1.1 Griffith Criterion

The classic example of Griffith's work (1920) is the analysis of a single through-the-thickness crack in an infinite body subjected to a uniaxial load σ_o applied in the

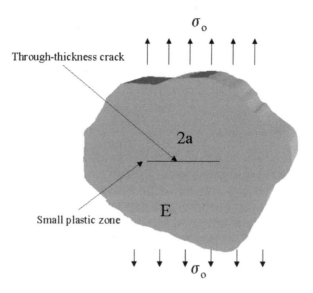

Figure 2.3   Single crack in an infinite body subjected to far-field loading

y-direction. A crack of length $2a$ lies along the x-direction or transverse to the loading direction as shown in Figure 2.3. The stress field along the crack plane is given by

$$\sigma_{yy} = \sigma_{xx} + \sigma_o = \frac{\sigma_o}{\sqrt{\frac{x^2}{a^2} - 1}} \left(\frac{2x}{a} + \sqrt{\frac{x^2}{a^2} - 1} \right)^{-1} + \sigma_o \tag{2.1}$$

and

$$\sigma_{xy} = 0 \tag{2.2}$$

where a is the crack half length and the origin of the $x - y - z$ coordinate system is the center of the crack. When the crack extends from a to $a + \delta a$, where δa is the infinitesimal crack extension, strain energy is released. The amount released must be greater than the amount absorbed to create new fracture surfaces. Griffith postulated that a crack can propagate when the strain energy released by crack growth is equal to the surface energy required to create new crack surfaces. Minimizing the total energy of the body with respect to a gives the following criterion for fracture

$$G_{Ic} = R_o \tag{2.3}$$

where G_{Ic} is the critical value of the strain energy release rate and R_o is the intrinsic resistance to fracture. When the above criterion, (equation 2.3), is achieved, the system is considered unstable. In other words, when $G_{Ic} < R_o$, the crack does not grow, and when $G_{Ic} \geq R_o$ the crack extends in the plane of the initial crack. The critical applied stress σ_c required to propagate the crack in an isotropic brittle material is given by

$$\sigma_c^2 = \frac{2E\gamma}{\pi a} \tag{2.4}$$

where E is the Young's modulus and γ is the surface energy per unit area.

The displacement of the crack surfaces can be uniquely distilled into three distinct modes. The first mode, opening Mode I, is associated with the separation of the crack faces in the direction normal to the crack direction. Mode II is associated with an in-plane shearing displacement of the crack where the shear loading is applied in the direction of the crack, resulting in in-plane relative sliding of the crack faces. The final mode, Mode III, represents tearing at the crack tip due to an out-of-plane shear displacement, causing sliding in the direction parallel to the crack front. Analogous expressions for Modes II and III have been developed (Broek, 1986) and are summarized in Williams and Beyerlein (2004).

2.5.1.2 K-Field Analysis

It is well known that the Griffth's elastic stress field contains a singularity at the crack tip, i.e. when $x = a$ in equation 2.1. Irwin (1957) and Orowan (1955) independently extended Griffith's concept by introducing a two-zone model (Figure 2.4). They considered the inelastic effects to be confined to a small inner zone. At distances greater than the inner (inelastic) zone size, but still smaller than the crack size, lies the outer zone. The singular elastic stress field obtained by Griffith still prevails in the outer zone, yet based on the outer zone size assumption, can be further simplified by considering $x \ll a$ and keeping only the leading order terms of this stress field. From this analysis evolved the stress intensity parameter, K, and the concept of the K-field for the outer zone. Inelasticity in the inner zone is assumed not to influence the elastic K-field in the outer zone, and accordingly does not alter the fracture criteria. Applying this assumption, Irwin effectively related the energy release rate G_I to the crack tip K-field

$$G_I = \frac{K_I^2}{E'} \tag{2.5}$$

which defines the stress intensity factor $K_I = \sigma_o \sqrt{\pi a}$ for Mode I. For plane stress conditions $E' = E$ where E is the Young's modulus. For plane strain E' is replaced with $E/(1 - \nu^2)$, where ν is the isotropic material Poisson's ratio. For a straight $2a$ crack, the stress intensity factors in the other two modes take on the same form, that is

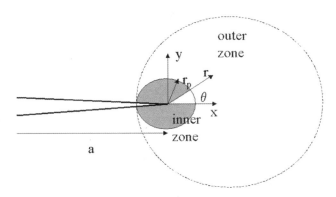

Figure 2.4 Two-zone model around the tip of a single Mode I crack in an infinite body subjected to a uniform far field loading state

$K_{II} = \sigma_{II}\sqrt{\pi a}$ and $K_{III} = \sigma_{III}\sqrt{\pi a}$ where σ_{II} and σ_{III} are the uniform applied stresses, but in Modes II and III, respectively.

Some notable efforts to remove the crack tip singularity were performed by Baranblatt (1962) and Dugdale (1960), who independently developed similar models to include cohesive forces distributed over a finite zone at the crack tip. The cohesive forces, acting along the crack plane, represent atomic attraction forces in the Barenblatt model or plastic yielding in the Dugdale model. If these cohesive forces are generally represented by a nonlinear function $p(u)$, where $2u$ is the total crack opening displacement (see Figure 2.5), then the critical crack resistance energy is

$$R_{CZ} = \int_0^\delta p(u)\mathrm{d}u \tag{2.6}$$

where δ is a critical, nonzero value of u, at which point, the cohesive stress vanishes, i.e. $p(\delta) = 0$. It is significant to point out that only the total area under $p(u)$ is important. The Barenblatt–Dugdale models retain the small scale zone assumption: the cohesive zone is much smaller than the crack size and the form of the cohesive forces do not depend on the remote loading, i.e. $p(u)$ is a material property. If $p(u)$ is associated with atomic cohesive forces, then $R_{CZ} = R_o$, the Dupré work of adhesion, and the cohesive zone theory reduces to the Griffith failure criterion (equation 2.3).

For greater detail on the theory of fracture mechanics see Dieter (1986), Anderson (1995), or Karihaloo and Xiao (2003). Also, a comprehensive reference list of current work is given in Karihaloo and Xiao (2003).

2.5.1.3 *Application to Composites*

The usefulness and validity of applying CFM to composites largely depends on the associated length scale and material behavior. However the regimes of validity are few, suggesting that for a number of reasons CFM is not a promising approach to predicting composite fracture behavior. At the macrolevel, a composite is highly anisotropic and, in some cases, heterogeneous. At the microstructural level inclusions, matrix, and interfaces with vastly different properties are revealed. Composite failure involves relatively large-scale dissipative mechanisms, such as matrix yielding, fiber pull-out, and interface debonding, that can occur around crack tips to an extent larger than the

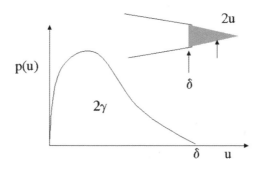

Figure 2.5   Generic traction-displacement jump response curve for a cohesive zone model

cracks themselves. These large scale nonlinear events will significantly impact the stress fields at distances far from the crack tip. This inherent response clearly violates the small scale process zone assumption of CFM or the reversible assumption of the J-integral concept of Rice (1968) (see Williams and Beyerlein, 2004 for a discussion of the J-integral approach). If the process zones are in the mm to cm range, this implies K_{Ic} fracture toughness test specimens need to be prohibitively large, of the order of meters to tens of meters in size (Argon, 2001). Cracks in composites are often small, of the order of one or a few fiber diameters, and tend to coalesce in failure. For instance, composite failure in fiber composites, unidirectional or laminates, involves stochastic random fiber fracture, or interfacial debonds, which eventually accumulate in weaker regions forming macroscopic cracks. Larger crack sizes overstress larger volumes of material, resulting in an additional statistical size effect and influence on the mode of crack propagation. Failure in composites is highly anisotropic. Even when subsequent fractures extend from a crack tip, the propagation does not always occur self-similarly, that is, the most likely position of fracture is not always adjacent to the crack tip or the fracture plane. Also failure planes are rather tortuous and follow preferred weak planes dictated by the composite microstructure. Lastly in CFM, there is a one-to-one relationship between the strength ratio and toughness ratio of the cracked material to the uncracked material. In composites there is no such relationship. In fact, the strength of unidirectional and laminate composites is found not to scale with the initial crack size, see Zweben (1973), Beyerlein and Phoenix (1997a, b) and references therein, making it impossible to determine a value of K_{Ic}.

Issues associated with classical fracture mechanics approaches are as follows:

1. The analyses must assume initial crack size, location, and orientation in the structure.
2. The near crack tip K-field is dependent upon the structural geometry and the boundary conditions only through a stress intensity factor K.
3. Extending these analytical CFM tools to a large number of cracks or cracks at the level of the microstructure is challenging.
4. CFM based results are available for different configuration of cracks.
5. Though not discussed, the J-integral (Rice, 1968) is a powerful extension to Griffith's fracture criterion for nonlinear material response, although this response must be reversible.
6. CFM has proven to be an effective approach for analyzing fracture in anisotropic elastic structures subjected to simple loading conditions. (see Williams and Beyerlein, (2004) for more detail).

2.5.2 Strength and Size Effects in Brittle Materials and Reinforcement Fibers: Weibull Model

It is widely accepted and confirmed by experiments that the tensile strength of brittle solids, such as ceramics and concrete, and brittle reinforcement fibers, such as carbon, graphite, glass, boron, and glass fibers, are not deterministic, but statistically varying. In addition, they exhibit a size effect, that is, under the same stress states larger specimens (or longer fibers) fail at lower stresses than smaller ones (or shorter fibers). The Weibull distribution (Weibull, 1951) has been the empirical model of choice to describe this strength variability and size effect and is based on the weakest link concept, i.e. a material is only as strong as its weakest link.

Following the weakest link concept, the tensile strength of individual fibers S randomly sampled and tested at gauge length l_0 are assumed to follow a Weibull distribution (Weibull, 1951).

$$F(\sigma) = 1 - exp\left(-\frac{l}{l_0}\left(\frac{\sigma}{\sigma_0}\right)^p\right) \qquad \sigma \geq 0 \qquad (2.7)$$

where $F(\sigma)$ is the probability that the fiber strength is less than σ, i.e. $F(\sigma) = P(S \leq \sigma)$, σ_0 is the Weibull scale parameter and ρ the Weibull shape parameter or modulus. The mean is $\sigma_0\Gamma(1 + 1/\rho)$ and its standard deviation is $\sigma_0(\Gamma(1 + 2/\rho) - \Gamma(1 + 1/\rho)^2)^{1/2}$, where $\Gamma()$ is the Gamma function. The mean is slightly smaller than the scale parameter σ_0 and a useful rule of thumb for the coefficient of variation CV (ratio of the standard deviation to the mean) is $1.2/\rho$. The fiber scale parameter σ_R decays with its length l, according to

$$\sigma_R = \sigma_0\left(\frac{l}{l_0}\right)^{-\frac{1}{\rho}} \qquad (2.8)$$

That is, the higher the variability in fiber strength (or the lower ρ), the stronger the size effect. The ability and accuracy of this scaling law in forecasting the strength of fibers from laboratory tests is crucial as most often one wants fiber strengths at lengths much longer or much shorter than can be tested. In addition, though Weibull fiber strengths can be rapidly obtained from laboratory tests, see for example Beyerlein and Phoenix (1996), one must perform a sufficient number of repeated tests to obtain accurate Weibull parameters (at least 50 to 100 at one length). As a rule of thumb, it is recommended that 10^n tests be performed at a given length if the probability of failure at $10^{-(n-1)}$ is desired.

Issues associated with Weibull models for brittle materials and reinforcement fibers are as follows:

1. Weibull theory for strength applies best to brittle materials, such as ceramics and reinforcement fibers, as opposed to metals, metal wires, and rope.
2. Though a standard for fibers, the Weibull fiber model is not the only possible model.
3. Weibull theory tends to over predict the size effect actually measured in materials.

2.5.3 Phenomenological Failure Theories

Phenomenological failure theories attempt to predict when a material will fail based on some simple macroscopic criteria.

The simplest of failure theories are the maximum stress or maximum strain theories. These theories state that failure occurs when the stress/strain in a given direction reaches the failure stress/strain of the material in that direction. These types of criteria do not exhibit any coupling between the different failure modes since failure in some direction occurs regardless of the values of the stress/strain field in the other directions. These types of theory are capable of accounting for differences in the tension and compression failure of a material.

The Tensor Polynomial Failure Criteria, also called the Tsai–Wu failure criteria (Tsai and Wu, 1971), have provided the basis for more advanced phenomenological failure theories. This theory defines a scalar function $f(\sigma)$ and considers failure to occur when

$$f(\sigma) = \boldsymbol{F} : \sigma + \sigma : \hat{\boldsymbol{F}} : \sigma = 1 \tag{2.9}$$

where $\boldsymbol{F}$ and $\hat{\boldsymbol{F}}$ are second-order and fourth-order tensors, respectively, related to material strength. The tensor F accounts for differences in the tension and compression response of the normal deformation modes. Mathematically these terms represent translation effects of $f(\sigma)$. The tensor $\hat{\boldsymbol{F}}$ introduces coupling effects between the different deformation modes. These terms result in $f(\sigma)$ having an ellipsoidal shape in stress space. This theory is inherently anisotropic in nature.

A recent round-robin study of various advanced failure theories (Hinton and coworkers, 1998, 2002), has shown that while many failure criteria are capable of predicting some aspects (not always the same ones) of material failure none of the currently available theories is capable of predicting the entire range of failure behavior of composite systems.

For a more detailed description of phenomenological failure models see Herakovich (1998), Mroz (2003), or the works referenced by Hinton and coworkers (1998, 2002).

Issues associated with failure theories are:

1. Experimental determination of some strength parameters for mixed-mode loading conditions in more advanced failure theories can prove difficult.
2. Phenomenological failure theories do not, in general, generate both constitutive (stress–strain) predictions and failure predictions.
3. These theories are typically not rate or history dependent.
4. The basis for these models is empirical.
5. The current state-of-the art is not robust enough to be reliable for a broad range of composite materials subjected to generalized loading states (see Hinton and coworkers, 1998, 2002).
6. Frequently these types of theories, when applied to unidirectional composites, assume a state of plane stress. In practice this assumption may or may not be appropriate.

2.6 PHENOMENOLOGICAL CONSTITUTIVE MODELING

One damage modeling approach incorporates the influence of the damage mechanisms directly into the macroscopic (phenomenological) material constitutive relations. Thus, these approaches are inherently continuum level theories. The development of such constitutive relations rests on the use of the first and second laws of thermodynamics

$$\rho \dot{e} = \sigma : \dot{\boldsymbol{\epsilon}} + \rho r - \nabla \cdot q , \ \rho T \dot{s} - \rho r + T \nabla \cdot \left(\frac{q}{T}\right) \geq 0 \tag{2.10}$$

where ρ is the density, e is the specific internal energy, σ is the stress tensor, $\dot{\epsilon}$ is the strain rate tensor, r represents a source term, q is the thermal flux, T is the absolute temperature, and s is the specific entropy.

The concepts of material state and the associated state variables are central to the development of thermodynamically derived constitutive relations. At every instant a material exhibits an internal structure. Two types of state variables (observable and internal) can be used to describe this state. The observable state variables (OSVs) are the strain (ϵ) or the stress (σ) state and the temperature T or the specific entropy s state. The strain and stress states are conjugate states as are the temperature and entropy states. The OSVs are sufficient to completely describe a material's elastic state and behavior. When a material exhibits unrecoverable, nonlinear (history-dependent) behavior it is necessary to introduce history-dependent internal state variables (ISVs) to describe completely a material's evolving deviation from the elastic internal state and behavior.

The theoretical framework is also based on the concept of the existence of a convex, single valued, thermodynamic potential, that is a function of the OSVs and ISVs which encloses their origin (Malvern, 1969). Most thermodynamically-based constitutive theory formulations employ either the Helmholtz free energy Ψ or the Gibb's free energy g as the basic potential. The following discussion will center on the use of the Helmholtz free energy Ψ.

Combining the above concepts with the conservation principles (equation 2.10), functional forms for the constitutive relations and an associated constraint equation that governs how material deformation proceeds can be obtained. For example, a Helmholtz free energy-based analysis gives the following functional constitutive relations

$$\sigma = \rho \frac{\partial \Psi}{\partial \epsilon}, \quad s = -\frac{\partial \Psi}{\partial T}, \quad \tau = \rho \frac{\partial \Psi}{\partial v} \tag{2.11}$$

and associated thermodynamic constraint equation

$$-\rho \frac{\partial \Psi}{\partial v} \vert \dot{v} - T^{-1} q \cdot \nabla T \geq 0 \tag{2.12}$$

In the above equations the terms v are the independent, kinematic ISVs while the τ (conjugate to the v) are the dependent, force-like ISVs.

In general, ISVs cannot be directly observed or measured and it becomes necessary to develop evolution equations for these terms by satisfying the thermodynamic constraint equation (equation 2.12). This is usually done through the use of a 'dissipation potential' Φ, considered to be a function of the OSVs and the ISVs. Often it is convenient to satisfy equation (2.12) in a strong sense by requiring $\Phi_m \geq 0$ and $\Phi_T \geq 0$, where Φ_m and Φ_T are mechanical and thermal effects, respectively. For rate-dependent material behavior, the functional form for the ISV evolution equations are

$$\dot{v} = -\frac{\partial \Phi_m}{\partial \tau} \tag{2.13}$$

The above discussion sets the stage for the thermodynamically-based development of different types of phenomenological constitutive relations. In order to utilize this framework it is necessary to postulate explicit forms for the different potentials. Often the basic thermodynamic potential is expressed as a second order Taylor series expansion in terms of the independent state variables. A simple example is given by the Helmholtz potential for thermoelasticity

$$\rho\Psi = \frac{1}{2}\epsilon : \ C : \ \epsilon + \Delta T\boldsymbol{\beta} : \ \epsilon - \frac{C_\epsilon}{2T_o}\Delta T^2 \qquad (2.14)$$

where $\Delta T = T - T_o$, T_o is the material reference temperature, c_ϵ is the specific heat at constant strain, $\boldsymbol{\beta}$ is a second-order tensor related the thermal expansion behavior, and C is the fourth-order stiffness tensor. Using equation (2.14) in equation (2.12) gives the classical Hookean constitutive relations.

Continuum damage mechanics (CDM) focuses on incorporating microfracture effects into the above framework. The simplest forms of CDM are based on the assumption that microfracture-induced damage can be represented in terms of scalar state variables. For example, if a single scalar damage variable D is assumed the Helmholtz free energy for thermoelasticity with damage can be written as

$$\rho\Psi = \frac{1}{2}(1 - D)[\epsilon : \ C : \ \epsilon + \Delta T\boldsymbol{\beta} : \ \epsilon^e] - \frac{c_\epsilon}{2T_o}\Delta T^2 \qquad (2.15)$$

For more detail about the development of thermodynamically based phenomenological constitutive relations see Coleman and Gurtin (1967), Lubliner (1990), Lemaitre and Chaboche (1994), or Chaboche (2003).

Issues associated with phenomenological constitutive modeling are:

1. No generally accepted set of ISVs or their evolution laws exists. Hence, there is no generally accepted set of constitutive relations for different classes of materials.
2. Experimental determination of some model parameters in more advanced phenomenological constitutive theories can prove difficult as can the physical interpretation of these parameters.
3. There is a certain 'art' in specifying the potentials in a given formulation as this approach requires some intuitive insight into what constitutes appropriate forms for these potentials.
4. These approaches can generate predictions for both the material constitutive behavior, as well as the material failure.
5. Developing phenomenological constitutive relations for anisotropic materials, particularly in the nonlinear response regime, becomes a more complex task than is the case for isotropic materials.

2.7 MICROMECHANICAL MODELING OF MATERIALS

At some length scale, all materials are heterogeneous. The previous approaches to modeling damage and behavior in materials represent continuum level analyses. Alternatively, micromechanical theories attempt to predict bulk material behavior based on a direct analysis of the material microstructure, the response characteristics of the individual phases, and the interfacial properties.

There are basically two types of micromechanical analysis. The first type, homogenization, attempts to bridge the length scales associated with a material by predicting the bulk stress–strain behavior based on a suitable averaging of the microstructural response. The other type, direct micromechanical analysis (DMA), directly links the local failure events to the response of a relatively detailed model of the microstructure without necessarily defining the bulk behavior of an 'equivalent' homogeneous material. The examples of DMA approaches presented in this chapter focus on statistical failure modeling of composites.

2.7.1 *Homogenization-Based Constitutive Modeling*

To make the connection with traditional concepts of material response, it is necessary to introduce the concept of mean and fluctuating fields, i.e. $f = \bar{f} + f'$, where f is some generic local field (such as the displacement, strain, or stress field), the mean (volume averaged) component of the field is defined by

$$\bar{f} = V^{-1} \int f \mathrm{d}V \tag{2.16}$$

and f' is the fluctuating component of the field where $\bar{f'} = 0$. The appropriate definition of the volume V in the above definitions will be discussed later.

Consistent with classical definitions of material behavior, homogenization theories (typically) assume that the bulk material response is driven by the existence of uniform bulk (macroscopic) strain or stress fields induced by one of the applied loading states

$$u = \bar{\epsilon} \cdot x , \quad t = \sigma \cdot n \tag{2.17}$$

Homogenization approaches also utilize the fundamental concept of a representative volume element (RVE). An RVE incorporates all of the pertinent aspects of the material response and geometry at the microstructural level. This generic definition for an RVE sets the scope of the associated micromechanical analysis as it determines how much information must be incorporated into the analysis. Furthermore, the characteristic length scale of the RVE must be sufficiently smaller than the characteristic structural length scale so that the mean fields can be considered to be constant within the RVE. This point implies that an RVE can only exist in situations where the field gradients for the mean fields change slowly (as is often the case away from material boundaries). A consistently defined RVE behaves the same under an applied macroscopic strain field or the equivalent applied macroscopic stress field.

Realistic, simplifying assumptions concerning microstructural arrangements are desirable since they can lead to enhanced efficiency in the analysis. A frequently used assumption is that of a periodic microstructure, which is a microstructure that can be translated in space in increments of the RVE length to obtain the entire composite microstructure. A periodic microstructure, in conjunction with the applied uniform bulk stress/strain field (equation 2.17) implies, that the local fluctuating fields are also periodic. Under these circumstances, the analysis of the RVE can be reduced to the analysis of a small number of inclusions required to adequately represent the repeating microstructure, see Figures 2.6 and 2.7 for examples.

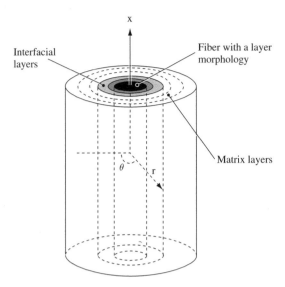

Figure 2.6   Concentric cylinder assemblage model unit cell indicating the types of microstructural detail that can be accommodated

The requirement of the existence of an RVE implies the need to satisfy a boundary value problem (BVP). The BVP governing equations are the principle of the conservation of linear and angular momentum specialized to the equilibrium case, the continuity of the displacement and tractions across interfaces between constituents, the constitutive relations for the constituents, and the appropriate boundary conditions.

There are three types of potential boundary conditions (BCs) that can be imposed within an homogenization analysis. The first two types of BCs are implied by the previously discussed uniform macroscopic fields (equation 2.17) and are simply the imposition of macroscopic deformation or stress field given by equation (2.17) directly on the boundary of the RVE. The constraint that the behavior of the RVE be the same under the different types of homogeneous boundary conditions can, typically, only be achieved in the limit as the size of the considered volume increases indefinitely for these BCs. For a volume of the microstructure that does not satisfy this behavior coincidence criteria, the response under the homogeneous displacement BC represents an upper bound estimate for the bulk behavior, while the response under the homogeneous traction BC represents a lower bound estimate. The third type of BC is based on imposing the homogeneous BCs in conjunction with the assumption of periodicity of the local fields. In this case, the homogeneous BCs are not satisfied on the boundary of the RVE but rather are satisfied as a constraint on the volume-averaged value of the local fields, see equation 2.16.

The solution of the BVP associated with a given RVE is used to define the effective constitutive relations of a composite. The first step in this process is to ensure equivalency between the applied bulk field and its local equivalent field by solving the BVP. Next the bulk response to the applied bulk field is obtained by calculating the volume average of the local response field using equation (2.16).

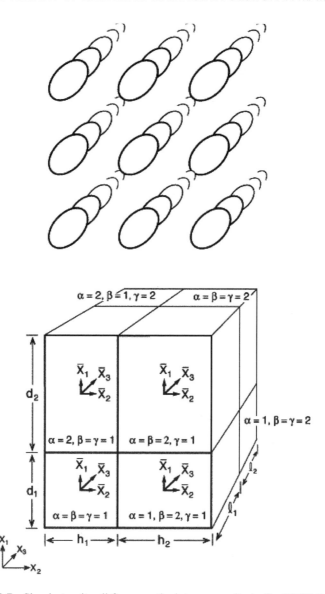

Figure 2.7 Simplest unit cell for a particulate composite in the GMC theory

Much of the initial foundation for homogenization theories was laid down by Hill (1963, 1964a,b,c, 1965, 1967). Hill recognized that the BVP outlined above for elastic composite materials resulted in solutions for the local strains of the form

$$\epsilon' = A' : \bar{\epsilon} \tag{2.18}$$

where A' is a fourth order tensor called the fluctuating mechanical strain concentration tensor. These (mechanical) concentration tensors are functions of the composite microstructure and the elastic properties of the different phases. Using equation 2.18 in equation 2.16, the following exact relations for the effective or bulk constitutive relations for a heterogeneous material can be obtained

$$C^{eff} = \overline{C} + \overline{C_i : \hat{A}_i} \qquad (2.19)$$

where $\hat{A}_i$ is the average fluctuating mechanical strain concentration tensor within phase i (see Williams, 2004 for additional details). Equation (2.19) shows that only information about the average phase response is required to obtain accurate estimates for the effective elastic properties of a heterogeneous material. It is noted that the effective properties can be anisotropic even when the component phases are isotropic due to geometric effects at the microstructural (RVE) level.

The above discussion has focused on the situation where all material behavior is elastic. If the individual phases exhibit history-dependent deformations, then the BVP outlined above has general solutions of the form

$$\epsilon' = A' : \overline{\epsilon} + \boldsymbol{d}' : \mu' \qquad (2.20)$$

where the eigenstrains μ represent history-dependent (damage) effects in the phases (see Williams, 2004 for details). The fourth-order transformation field concentration tensors $\boldsymbol{d}'$ are functions of the microgeometry of the RVE and the elastic properties of the phases. The relations for the effective elastic properties (equation 2.19) continue to hold. It can be shown that the effective eigenstress, i.e. the bulk history-dependent behavior, is given by

$$\lambda^{eff} = \overline{\lambda} + \overline{C' : \boldsymbol{d} : C^{-1} : \lambda} \qquad (2.21)$$

An important point to note is that, unlike the effective elastic properties of the composite, the determination of the effective eigenfields requires detailed information about the local (or fluctuating) eigenfields. Thus, obtaining accurate estimates for the effective eigenfields represents a substantially more difficult problem than the corresponding elastic analysis.

In summary, homogenization theories attempt to accurately predict the different types of concentration tensors (A' and $\boldsymbol{d}'$). Particular homogenization theories are discussed below. The different theories are separated into classes, such that the theories within each class have a common theme. The consequences of these themes is discussed for each class.

2.7.1.1 Simplified Homogenization Models

The dominant characteristic of the models in this section is that they do not attempt to satisfy the BVP outlined above but rather impose simple restrictions on the concentration tensors. These types of models represent the simplest forms of homogenization theories.

Two of the earliest attempts to develop homogenization theories are represented by the Voigt and Reuss models. The Voigt model assumes that all phases see the applied average strain $\overline{\epsilon}$, i.e. $A'_i = I$. This assumption satisfies the continuity of the displacements equations but violates the equilibrium constraints. The Reuss model, by assuming that all phases see the applied mean stress, satisfies equilibrium but not displacement continuity. Obviously, in the Voight/Reuss models the transformation concentration tensors are identically zero. These estimates for the effective properties represent absolute upper (Voigt model) and lower (Reuss model) bounds on the bulk elastic response. Various other simplified micromechanics theories, such as the vanishing fiber diameter model (Dvorak and Bahei-El-Din, 1982) or rule-of-mixtures type models, utilize the concepts embodied by the Voigt/Reuss models.

In general, this class of models gives only crude estimates for the effective elastic properties of a composite system. Therefore, the use of these models for damage modeling can result in highly inaccurate predictions for the bulk behavior.

2.7.1.2 Average Field Models

Average field models consider only the influence of the local average fields in the different phases on the local and global behavior. These theories can provide accurate estimates of the effective elastic properties of a composite.

The basis for most theories in this class is the pioneering work by Eshelby (1957). Eshelby considered the influence of a single inclusion in a infinite matrix of linear elastic material where the matrix was subjected to far-field loading according to equation (2.17). The presence of the inclusion results in disturbances, denoted by σ^d and ϵ^d, of the average fields $\bar{\sigma}$ and $\bar{\epsilon}$. Introducing the concept of a reference material and utilizing a Green's function-based analysis, Eshelby was able to show

$$\epsilon^d = S^E : \epsilon^*$$

$$A_i^E = (L_m - L_i)^{-1} : L_m : \left[(L_m - L_i)^{-1} : L_m - S^E \right]^{-1}$$
$$= [I + S^E : L_m^{-1} : (L_i - L_m)]^{-1} \quad \quad (2.22)$$

where ϵ^* is the eigenstrain that must be introduced to account for the differences in material properties and S^E is the so-called fourth-order Eshelby tensor. There are several important properties of the Eshelby tensor: (1) for a generalized ellipsoidal inclusion S^E is constant and, hence, all fields within the inclusion are constant; (2) S^E exhibits the minor symmetries $S_{ijkl}^E = S_{jikl}^E = S_{ijlk}^E$ but not major symmetry ($S_{ijkl}^E \neq S_{klij}^E$); (3) the tensor only depends on the elastic properties of the matrix; (4) if the matrix is isotropic it only depends on the Poisson's ratio of the matrix; (5) it is a function of the geometric properties of the inclusion; (6) for an ellipsoidal inclusion there is no coupling between the shears and the normal effects. Eshelby's analysis represents a dilute (low inclusion volume fraction) solution. The above estimate for the concentration tensor can be utilized in the previous results for the effective elastic properties in order to obtain an estimate for these properties. It is noted that Eshelby's analysis is based on linear elastic behavior for the matrix (comparison) material.

The Mori-Tanaka (MT) theory (Benveniste, 1987) represents an extension of the Eshelby analysis. The central assumption of the MT theory for a two-phase composite is that the average strains in the inclusions are related to the average matrix response through the Eshelby concentration tensor, i.e.

$$\epsilon_i = A_i^E : \epsilon_m \quad \quad (2.23)$$

where A_i^E is the concentration tensor for the inclusion obtained from the Eshelby analysis (equation 2.22). Utilizing this assumption the following forms for the matrix and inclusion concentration tensors can obtained

$$A_m = (c_m I + c_i A_i^E)^{-1}, \ A_i = A_i^E \ : \ (c_m I + c_i A_i^E)^{-1} \tag{2.24}$$

Utilizing equation (2.24) in equation (2.19) gives estimates for the effective properties (equation 2.19). Benveniste (1987) showed that the effective stiffness and compliance tensors obtained using the MT approach are each other's inverse. The MT theory has been proven analytically to represent a nondilute solution for the effective behavior in some (but not all) cases. In practice, many researchers utilize the MT approach regardless of the volume fraction of the inclusions.

While average field theories can accurately predict the effective elastic properties of heterogeneous materials, they, in general, give overly stiff predictions for the damage behavior of such materials. Exceptions to this statement are the cases where self-consistent schemes are utilized to model the response of polycrystalline materials such as a metals. Due to the low ratio of anisotropy within the individual crystals such approaches have been very successful in modeling the response of metals and even ceramics (see Kocks *et al.* 1998).

One issue associated with the use of Eshelby-based, self-consistent, schemes in modeling history-dependent material behavior is the necessary and often complex step of linearizing the constitutive relations of the constituent phases. A second issue that can arise in using these types of models is that there is often no geometric coupling between the local fields, i.e. an applied far-field normal loading does not induce shearing locally and vice versa. This lack has an impact on the evolution of the local nonlinearities and hence on the bulk nonlinear behavior. Again for materials with relatively low anisotropy ratios in the material properties, this last consideration may not be significant.

2.7.1.3 Higher-Order Homogenization Theories

This class of homogenization theory has the common ability to predict the spatially varying fluctuations in the various fields about the phase average responses. This class of theory can provide accurate estimates for both the elastic and inelastic behavior of composites. These abilities arise from the attempts of these models to solve the governing equations of continuum mechanics directly which in turn gives estimates for the spatially varying fields and it is in this sense that they are referred to as 'higher-order' theories.

Some of the simplest models of this class are the concentric cylinder assemblage (CCA) and concentric sphere assemblage (CSA) models. Both of these approaches consider idealized microstructures composed of a single inclusion (either a continuous fiber or a spherical particle, respectively) embedded in an annulus of matrix material (Figure 2.6), with the outer boundary of the resulting layered structure being subjected to equation 2.17. These approaches cannot consider periodic boundary conditions. The microstructural idealization arises due to the assumption that the composite volume can be completely filled by an infinite number of these simple microstructures by varying the the outer radius of the assemblage down to an infinitesimal size, while maintaining constant volume fractions of phases.

As an instructive example, the basic equations for the CCA model for axial shear loading for inelastic behavior of the phases are presented (Williams and Pindera, 1997). The geometry of the problem consists of a fiber aligned along the x_1 axis for the polar coordinate system x_1, r, and θ, (Figure 2.6). The assemblage can consist of an arbitrary number of layers, denoted by K, of either fiber or matrix material. Individual layers are denoted by k. The layers must have at least orthotropic symmetry. The external BCs for this problem are

$$u_1 = \bar{\epsilon}_{12} r_K \cos\theta \ , \ u_r = \bar{\epsilon}_{12} x_1 \cos\theta \ , \ u_\theta = -\bar{\epsilon}_{12} x_1 \sin\theta \qquad (2.25)$$

The interfacial continuity conditions between layers must also be satisfied where it must be recognized that these conditions are functions of the angular coordinate θ. Furthermore, the displacements are required to be bounded at the core, $r=0$. Based on these constraints the displacement field within a layer k is given by

$$u_1^k(r,\theta) = \phi^k(r,\theta) - \bar{\epsilon}_{12} r \cos\theta \ , \ u_r^k(r,\theta) = \bar{\epsilon}_{12} x_1 \cos\theta \ , \ u_\theta = -\bar{\epsilon}_{12} x_1 \sin\theta \qquad (2.26)$$

The term ϕ^k represents the fluctuating effects in the axial displacement component induced by the presence of the microstructure. Substituting the above form for the displacement field into the equilibrium equation gives the governing differential equation

$$\frac{1}{r}\frac{\partial}{\partial r}\left(r\frac{\partial\phi^k}{\partial r}\right) + \frac{\eta}{r}\frac{\partial^2\phi^k}{\partial\theta^2} = e \qquad (2.27)$$

where $\eta = C_{1\theta1\theta}/C_{1r1r} = G_{1\theta}/G_{1r}$ is a measure of the anisotropy of the layer, and $e = 2\nabla \cdot \epsilon^I$. The development and evolution of the inelastic strains ϵ^I are determined through the use of a submodel for the material behavior at a point within each layer. The solution equation (2.27) can be obtained using Fourier series expansions for ϕ^k and e and subsequently applying standard solution techniques for the ordinary differential equations. The final solution form can be cast in the functional form $Cu = F\bar{\epsilon} + f$ where the coefficient matrices C and F are known functions of the material and geometric properties for the different layers, the forcing vector f is a function of the layer properties and the inelastic strain distributions within the layers, and u is the vector of unknown.

The CCA model discussed above is based on the analysis of rather specialized microstructures. More general and flexible approaches to modeling the behavior of composite materials are the method of cells (MOC) (Aboudi, 1991) and the generalized method of cells (GMC) (Paley and Aboudi 1992, Aboudi, 1995). In these theories the composite microstructure is represented by two- or three-dimensional (2D or 3D, respectively) RVEs discretized into an array of rectangles or rectangular parallelepipeds. Each of these subregions, called a subcell, is considered to be composed of a single phase (Figure 2.7). An arbitrary number of phases can be considered. The number of subcells in each direction is arbitrary. The 3D GMC analysis starts with a linear expansion for the displacement field within each subcell given by

$$u_i^{(\alpha,\beta,\gamma)} = \bar{\epsilon}_{ij} x_j + w_i + y_1\phi_i^{(\alpha,\beta,\gamma)} + y_2\xi_i^{(\alpha,\beta,\gamma)} + y_3\psi_i^{(\alpha,\beta,\gamma)} \qquad (2.28)$$

where α, β, and γ denote the subcell in the x_1, x_2, and x_3 directions and the y_i denote local coordinates within each subcell. The terms $\phi_i^{(\alpha,\beta,\gamma)}$, $\xi_i^{(\alpha,\beta,\gamma)}$, and $\psi_i^{(\alpha,\beta,\gamma)}$ denote fluctuating displacement effects induced by the presence of the microstructure. Based on this form for the displacement field, the subcell strain field is the sum of the applied average strains and fluctuating strains $\mu_{ij}^{(\alpha,\beta,\gamma)}$, i.e. $\epsilon_{ij}^{(\alpha,\beta,\gamma)} = \bar{\epsilon}_{ij} + \mu_{ij}^{(\alpha,\beta,\gamma)}$ where $\mu_{ij}^{(\alpha,\beta,\gamma)}$ is a function of $\phi_i^{(\alpha,\beta,\gamma)}, \xi_i^{(\alpha,\beta,\gamma)}$, and $\psi_i^{(\alpha,\beta,\gamma)}$. The strains and, hence, the stresses within each subcell are spatially uniform. This fact implies that the equilibrium equations within each subcell are satisfied identically. Therefore, it remains to satisfy the interfacial constraints between the subcells. The displacement and traction continuity conditions are satisfied in an integral (average) sense across the interfaces of the subcells, as well as between RVEs, i.e.

$$\int_A u_i^- \, \mathrm{d}S = \int_A u_i^+ \mathrm{d}S, \quad \int_A \sigma_i^- \, \mathrm{d}S = \int_A \sigma_i^+ \mathrm{d}S \qquad (2.29)$$

where $\int_A$ denotes the integral across the surfaces of a subcell face, $u_i^\pm$ denotes the displacement components on either side of an interface, and $\sigma_i^\pm$ denotes the stress components on either side of an interface. The stresses in the traction continuity equations are rewritten in terms of the fundamental kinematic unknowns $\phi_i^{(\alpha,\beta,\gamma)}, \xi_i^{(\alpha,\beta,\gamma)}$, and $\psi_i^{(\alpha,\beta,\gamma)}$ and any damage effects (through the use of constitutive models that describe the material behavior within each subcell). A final set of governing equations of the form $C\mu = F\bar{\epsilon} + f$ can be obtained. The coefficient matrices C and F are a known functions of the material and geometric properties for the different layers, the forcing vector f is a function of the layer properties and the inelastic strains, and μ is the vector of unknown strains in all of the subcells. This approach is formulated independently of any particular set of constitutive assumptions concerning the behavior of the component phases and hence can incorporate any desired set of such relations. The MOC and GMC approaches have been shown to provide accurate estimates for the elastic and inelastic behavior of multiphase composites.

Various other approaches in this class of homogenization theory include Green's-function-based theories (and associated theories that employ Fourier series techniques), asymptotic-expansion-based theories, and bounding analyses. For greater detail about different higher order homogenization approaches see Christensen (1979), Bensoussan *et al.* (1978), Aboudi (1991), Nemat-Nasser and Hori (1993), and Torquato (2001).

Issues associated with homogenization theories are:

1. These theories provide information about the local fields and failure processes in a heterogeneous material.
2. At times the information about the behavior of the constituents and the interfaces required by these models can be difficult to obtain. However, once models for the behavior of the constituents and the interfaces have been specified the only variables are the descriptions of the material microstructure.
3. These types of theory can naturally account for anisotropic effects.

2.7.2 Statistical Analyses for Fiber Composites

This section focuses on the development of microstructural analysis tools for determining the statistical strength and toughness of composite materials. These microstructural models appreciate features and failure mechanisms at the length scale of the individual phases in the composite.

2.7.2.1 Load Sharing Among Fractures

The dominant failure mechanisms in composites are fiber and matrix failure. Even an isolated fracture in a large composite does not go unnoticed. The load released by a fracture or fractures gets redistributed and overstresses the surrounding intact fibers and matrix material. Knowing the resulting internal stresses or strains is the first step to predicting composite strength and toughness (as well as other mechanical properties, such as fatigue strength). In this section, we will review analytical stress transfer models which attempt to solve this problem, as opposed to purely numerical schemes, such as finite element and finite difference.

The first composite strength models utilized idealized load sharing rules. They are referred to as 'rules', because stress states can be quickly calculated and they have very few mechanics considerations except equilibrium. The two most well known rules are the idealized local load sharing (ILLS) rule and equal load sharing (ELS) rule. These rules represent the two extremes in the load sharing spectrum.

In ELS, a finite number of fibers needs to be considered. Let σ be the axial tensile stress applied to all fibers. If i out of n fibers are broken on any plane, then the stress on the $(n-i)$ surviving fibers is equal and is $K_i\sigma$, where

$$K_i = \frac{n}{n-i} \qquad i = 1, 2, 3, \ldots, n-1 \qquad (2.30)$$

This equation states that all fibers equally share the load from the broken fibers. The ELS rule best applies to a dry bundle, where there is no matrix or interface friction. In ELS, composite size n does matter.

In the other extreme, the ILLS model, the fiber packing arrangement becomes important, but not the total number of fibers n or composite size. For planer (2D-type) composite ILLS states that the load of a failed fiber is redistributed equally only onto its two nearest neighbors. Thus an intact fiber adjacent to r breaks is $K_r\sigma$, where K_r is

$$K_r = 1 + \frac{r}{2} \qquad r = 1, 2, 3, \ldots \qquad (2.31)$$

In 3D, early ILLS versions (Smith et al., 1983) were not analytically convenient, nearly defeating the purpose of defining a rule. In Phoenix and Beyerlein (2001), a simple rule is introduced. This 3-D analysis considers a tight round cluster of *dimensionless* diameter D. The number of broken fibers in this cluster is approximately $\sim \frac{\pi D^2}{4}$ and the load is distributed equally on the nearest ring of intact fibers, $\sim \pi D$ in number. The stress concentration factor on these fibers is approximately

$$K_r = 1 + \frac{D}{4} \qquad (2.32)$$

Though the ELS and ILLS rules are highly simplified, they have been very successful in giving insight into statistical failure processes because they concentrate on probable sequences of fiber fracture progression rather than complicated computation of stress concentration factors (See Phoenix and Beyerlein, 2001, for review, and Mahesh *et al.*, 2002).

Another set of stress transfer models which incorporates more mechanics considerations than idealized load sharing rules are shear lag models. When one refers to elastic shear lag models for composites two versions come to mind: the single fiber shear lag model of Cox (1952) and the multiple fiber shear lag model of Hedgepeth (1961). The common features of these models are that each fiber is considered to be one-dimensional and capable of sustaining only axial stresses along its length and that the matrix serves to only transfer load from a fiber discontinuity, such as a fiber end or a fiber fracture, to the rest of the fiber or to other fibers. In the Hedgepeth model, interactions between multiple fibers are considered and therefore this model and its many extensions have more potential for determining the strength of fiber composites than the single fiber approach.

Figure 2.8(a) shows the problem considered by Hedgepeth, a unidirectional fiber lamina, containing an infinite number of infinitely long fibers, is loaded in tension parallel to the fiber direction and has a central transverse crack of r contiguous fiber fractures. The center fiber is numbered $n = 0$ and the fibers to the right are numbered $n = 1, 2, ..., \infty$ and those to the left, $n = -1, -2, ..., -\infty$. The matrix 'bay' to the right of fiber n is denoted matrix bay n. Likewise, the fiber coordinate axis is $+x$ above the crack and $-x$ below the crack. The fiber and matrix are both linear elastic. In accordance with the shear lag assumptions, the fiber deforms only in simple tension or compression and the matrix deforms only in shear, transmitting load from the broken fibers to the intact fibers. The Young modulus of the fibers is E_f and G is the effective

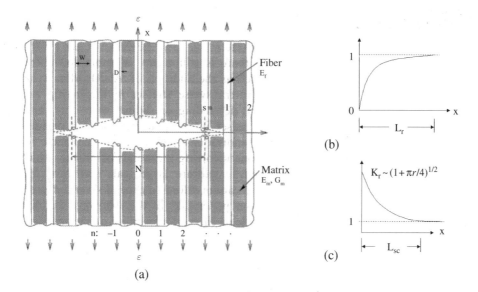

Figure 2.8   (a) Two-dimensional unidirectional composite containing a central crack of r contiguous fiber breaks along x. Typical shear lag profiles of stress concentration along (b) a broken fiber and (c) a neighboring intact fiber near a fiber break. (a) Taken from Beyerlein and Landis (1999)

shear modulus of the matrix. Also w is the surface-to-surface fiber spacing, h is the effective matrix thickness or the out-of-plane length of load transfer which is on the order of the fiber diameter or lamina thickness, and A_f is the cross-sectional area of the fiber. These assumptions simplify the analysis considerably as the equilibrium equations become decoupled and therefore the stresses found in any given fiber or matrix region depend only on the fiber axial coordinate x.

Computation is needed to solve the Hedgepeth stress field for any x, n, and r (Hedgepeth, 1961). Fortunately, there are some analytical expressions for certain cases. For instance, Hedgepeth is most quoted for his result on the stress concentration on the fiber immediately adjacent to the last broken fiber of a row of r breaks, denoted K_r

$$K_r = \frac{4 \cdot 6 \cdots (2r + 2)}{3 \cdot 5 \cdots (2r + 1)} \quad r = 1, 2, 3, \ldots \tag{2.33}$$

Later Hikami and Chou (1990) obtained in closed form the stress concentration factor, $K_{r,s}$, on fiber s ahead of the last broken fiber $n = r$ along the crack plane.

$$K_{r,s} = (r + 2s - 1)\frac{(2s)(2s + 2) \ldots (2s + 2r - 2)}{(2s - 1)(2s + 1) \ldots (2s + 2r - 1)} \quad r, s = 1, 2, 3, \ldots \tag{2.34}$$

Beyerlein et al. (1996) derived the following approximation to equation (2.33):

$$K_r \simeq \sqrt{1 + \frac{\pi r}{4}} \tag{2.35}$$

which has less than 0.2 % error for $r = 2$, an error that decays rapidly as r increases, as well as the following approximation to equation (2.34):

$$K_{r,s} \simeq \sqrt{1 + \frac{\pi r}{4}\left(\frac{s}{2s - 1}\right)}\sqrt{\frac{1}{1 + \frac{\pi(s-1)}{4}}} \tag{2.36}$$

A simpler but less accurate version of equation 2.36 is:

$$K_{r,s} \simeq \sqrt{\frac{1 + \frac{\pi r}{4}}{1 + \pi(s - 1)}} \quad s = 1, 2, 3, 4, \ldots \tag{2.37}$$

which differs from equation (2.36) by only 1.5 % for $s = 2$.

Hedgepeth and Van Dyke (1967) were the first to extend the Hedgepeth (1961) analysis to a three-dimensional packing of fibers. In Phoenix and Beyerlein (2001), analogs of equations (2.35) and (2.37) were also derived for the 3-D fiber packing arrangements. In 3-D, fiber break clusters form rather noncircular and amorphous shapes, particularly as the variability in fiber strength, ρ, decreases. Regardless, such clusters are idealized as circular with diameter Dd, where d is the center-to-center fiber spacing and thus D is a dimensionless diameter. For the ring of fibers immediately surrounding this circular cluster of r breaks, a stress concentration factor is

$$K_r \simeq \sqrt{1 + \frac{D}{\pi}} \tag{2.38}$$

Note that equation 2.38 is an approximation for the effective average stress concentration, which ignores fluctuations in stress around the cluster periphery. For a fiber a distance sd from the edge of the cluster, where s is the number of effective fiber spacings d and where $d = w' + h'$, the stress concentration in the plane of the cluster r is

$$K_{r,s} \simeq \sqrt{\frac{\frac{1+D}{\pi}}{1 + \pi(s-1)}} = \frac{K_r}{\sqrt{1 + \pi(s-1)}} \qquad s \geq 0, \; D \gg s \qquad (2.39)$$

Shear lag models are generally believed to be adequate when the fiber carries most of the load, i.e when fiber volume fraction V_f of the composite is high or when the fiber and matrix moduli are not comparable. A general rule of thumb is $E_f V_f / E_m (1 - V_f) \gg 1$, which applies to nearly all PMCs. Having said this, the full 3-D finite element calculations, performed by Reedy (1984) on a planar composite lamina with $V_f = 0.5$ were in excellent agreement with shear lag predictions, even when extensive yielding was included.

One of the advantages of the shear lag model is that it shows promise in modeling the essential physics of extensional deformation of the matrix, as well as inelastic matrix and interface deformations, random fiber fractures, fiber pull-out and crack propagation (Phoenix and Beyerlein, 2001; Williams and Beyerlein, 2004). The reader is referred to Chou (1992), Phoenix and Beyerlein (2001), and Beyerlein *et al.* (1996), which give comprehensive summaries and highlights of the many variations of shear lag analyses to date and include extensive references.

2.7.2.2 Composite Fracture Mechanics: Connecting LEFM and Shear Lag Modeling

As fiber fractures accumulate into a large macrocrack or as laminate plies tend to delaminate, it is certainly not surprising that many have turned to the field of fracture mechanics to understand composite failure. A brief overview of LEFM and issues surrounding its application to composites have already been given in Section 5. Here an application of LEFM for modeling composites is presented.

To make comparisons between shear lag models and LEFM in Beyerlein *et al.* (1996) and Beyerlein and Landis (1999), a few continuum length scales need to be related to those of the composite microstructure. Phoenix and Beyerlein (2001) showed that the continuum crack length $2a$ is best related to the total number of fiber breaks contained in the initial crack r by $2a = (r + 1 - \frac{2}{\pi})d$ and the distance from the crack tip along the crack plane $\bar{r}$ by $\bar{r} = (s - 1 + \frac{1}{\pi})d$, where d is the center-to-center spacing and s is the center positions of the fibers ahead of the crack tip fiber. Consistent with shear lag assumptions, the tensile carrying capability of the matrix is neglected and the effective tensile stress becomes $V_f \sigma_f$, where σ_f is the fiber stress. In a composite, the tensile stress distribution along the crack plane decays according to

$$V_f \sigma_f \simeq \frac{K_I}{\sqrt{2\pi r}} \qquad (2.40)$$

Using the composite definitions for $\bar{r}$ and $2a$ above one can obtain

$$K_I \simeq V_f \sigma_f \sqrt{\pi \left(a + \frac{d}{2} \right)} \tag{2.41}$$

where K_I is the mode I stress intensity factor for a crack in a continuum composite, which compares well to $\sigma \sqrt{\pi a}$ for a continuum. Thus the discrete shear lag model exhibits the characteristic square root of $\bar{r}$ decay of classical linear elastic solutions.

In order to investigate the implications of a crack on composite strength and toughness, we now consider the strength of the fibers. If the fiber strength is deterministic (i.e. $\rho = \infty$), then the strength of the cracked composite defined by the failure of the fiber with the highest stress concentration is

$$\sigma_r = \frac{V_f \sigma_s}{\sqrt{1 + \frac{\pi r}{4}}} \tag{2.42}$$

and the critical Mode I stress intensity factor is

$$K_{Ic}^{\infty} = \sigma_r \sqrt{\pi \left(a + \left(\frac{3}{\pi} - \frac{1}{2} \right) d \right)} \tag{2.43}$$

If the fiber strength is not deterministic but is, instead, statistical, following, for instance a Weibull distribution (Weibull, 1951), with ($2 < \rho < 20$), then additional considerations need to be made. Assume that an initially uncracked composite (i.e., $r=0$) fails when a cluster of fiber breaks forms and reaches a critical size k, meaning there are k fiber breaks contained in the cluster. If the size of the initial crack r is smaller than the critical cluster size k, i.e. $r < k$, then the initial crack does not change the strength of the composite from that of an initially uncracked composite (Beyerlein and Phoenix 1997a, b). However, if $r > k$, then the composite strength distribution depends on the initial crack size r. Beyerlein and Phoenix (1997b) derived the strength distribution of an initially cracked composite, allowing for crack extension from either end of the crack to the point of instability. The reader is referred to Beyerlein and Phoenix (1997b) for the form of the distribution, though it is mentioned that the distribution is non-Weibull and similar in form to that for an uncracked composite. The analogous composite strength to the deterministic value in equation 2.42 or formally, the stress $\sigma(r, P)$ to cause failure at probability level P with an r cluster is

$$\sigma(r, P) = \sigma_r \left(\frac{1}{2} \ln \left(\frac{r + \frac{4}{\pi}}{\frac{\rho}{2} \ln \left(\frac{1}{P} \right)} \right) \right)^{\frac{1}{\rho}} \tag{2.44}$$

The critical Mode I stress intensity factor becomes

$$K_{Ic} \simeq K_{Ic}^{\infty} \left(\frac{1}{2} \ln \left(\frac{\frac{2a}{d} + \frac{(\sigma - \pi)}{\pi}}{\frac{\rho}{2} \ln \left(\frac{1}{P} \right)} \right) \right)^{\frac{1}{\rho}} \tag{2.45}$$

Thus the fracture toughness of a composite lamina with Weibull fibers depends on the modulus ρ (see equation (2.7)), failure probability of interest P, and a. The dependence on a suggests that the so-called R-curve toughening prevails and is due to the variability in fiber strength as the occasional relatively strong fibers encountered by the growing crack can pin the crack (Beyerlein and Phoenix 1997a, b).

2.7.2.3 Probability Models for Fiber Composite Strength

In this section, analytical and numerical probability models for composite strength, which account for the complexities in modeling the stochastic progression of fiber fractures up to final failure are reviewed. In principle, these models provide a statistical distribution for composite strength, say $P(\sigma)$, which is the probability of failure at a given stress σ or the probability that the strength of the composite is less than or equal to σ. The parameters of the distribution will depend on the five considerations discussed in Section 2, namely the types of damage and their interactions, loading conditions, microstructure and the composite material system. Furthermore, if the strength distribution depends on the composite volume, then the model predicts a size effect as well.

Some of the first statistical models developed used the load sharing rules (ELS and ILLS) presented in Section 2.7.2.1. Idealized rules are still being developed and used in composite strength models. Some other versions are tapered load sharing (TLS) (Phoenix and Beyerlein, 2000; Mahesh and Phoenix, 2003) and global load sharing (GLS) (Curtin, 1991a, b; Hui *et al.*, 1995) and elastic global load sharing (GLSE) (Phoenix and Beyerlein, 2001). In the TLS rule, the first and second nearest neighbors share the load (Phoenix and Beyerlein, 2000). Though a seemingly trivial extension of the ILLS rules the alterations in the failure progression due to TLS are nontrivial. Theories for composite strength developed under TLS can be found in Phoenix and Beyerlein (2000) and Mahesh and Phoenix (2003). Unlike the other rules introduced here, GLS and GLSE consider the longitudinal unloading of stress from the fiber fracture via slip (in GLS) or elastic matrix deformation (in GLSE) yet load transfer from fiber to fiber still occurs in an ELS-like fashion transverse to fiber direction. GLS is used frequently in the study of ceramic matrix composites with very weak interfaces. In such composites, fiber fragmentation is the prevalent failure mode and thus load transfer along the fiber is important more so than that from fiber to fiber. Several composite strength theories have been developed using GLS or GLSE and they are reviewed and their predictive capabilities are compared in Phoenix and Beyerlein (2001).

Statistical models using ELS or GLS for load transfer best apply to composites, which display a 'dispersed' or 'ductile-like' failure, whereas those using LLS type models (e.g. ILLS, ILS) best apply to composites, which display a 'brittle-like' failure.

Early statistical models, quite sophisticated in formulation, utilized the chain of bundles concept first introduced by Rosen (1964) where a composite is viewed as a chain of m independent and identically distributed bundles. Even to date, this chain of bundles model has seen considerable success in modeling composite strength and size effect, for all models of load transfer, assuming ELS (Rosen, 1964; Smith, 1982; Harlow and Phoenix, 1978a, b), GLS (Phoenix and Raj, 1992; Phoenix *et al.*, 1997, Phoenix and Beyerlein, 2001), LLS type models (e.g. ILLS shear lag) (Batdorf, 1982; Phoenix and Beyerlein, 2001; Phoenix and Smith, 1983; Smith, 1980, 1983; Zweben and Rosen, 1970), and Hedgepeth (1961), and Hedgepeth and Van Dyke (HVD) shear lag type load sharing (Beyerlein and Phoenix 1997a, b; Mahesh *et al.*, 2002). A comprehensive review dedicated to these sorts of models is found in Phoenix and Beyerlein (2001). A review of some of these models is given starting with the earliest one by Rosen (1964), who used for the strength of individual bundles the classic result of Daniels (1945) assuming ELS and ending with the most recent ones utilizing LLS.

2.7.2.4 Analytical Modeling under Dispersed Type Loading Sharing

Let G_n be the strength of the bundle with n fibers under ELS. Daniel's classic result (1945) is that as the bundle size n increases, G_n follows a Gaussian (normal) distribution with mean μ and standard deviation γ_n; the proof of which is difficult. In his result, μ was independent of n while γ_n was a function of n. Smith (1982) derived a correction term to μ which depends on n, resulting in a new mean, μ_n. The resulting Gaussian approximation for the bundle strength distribution function can be written as

$$G_n(\sigma) = \Phi\left(\frac{\sigma - \mu_n}{\gamma_n}\right) \quad \sigma \geq 0 \tag{2.46}$$

where

$$\Phi(z) = \frac{1}{\sqrt{2\pi}} \int_{-\infty}^{z} exp\left(-\frac{t^2}{2}\right) dt \tag{2.47}$$

For Weibull fibers, μ_n and γ_n depend on n, ρ, and the Weibull scale parameter. Several corrections and improved expressions for μ_n and γ_n have been introduced since Daniel (1945) and Smith (1982). The reader is referred to Phoenix and Beyerlein (2001) for these modified forms of μ_n and γ_n. Generally, as ρ and n increase, μ_n and γ_n decrease.

For a composite viewed as a chain of m bundles with overall length $L = m\delta_R$, the strength of this chain is equal to that of its weakest bundle (Rosen, 1964). Stated another way, the chain survives if and only if all the bundles survive the applied stress. The probability of survival of the bundle is $1 - G_n(\sigma)$ and that of all m bundles is $(1 - G_n(\sigma))^m$. Thus the probability of chain failure at stress s is

$$H_{m,n} = 1 - (1 - G_n(\sigma))^m \tag{2.48}$$

forecasting a size effect (or dependency on n and m) in composite strength. Asymptotic analysis for large n yields a double exponential (Smith, 1982).

$$H_{m,n} = 1 - exp\left(-exp\left(\frac{\sigma - b*}{a*}\right)\right) \tag{2.49}$$

where

$$a* = \frac{\gamma_n}{\sqrt{2ln(m)}}$$

$$b* = \mu_n - \gamma_n\left(\sqrt{2ln(m)} - \frac{ln(ln(m)) + ln(4\pi)}{\sqrt{8ln(m)}}\right)$$

$$\rho* = \frac{b*}{a*}$$

and where the parameters of these distributions depend additionally on m. Though geometry and load sharing are idealized, equation (2.46) and (2.49) do, however, illustrate one important fact: as composite failure is not a linear process, the distribution of composite strength is not generally equal to the distribution for fiber strength.

2.7.2.5 Analytical Modeling under Localized Load Sharing (LLS)

Like the models described in the previous section, the chain of bundles concept is also applied in modeling composite strength under localized type load sharing (LLS); however, in determining individual bundle strength some choices need to be made. First there is a choice of LLS models; such as ideal localized load sharing (ILLS) rule and shear lag models (see 2.7.2.1), including extensions for nonlinear matrix and interface deformation. In using these types of LLS schemes, models seek to represent composites which tend to fail close to the localized failure limit. With the stress transfer model selected, one can either model the failure process (a) computationally, by numerically calculating stress transfer and nucleating new fractures at each step, or (b) analytically using stress transfer rules and idealizing the fracture process. For (a), simulation techniques have often used Monte Carlo methods (Landis *et al.*, 1999a; Beyerlein and Phoenix, 1998a; Mahesh *et al.*, 1999). General descriptions of the algorithmns and other references are given these works. Second, modeling bundle strength under LLS involves making certain assumptions about how the fractures propagate to failure in a single bundle of fibers n, whether in-plane or out-of-plane, consecutive or random, etc. Typically simulations make fewer assumptions on when and where fiber fractures occur than probability models and provide guidance and validation for analytical probability models (Phoenix and Beyerlein, 2001; Beyerlein and Phoenix, 1997a, b; Mahesh *et al.*, 2002; Ibnabdeljalil and Curtin, 1996).

In the localized failure limit (correlated, brittle-like), failure begins with random fractures, some of which eventually grow into larger clusters and only one of which will culminate into a critical cluster of size k in a statistically weak region of the composite. The concept of critical cluster size k is interpreted as follows: the composite does not fail if it contains a cluster of size less than k, the failure stress is the minimum stress at which a cluster grows to a size k, and if increased an infinitesimal amount above this minimum stress, there will be only one k cluster which will become unstable and fail the composite. Composites with lower failure stress, more volume, or more variability in fiber strength (ρ decreases) tend to require larger k values.

As an example, asymptotic analyses for large composites are presented. In this limit, Harlow and Phoenix (1978a, b) proposed that H_{mn} will have the following form

$$H_{m,n} = 1 - exp(-mnW(\sigma)) \quad \sigma \geq 0 \qquad (2.50)$$

where $W(\sigma)$ is a characteristic distribution function. Contrary to popular belief, $W(\sigma)$ is not of the Weibull type. For a planar composite, Beyerlein and Phoenix (1997a, b) proposed that $W(\sigma)$ follows $W(\sigma) = W^*(\sigma)\Psi_2(\sigma)$, where $W^*(\sigma)$ is the component representing formation of a critical cluster and $1 - \Psi_2(\sigma)$ is the probability that a cluster achieves a critical size, and stalls due to a local increase in fracture resistance. Figure 2.9 compares results of this probability model with Monte Carlo simulation results for various values of ρ. These distributions are plotted on Weibull coordinates for $H_{m,n}$ vs applied stress normalized by σ_δ

$$\sigma_\delta = \sigma_l \left(\frac{\delta}{l}\right)^{-\frac{1}{\rho}} \tag{2.51}$$

Weibull coordinates are scaled such that a Weibull distribution would plot as a straight line.

The Monte Carlo simulation results in Figure 2.9 reveal a few interesting facts. First, judging from their non-linear form on Weibull coordinates, the composite distribution is not Weibull, though the fiber strengths are Weibull. Second, the variability in composite strength is much less than that of the fiber strength, and the composite is much weaker than a fiber element of the same length. Lastly, and perhaps nonintuitively, the variability in fiber strength affects the mean composite strength much more than it does the variability in composite strength.

The agreement between the simulations (dots) and the model utilizing $W(\sigma) = W^*(\sigma)\Psi_2(\sigma)$ in equation (2.50) (solid lines) is excellent even for $\rho = 3$ (large variability in fiber strength). The dashed lines are the model utilizing $W(\sigma) = W^*(s)$ only which provides a conservative estimate for all values of ρ. The comparison suggests that $\Psi_2(\sigma)$, the probability representing toughening by fiber strength variability only has significance when $\rho < 5$.

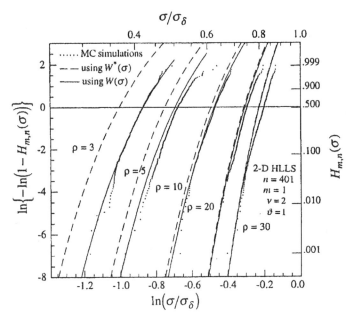

Figure 2.9 Distributions for composite strength under Hedgepeth shear lag load sharing in a planar unidirectional fiber composite. Analytical probability models for $W(b)$ and $W^*(b)$ from Beyerlein and Phoenix (1997b) are compared to Monte Carlo simulation results. Taken from Phoenix and Beyerlein (2001)

This method was also used to develop the distribution function for the strength and creep lifetimes of a 3-D composite (Phoenix and Beyerlein, 2001; Mahesh *et al.*, 2002; Mahesh and Phoenix, 2004). The analytical results compared favorably with distributions obtained from Monte Carlo simulations.

2.7.2.6 Ductile–Brittle Transitions in Fiber Composite Failure

Many concepts and issues in modeling failure of composites can be seen in simple unidirectional continuous fiber composites. Two extreme types of failure have evolved, namely, brittle-like and ductile-like. These extremes have also been variously called the localized vs dispersed failure limits (Phoenix and Beyerlein, 2001) or the concentrated vs dilute fiber limits (Argon, 2001). Based on the discussions in this chapter, it is now possible to extend this list of sources for material failure behaviors from just simple material considerations to modeling and microstructural parameters. Ductile fracture results from (i) a weak fiber-matrix interface, (ii) a matrix with a low yield strength compared to its stiffness, (iii) fibers with large variability in strength, (iv) a three dimensional load sharing geometry (as opposed to a planar lamina or tape), and (v) small volume composites. Brittle fracture results from generally the opposite of (i)–(v). Naturally most composite failure modes fall between these two extremes. These lists suggest that the fundamental basis for failure mode is the volume of overstressed material relative to the volume of damage.

Issues associated with statistical analyses for composites are

1. Approaches discussed here have only been applied to unidirectional fiber composites, in which loading is parallel to the fiber axis and monotonic.
2. As composite failure is not a linear process, the distribution of composite strength is not generally equal to the distribution for fiber strength.
3. The Monte Carlo simulation results reveal that (i) the composite distribution is not Weibull, though the fiber strengths are Weibull, (ii) the variability in composite strength is much less than that of the fiber strength, and the composite is much weaker than a fiber element of the same length, and (iii) the variability in fiber strength affects the mean composite strength much more than it does the variability in composite strength equal to the distribution for fiber strength.
4. Analytical probability distributions have been developed which compare favorably to Monte Carlo simulations. These distributions are useful in reliability calculations and provide scaling rules to relate the strength of composites of different sizes.
5. There are no size limits in these scaling rules; however, they have not been validated by Monte Carlo simulations beyond a few thousand fibers and by experimental data.
6. In statistical composite modeling, the damage mechanisms and stress analyses must be determined *a priori* and tend to be specific to ceramic matrix, metal matrix or polymer matrix composites.

2.8 SUMMARY

Accurate material models are indespensible in the development and failure analysis of advanced structures as well as the development of structural health monitoring schemes. There are many issues and considerations that drive the choice and development of what constitutes an appropriate material for a given application. The aim of this chapter was to expose the reader to the general issues and considerations that need to be addressed in making this choice. In doing so, the characteristics, strengths, and weaknesses of material models from a number of different classes of theory, both continuum and micromechanical in nature, were reviewed. The examples covered in this chapter generate by no means a comprehensive list of the models but serve to emphasize certain issues. Some important problems that were not directly addressed are high temperature environments, inertial effects, fatigue, fracture using cohesive zone models, and lifetime predictions under constant or cyclic loading. Several broad classes of models not covered were those applying to length scales lower than a micron, such as dislocation modeling, molecular dynamics, atomistics, and molecular modeling of polymers, to name a few. Nonetheless, it is hoped that the resulting work has provided the reader with some appreciation of both the scope of current material models available, as well as a starting point for choosing and developing models for particular applications.

Some final points deserve mentioning. There is a general concensus that there is no unified model for all applications and materials. In fact, it may be most appropriate to utilize two (or more) models in tandem for the multiscale analysis of a given material/ structure system, in using one model to 'guide' another, or in using different approaches at the same length scale. Finally, as outlined above every model has strengths and weaknesses and no one currently available material model may be able to predict all of the phenomena experimentally observed in the material/structural responses. In this case, it may be necessary to employ a suite of models to be able to confidently quantify all aspects of the material/structural behavior.

Despite the fact that nearly all analytical models are idealized compared to the actual material, they do provide guidance in experimental design, insight for invention of new and better materials, and can serve as building-block models in developing large-scale numerical models.

Finally, no model can be used for structural design without some experimental input. In fact, experimentation is crucial for model development, validation, and implementation.

In the end, it is important to remember that models are models and always introduce some level of assumption and/or phenomenology. Model development is an evolutionary process. Because models play a critical role in structural design, their accuracy and applicability are constantly being questioned. As new materials, applications, demands, and approaches blossom, earlier models become unsuitable, newer ones are developed, and improved versions will inevitably be requested.

REFERENCES

Aboudi, J. (1991) *Mechanics of Composite Materials: A Unified Micromechanics Approach*, Elsevier, New York.

Aboudi, J. (1995) 'Micromechanical analysis of thermo-inelastic multiphase short-fiber composites', *Comp. Eng.*, **5**, 839–850.

Anderson, T.L. (1995) *Fracture Mechanics: Fundamentals and Applications*, Second Edition, CRC Press, New York.

Argon, A.S. (2001) Chapter 24 in *Comprehensive Composites*, Chou, T.-W. and Zweben, C. (Eds), Elsevier, New York, pp. 763–802.

Baranblatt, G.I. (1962) 'The mathematical theory of equilibrium cracks in brittle fracture', *Adv. Appl. Mech.*, **7**, 55–129.

Batdorf, S.B. (1982) 'Statistical treatment of stress concentration factors in damaged composites', *J. Reinf. Plastic Comp.*, **1**, 153–164.

Bensoussan, A., Lions, J.-L. and Papanicolaou, G. (1978) *Asymptotic Analysis for Periodic Structures*, North-Holland Publishing Company, New York.

Benveniste, Y. (1987) 'A new approach to the application of Mori-Tanaka's theory in composite materials', *Mech. Matl.*, **6**, 147–157.

Beyerlein, I.J., (2000) 'Stress fields around cracks with a viscous matrix and discontinuous fiber bridging', *Comp. Sci. Tech.*, **60**, 2309.

Beyerlein, I.J. and Landis, C.M. (1999), 'Shear-lag model for failure simulations of unidirectional fiber composites including matrix stiffness', *Mech. Mater.*, **31**, 331–350.

Beyerlein, I.J. and Phoenix, S.L. (1996) 'Statistics for the strength and size effects of microcomposites with four carbon fibers in epoxy resin', *Comp. Sci. Technol.*, **56**, 75–92.

Beyerlein, I.J. and Phoenix, S.L. (1997a) 'Stress profiles and energy release rates around fiber breaks in a lamina with propagating zones of matrix yielding and debonding', *Comp. Sci. Tech.*, **57**, 869–885.

Beyerlein, I.J. and Phoenix, S.L. (1997b) 'Statistics of fracture for an elastic notched composite lamina containing Weibull fibers – Part I: Features from Monte Carlo simulation', *Engng Fract. Mech.*, **57**, 241–265.

Beyerlein, I.J. and Phoenix, S.L. (1997c) 'Statistics of fracture for an elastic notched composite lamina containing Weibull fibers – Part II: Probability models of crack growth', *Engng Fract. Mech.*, **57**, 267–299.

Beyerlein, I.J., Phoenix, S.L. and Sastry, A.M. (1996) 'Comparison of shear-lag theory and continuum fracture mechanics for modeling fiber and matrix stresses in an elastic cracked composite lamina', *Int. J. Solids Struct.*, **33**, 2543–2574.

Beyerlein, I.J., Phoenix, S.L. and Raj, R. (1998a) 'Time evolution of stress distributions around arbitrary arrays of fiber breaks in a composite with a creeping viscoelastic matrix', *Int. J. Solids Struct.*, **35**, 3177.

Beyerlein, I.J., Amer, M., Schadler, L.S. and Phoenix, S.L. (1998b) 'New methodology for determining in situ fiber, matrix, and interface stresses in damaged multifiber composites', *Sci. Engng Compos. Mater.*, **7**, 151.

Beyerlein, I.J., Zhou, C.H., Schadler, L.S. (2003) 'A time dependent micromechanical fiber composite model for inelastic growth in viscoelastic matrices', *Int. J. Solids Struct.*, **40**, 2171–2194.

Broek, D. (1986) *Elementary Engineering Fracture Mechanics*, Fourth revised edition, Kluwer, London.

Chaboche, J.L. (2003) '*Damage Mechanics*' in *Comprehensive Structural Integrity*, Milne, I., Ritchie, R.O., and Karihaloo, B. (Eds), Elsevier-Pergamon Press, New York.

Chou, T.-W. (1992) *Microstructural Design of Fiber Composites*, Cambridge Solid State Science Series, R.W. Cahn, E.A. Davis and I.M. Ward (Eds), Cambridge University Press, New York.

Christensen, R.M. (1979) '*Mechanics of Composite Materials*', John Wiley & Sons, New York.

Coleman, B.D. and Gurtin, M.E. (1967) 'Thermodynamics with internal state variables', *J. Chem. Phys.*, **47**, 597–613.

Cox, H.L., (1952) 'The elasticity and strength of paper and other fibrous materials', *Br. J. Appl. Phys*, **3**, 72–79.

Curtin, W.A., (1991a) 'Theory of mechanical properties of ceramic matrix composites', *J. Am. Ceram. Soc.*, **74**, 2837–2845.

Curtin, W.A., (1991b) 'Exact theory of fibre fragmentation in a single-filament composite', *J. Mater. Sci.*, **26**, 5239–5253.

Daniels, H.E., (1945) 'The statistical theory of the strength of bundles of threads I', *Proc. R. Soc. Lond. A*, **183**, 405–435.

Dieter, G.E. (1986) *Mechanical Metallurgy* McGraw-Hill Book Company, New York.

Dugdale, D.S. (1960) 'Yielding of steel sheets containing slits', *J. Mech. Phys. Sol.*, **8**, 100–104.

Dvorak, G.J. and Bahei-El-Din, Y.A. (1982) 'Plasticity analysis of fibrous composites', *J. Appl. Mech.*, **49**, 327–335.

Eringen, A.C. (1989) *Mechanics of Continua*, Robert E. Krieger Publishing Co., Melbourne, FL.

Eshelby, J.D. (1957) 'The determination of the elastic field of an ellipsoidal inclusion, and related problems', *Proc. R. Soc. Lond. A*, **A241**, 376–396.

Griffith, A.A. (1920) 'The phenomena of rupture and flow in solids', *Phil. Trans. R. Soc. Lond. A*, **221**, 163–198.

Harlow, D.G. and Phoenix, S.L. (1978a) 'The chain-of-bundles probability model for the strength of fibrous materials: I', *J. Comp. Mater.*, **12**, 195–214.

Harlow, D.G. and Phoenix, S.L. (1978b) 'The chain-of-bundles probability model for the strength of fibrous materials: II', *J. Comp. Mater.*, **12**, 314–334.

Hedgepeth, J.M. (1961) *Stress Concentrations in Filamentary Structures*, NASA TN D-882.

Hedgepeth, J.M. and Van Dyke, P. (1967) 'Local stress concentrations in imperfect filament composites', *J. Comp. Mater.*, **1**, 294–309.

Herakovich, C.T. (1998) *Mechanics of Fibrous Composites*, John Wiley & Sons, New York.

Hikami, F. and Chou, T.-W. (1990) 'Explicit crack problem solutions of unidirectional composites: Elastic stress concentrations', *AIAA J.*, **28**, 499–505.

Hill, R. (1963) 'Elastic properties of reinforced solids: Some theoretical principles', *J. Mech. Phys. Solids*, **11**, 357–372.

Hill, R. (1964a) 'Theory of mechanical properties of fibre-strengthened materials: I Elastic behavior', *J. Mech. Phys. Solids*, **12**, 199–212.

Hill, R. (1964b) 'Theory of mechanical properties of fibre-strengthened materials: II Inelastic behavior', *J. Mech. Phys. Solids*, **12**, 213–218.

Hill, R. (1964c) 'Theory of mechanical properties of fibre-strengthened materials: III Self-consistent model', *J. Mech. Phys. Solids*, **13**, 189–198.

Hill, R. (1965) 'A self-consistent mechanics of composite materials', *J. Mech. Phys. Solids*, **13**, 213–222.

Hill, R. (1967) 'The essential structure of constitutive laws for metal composites and polycrystals', *J. Mech. Phys. Solids*, **15**, 79–95.

Hinton, M.J. and Soden, P.D. (1998) *Comp. Sci. Tech.*, **58** (entire volume).

Hinton, M.J., Kaddour, A.S. and Soden, P.D. (2002) *Comp. Sci. Tech.*, **62** (entire volume).

Hui, C.-Y, Phoenix, S.L., Ibnabdeljalil, M. and Smith, R.L. (1995) *J. Mech. Phys. Solids*, **43**, 1551–1585.

Ibnabdeljalil, M. and Curtin, W.A. (1996) 'Strength and reliability of fiber-reinforced composites: local load sharing and associated size effects', *Int. J. Solids Struct.*, **34**, 2649–2668.

Irwin, G.R. (1957) 'Analysis of stresses and strains near the end of a crack transversing a plate', *J. Appl. Mech.*, **24**, 361–364.

Karihaloo, B. and Xiao, Q.Z. (2003) 'Linear and nonlinear fracture mechanics', in *Comprehensive Structural Integrity*, Milne, I., Ritchie, R.O. and Karihaloo, B. (Eds), Elsevier-Pergamon Press, New York.

Kocks, U.F., Tome, C.N. and Wenk, H.-R. (1998) *Texture and Anisotropy: Perferred Orientations in Polycrystals and their Effect on Material Properties*, Cambridge University Press, Cambridge, UK.

Landis, C.M., Beyerlein, I.J. and McMeeking, R.M. (1999a) 'Micromechanical simulation of the failure of fiber reinforced composites', *J. Mech. Phys. Solids*, **48**, 621–648.

Lemaitre, J. and Chaboche, J.L. (1994) *Mechanics of Solid Materials*, Cambridge University Press, Cambridge, UK.

Levin, V.M. (1967) 'Thermal expansion coefficients of heterogeneous materials', *Mech. Solids*, **2**, 58–61.

Lubliner, J. (1990) *Plasticity Theory*, Macmillan Publishing Co., New York.

Mahesh, S. and Phoenix, S.L. (2003) 'Absence of a tough-brittle transition in the statistical fracture of unidirectional composite tapes under local load sharing', *Phys. Rev. E.*, accepted for publication.

Mahesh, S. and Phoenix, S.L. (2004) 'Lifetime distributions for unidirectional fibrous composites under creep-rupture loading'. *Int. J. Fract.*, submitted for publication.

Mahesh, S., Beyerlein, I.J. and Phoenix, S.L. (1999) 'Size and heterogeneity effects on the strength of fibrous composites', *Physica D*, **133**, 371–389.

Mahesh, S., Phoenix, S.L. and Beyerlein, I.J. (2002) 'Strength distributions and size effects for 2D and 3D composites with Weibull fibers in an elastic matrix', *Int. J. Fract.*, **115**, 41–85.

Malvern, L.E. (1969) *Introduction to the Mechanics of a Continuous Medium*, Prentice-Hall, Englewood Cliffs, New Jersey.

Mroz, Z. (2003) 'Strength Theories', in *Comprehensive Structural Integrity*, Milne, I., Ritchie, R.O. and Karihaloo, B. (Eds), Elsevier-Pergamon Press, New York.

Nemat-Nasser, S. and Hori, M. (1993) *Micromechanics: Overall Properties of Heterogeneous Materials*, North-Holland, New York.

Orowan, (1955) 'Energy criteria of fracture', *Weld. Res. Supp.*, **34**, 157.

Paley, M. and Aboudi, J. (1992) 'Micromechanical analysis of composites by the generalized cells model', *Mech. Matl.*, **14**, 127–139.

Phoenix, S.L. and Beyerlein, I.J. (2000) 'Distributions and size scalings for strength in a one-dimensional random lattice with load resdistribution in nearest and next-nearest neighbors', *Phys. Rev. E*, **62**, 1622–1645.

Phoenix, S.L. and Beyerlein, I.J. (2001) 'Statistical strength theory for fibrous composite materials', Chapter 19 in *Comprehensive Composites*, Chou, T.-W. and Zweben, C. (Eds), Elsevier, New York, pp. 559–639.

Phoenix, S.L. and Raj, R. (1992) 'Scalings in fracture probabilities for brittle-matrix fiberous composites', *Acta Metall. Mater.*, **40**, 2813–2828.

Phoenix, S.L. and Smith, R.L. (1983) 'A comparison of probabilistic techniques for the strength of fibrous materials under local load sharing among fibers', *Int. J. Solids Struct.*, **19**, 479–496.

Phoenix, S.L., Ibnabdeljalil, M. and Hui, C.-Y. (1997) 'Size effects in the distribution for strength of brittle matrix fibrous composites', *Int. J. Solids Struct.*, **34**, 545–568.

Reedy, E.D. (1984) 'Fiber stresses in cracked monolayer. Comparison of shear lag and 3-D finite element predictions', *J. Comp. Mater.*, **18**, 595–607.

Rice, J.R. (1968) 'A path independent integral and the approximate analysis of strain concentrations by notches and cracks', *J. Appl. Mech.*, **35**, 379–386.

Rosen, B.W. (1964) 'Tensile failure of fibrous composites', *AIAA J.*, **2**, 1985–1991.

Smith, R.L. (1980) 'A probability model for fibrous composites with local load sharing', *Proc. R. Soc. Lond. A*, **372**, 539–553.

Smith, R.L. (1982) 'The asymptotic distribution of the strength of a series-parallel system with equal load sharing', *Ann. Prob.*, **10**, 137–171.

Smith, R.L. (1983) 'Limit theorems and approximations for the reliability of load-sharing systems', *Adv. Appl. Prob.*, **15**, 304–330.

Smith, R.L. Phoenix, S.L., Greenfield, M.R., Henstenburg, R.B. and Pitt, R.E. (1983) 'Lower tail approximations for the probability of failure of 3-dimensional fibrous composites with hexagonal geometry', *Proc. R. Soc. Lond. A*, **388**, 353–391.

Torquato, S. (2001) *Random Heterogeneous Materials: Microstructure and Macroscopic Properties*, Springer-Verlag, New York.

Tsai, S.W. and Wu, E.M. (1971) 'A general theory of strength for anisotropic material', *J. Comp. Matl*, **5**, 58–80.

Van Dyke, P. and Hedgepeth, J.M. (1969) 'Stress concentrations from single-filament failures in composite materials', *Textile Res. J.*, **39**, 618–626.

Weibull, W. (1951) 'A statistical distribution function of wide applicability', *J. Appl. Mech.*, **18**, 293–297.

Westergaard, H.M. (1939) 'Bearing pressures and cracks', *J. Appl. Mech.*, **6**, 49–53.

Williams, T.O. and Pindera, M.J. (1997) 'An analytical model for the inelastic longitudinal shear response of metal matrix composites', *Int. J. Plasticity*, **13**, 261–289.

Williams, T.O. and Beyerlein, I.J. (2004) 'Modeling Damage Evolution in Materials: Concepts, Approaches, and Issues', *Los Alamos National Laboratory Report LA-14136* (oakhill@lanl.gov or irene@lanl.gov).

Williams, T.O. (2004) 'A general, stochastic transformation field analysis', submitted for publication.

Zweben, C. (1973) 'Fracture mechanics and composite materials: A critical analysis', *ASTM STP* **521**, 65–97.

Zweben, C. and Rosen, W.B. (1970) *J. Mech. Phys. Solids*, **18**, 189–206.

3

In Situ Observation of Damage Evolution and Fracture Toughness Measurement by SEM

Juan E. Perez Ipiña[1] and Alejandro A. Yawny[2]

[1]*Universidad Nacional del Comahue/CONICET, Buenos Aires, Argentina*
[2]*Centro Atómico Bariloche, Argentina*

3.1 OVERVIEW OF FRACTURE MECHANICS RELATED TO DAMAGE PROGNOSIS

From a damage prognosis point of view, *damage* in a structural and mechanical system will be defined as intentional or unintentional changes to the material and/or geometric properties of the system, including changes to the boundary conditions and system connectivity, which adversely affect the current or future performance of that system [1]. Some types of damage, although not all, are related to cracks. Fracture mechanics applies only to this last type of damage. There follows a very short overview of fracture mechanics; for those concerned with a deeper insight, there are excellent books available [2–4].

Fracture mechanics deals with stress fields, cracks and the resistance that the material opposes to the crack growth (Figure 3.1). When fracture mechanics is employed, an initial crack length, a_0, is assumed. This initial crack can exist because it has been allowed by design, has appeared during construction and accepted for operation, or has appeared during service and also accepted for operation. In cases where no crack is present, fracture mechanics is not the tool to be employed to analyze the structure health.

Damage Prognosis – For Aerospace, Civil and Mechanical Systems Edited by D.J. Inman, C.R. Farrar, V. Lopes Junior and V. Steffen Junior © 2005 John Wiley & Sons, Ltd

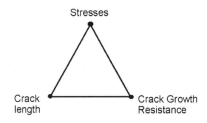

Figure 3.1   The fracture mechanics triangle

Complete failure, generally as a fracture, comes when the crack grows up to a critical crack length, a_C. Therefore, the component can go on working during certain period of time within which the crack is growing, as long as it remains less than a_C. This growth can occur through different mechanisms: fatigue, stress corrosion cracking, creep, etc. The relationship between the crack growth rate and the stress state together with the crack length can be experimentally determined, and then the structure's remaining life calculated as an estimation.

There are also thresholds related to these subcritical growth mechanisms that make the crack remain stationary, hence the damage in the structure does not increase. That is, there is a crack, there is a stress field, and also an environmental action or variation of stresses, and still there is no crack growth.

According to the conditions and the material's mechanical response, different parameters are used. When there is a contained plastic deformation, the so called linear elastic fracture mechanics (LEFM) is employed by using K_{IC} or G_{IC} as critical parameters. When the acting stress intensity factor, $K_I = f(\sigma, a)$, reaches K_{IC}, fracture occurs. If $a_0 < a_C$, then $K_I < K_{IC}$; and different relationships, for example the Paris' law in fatigue, link the acting SIF, K_I, to the crack growth rate. Thresholds having the most practical use are K_{ISCC} in stress corrosion cracking and ΔK_{th} in fatigue crack growth.

When linear elastic conditions are not met, especially when the surroundings of the crack tip present large plasticity, LEFM is no longer valid, so elastic plastic fracture mechanics, EPFM, has to be made use of. The most applied parameters come from the integral J theory and the British crack tip opening displacement. The equivalent to K_{IC} are J_C and δ_C respectively. Many structural materials present stable crack growth when they behave elastoplastically, then they can be characterized by means of resistance curves $J-R$ and ductile instability parameters as the tearing modulus, T. Another characterization is by initiation parameters: J_{IC}; δi. Creep crack growth occurs under elastoplastic conditions and the parameters used to relate the crack growth rate to stress and crack length are C^*; $C(t)$ or Ct.

In some special conditions (creep, hydrogen environmental, etc.), the resistance to crack growth can be gradually affected; furthermore, some external events, like fire, leading to microstructural changes can suddenly modify the fracture toughness. There are also other variables that affect this property, the most important being size limitations related to plane-stress/plane-strain conditions or constraint variation and, especially, the effects of temperature and strain rate, which many materials prove sensitive to.

Many materials such as ceramics, polymers and body-centered cubic metals present a transition from ductile to brittle behavior as temperature reduces or strain rate increases

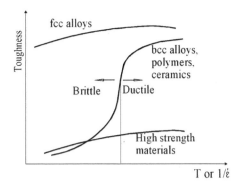

Figure 3.2   Ductile-to-brittle transition for different families of materials

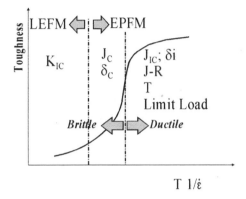

Figure 3.3   LEFM or EPFM parameters to be used in the regions of the ductile-to-brittle transition

(Figure 3.2). Then, at low temperature or high strain rate, they are brittle, that is for a given material and stress state, the critical crack length is small, while at higher temperature or quasi-static load application, a_C can be a lot larger. Additionally, LEFM or EPFM must be employed depending on temperature and strain rate. Figure 3.3 shows the fields of application of the above introduced parameters.

This is a very important issue in damage-tolerant structures because, depending on the conditions (temperature or strain rate), the fracture toughness and, of course, the critical crack length, a_0, can vary in materials sensitive to these variables. For example, the steel suspension components in a truck working in a tropical area for an oil service company will very likely lie on the upper-shelf of the ductile-to-brittle transition region, having the capacity to bear a large crack and its reliability not being affected. The same truck moved to an oil area in Alaska or southern Patagonia will be working within the transition itself or, indeed, on the lower shelf; and that crack being tolerable in the former will become unstable in the new environmental conditions. In the first case an EPFM analysis will be necessary, while in the second, LEFM will be the methodology to apply.

3.2 IN SITU *OBSERVATION OF DAMAGE EVOLUTION AND FRACTURE TOUGHNESS MEASUREMENT*

3.2.1 Introduction

The possibility to correlate the usual macroscopic fracture events, i.e. stable crack growth initiation, fatigue crack propagation, crack closure, etc., with the micromechanisms that are responsible for them represents an advantage in order to understand and improve their mechanical behavior. Most tests made for this purpose are carried out using macroscopic samples whilst the analysis of the mechanisms involved is performed in a second step. In this way, it is intended to correlate the microscopic evidence with the previously recorded mechanical features. A more direct correlation between both steps would be of great benefit and this can be achieved by performing *in situ* mechanical tests with concomitant mechanism observations.

However, microscope chamber sizes usually impose severe limitations on the maximum size of specimens to be tested. There are also a number of occasions where the use of specimens of reduced size is preferred when evaluating the mechanical properties (development of new materials, scarcity). All these facts generally lead to greater difficulties in fulfilling the standardized mechanical testing procedures (size limitations), and using specimens of reduced size requires understanding on the limits of applicability of the fracture toughness parameters, defined macroscopically.

In situ SEM techniques are being increasingly applied to study a great variety of materials-science related problems. In this relatively new development, the idea is that specially equipped SEMs can be used as microlaboratories [5]. In particular, micromechanisms operation associated with the mechanical behavior of different materials can be followed during the tests.

In the present work, however, the aim was to develop *in situ* experimental techniques to measure fracture toughness (K_{Ic}, J_{Ic}, *CTOD, R*-curves) in small samples while details of the crack blunting, initiation and propagation were simultaneously observed using scanning electron microscopy (SEM). A small load frame was employed to this purpose and *in situ* tests were performed with it in a Philips 515 scanning electron microscope with a small chamber. In this way a *P–V* record can be obtained from different types of tests in order to calculate any fracture mechanics parameter, determine the initiation and quantity of stable crack growth, measure directly the CTOD or Schwalbe's δ_5 [6] and to correlate any of these parameters with the observed micromechanisms. Several different materials were used in this work in order to illustrate the wide potential of the technique but no special emphasis was put on a detailed study of any particular material. Adequate references of thorough investigations on specific materials will be given in the text.

3.2.2 Material and Method

3.2.2.1 Description of the Loading Equipment

A small testing stage was developed and employed to perform conventional mechanical tests in small samples [7]. A photograph of the machine with a three-point bending specimen mounted in testing position is shown in Figure 3.4. The stage is a small

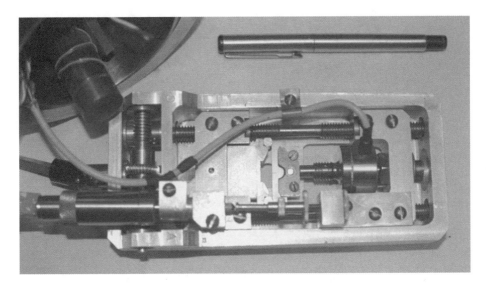

Figure 3.4 Small mechanical testing stage used inside the SEM camera. A thin, single-ended, notched laminate composite specimen is mounted together with antibuckling plates in a three-point bending test configuration

double-screw machine and for the purpose of all experiments presented in this work it was instrumented with a load cell HBM $\pm 2\,$KN and an inductive displacement transducer HBM WT1 $\pm 1\,$mm. The recommended maximum load for the testing stage is 500 N. The test temperature can be easily chosen in the range 243–343 K using a Peltier thermobattery. A thorough description of this stage can be found in Reference [7].

The stage is suitable to be used also in SEMs equipped with a bigger chamber and in light microscopes. Different types of fracture specimen can be loaded: tension SE(T) and DE(T), three-point bending SE(B) and compact tension specimens (CT).

Materials

The following materials were tested: Al – Si cast alloy; zirconium alloy (Zry-4); fiber – metal laminates (Arall®); soft martensitic stainless steel (SMSS), and shape-memory material (S-MM)

Specimen Preparation

- Samples geometries and dimensions:
    - Notched tensile SE(T) specimens were prepared in a spark erosion machine with nominal dimensions indicated in Figure 3.5.
    - Either precracked or notched bend SE(B) specimens were used with nominal dimensions shown in Figure 3.5 [8].
- Fatigue precracking:
  After cutting the mechanical notch, fatigue precracking was performed in most of the SE(B) specimens (Table 3.1).

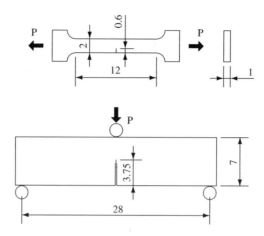

Figure 3.5   SE(T) and SE(B) specimens used in the present study

Table 3.1   Materials and specimen geometries tested

Material	Geometry	Precrack	Parameter
Al alloy	SE(T)/SE(B)	No/Yes	K/J
Zry-4	SE(B)	Yes	K/J
SMSS	SE(B)	Yes	$CTOD$
Arall 3 2/1	SE(B)	No/Yes	J
S-MM	SE(T)	No	—

Testing and Mechanisms Observation

The applied load (P) and the load point displacement (V) were recorded during the *in situ* tests. Specimens were loaded monotonously within regular load-point displacement intervals and images were recorded at the end of each of the intervals. Once the picture was obtained, the loading process was resumed. Besides the regular imaging sequence, any particular event observed in regions next to the crack tip and its wake was correlated with the corresponding point in the load-displacement (P vs. V) record. In addition, the initiation of the stable crack growth could be visually determined during the test and the corresponding point in the load/displacement record identified. In that way J_{IC} or δ_i could be evaluated. The evolution of the length of the crack during the stable growth period could be determined by direct measurement on the successive images and, by considering the images' homologous points in the P–V record, an R-curve could be derived. In the case of SMSS specimens, the crack length during the stable growth and the $CTOD$ values were both optically determined by measuring over the images obtained during the test.

The characteristics of the different tests, including material and test conditions, as well as the obtained fracture toughness parameters are summarized in Table 3.1.

3.2.3 *Results and Discussion*

3.2.3.1 *Fracture Toughness Obtained from* in situ *SEM Experiments*

Special effort was devoted to get the right conditions for the precracking procedure of tiny SE(B) specimens, having found it practicable. A razor blade pass over the mechanical notch was shown to enhance fatigue crack growth initiation. A number of cycles $N = 200\,000$ was used to obtain a fatigue crack length of approximately 0.5 mm in the SE(B) specimens of Al alloys. For Zry-4 and soft martensitic stainless steel (SMSS), similar growth lengths required between 30 000 and 60 000 cycles. In the case of Arall, notched and fatigue precracked ($N = 18\,000$) samples were tested and compared in order to study the influence of crack tip shielding mechanisms (fiber bridging) on the fracture toughness. Further details about the pre-cracking procedures can be found in References [9] to [11].

Table 3.2 summarizes the different toughness parameters calculated for materials and specimens considered in the present study. When available, results corresponding to fracture toughness evaluations in macro specimens were included for comparison. In some cases the calculated J values were converted to K equivalent ones, K_{JIC}, by applying equation 3.1[8]:

$$K_{JIC} = (J_{IC}E')^{1/2} \tag{3.1}$$

The P versus V record corresponding to a precracked SE(B) specimen of Al-Si cast alloy is presented in Figure 3.6(a). The plot depicts a clear nonlinear behavior, similar to that of almost all SE(B) specimens tested in this work. A significant difference between the load used to calculate the standardized K_Q, $P_{5\%}$, and the maximum load P_{max} can be seen clearly from Figuré 3.6(a). Whenever a specimen presents a nonlinear P–V record during a standard fracture test, this might generally be a consequence of plastic deformation, stable

Table 3.2 Examples of the fracture toughness parameters calculated from the *in situ* SEM experiments for the different materials and specimens. Toughness values obtained by testing macrospecimens were included for comparison

Material and specimen	J_{IC} [kJ/m^2]	J_i [kJ/m^2]	J_{Pmax} or $J_C{}^*$ [kJ/m^2]	K_{Pmax} [MPam$^{0.5}$]	K_{JIC} or $K_{Ji}{}^*$ [MPam$^{0.5}$]	δ_{max} [mm]	δ_i [mm]
Al-Si cast Alloy							
SE(B)	—	2.79	3.44	7.5	14.4* (E = 75 GPa)	—	—
Zry-4							
SE(B)	77.6	20.0	20.0	23.4	88/44.7*	—	—
CT25	69.0	—	—	57.6	83	—	—
SMSS							
SE(B)	—	—	—	—	—	0.062	0.08
CT25	—	—	—	—	—	0.170	—
Arall 3 2/1 SE(B)							
Pre-cracked	—	—	55.0*	—	—	—	—
Notched	—	—	65.5*	—	—	—	—

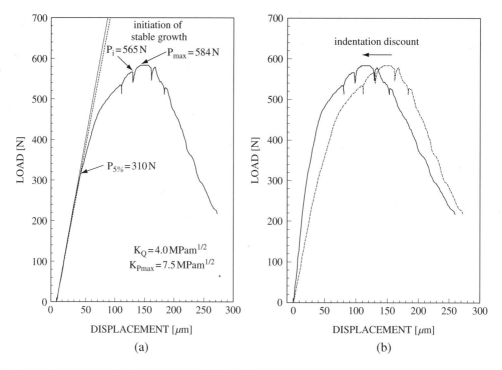

Figure 3.6   Test results for precracked SE(B) Al-Si cast alloy: (a) Load versus displacement record showing nonlinear behavior and differences between K_Q and K_{Pmax}. (b) Effect of indentation discount

crack growth, or combined influence of them both. Where these effects become relevant, an elastic – plastic methodology is necessary in order to determine initiation parameters (i. e. J_{IC}) or resistance curves (J–R curves). Figure 3.6(b) illustrates how the crude P–V plot represented in Figure 3.6(a) is modified after the displacements associated with the indentations were subtracted. The figure is self-explanatory about the importance of performing this correction procedure before any further analysis of the data is made.

Figure 3.7 shows a $J - \Delta a$ curve for Zry-4 SE(B) specimens. Photographs taken during the test have been included in the graph indicating the points they relate to. The amount of stable crack growth is easily obtained from the photographs and J values are calculated from the P–V record, making it possible to determine as many points on the R-curve as photographs have been taken.

$CTOD$ values can also be measured directly from the images, including the one where the initiation of slow stable crack growth is detected. This procedure is illustrated in Figure 3.8 where results for SMSS SE(B) specimens are presented. Figure 3.8 shows a series of images taken at different moments of the test, $CTOD$ and Δa values are indicated over one of these images.

In the analysis of the results obtained from the *in situ* tests, a significant difference between toughness parameters obtained by direct observation of the beginning of the stable crack growth and the equivalent toughness parameters but obtained after using 'engineering' procedures was found. The standardized methods define the conventional

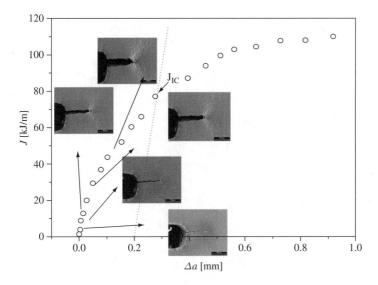

Figure 3.7 J_R curve for Zry-4 SE(B). The amount of stable crack growth is obtained from the micrographs and J values are calculated from the P–V record. The R-curve can be determined with as many points as images were taken

initiation point as the toughness corresponding to 0.2 mm of crack growth. These procedures gave values that were at least three-times higher than those obtained through direct observation of initiation on the specimen surface. For example, in Table 3.2 we can see that J_{IC} values in Zry-4 are 20.0 kJ/m² and 77.6 kJ/m² respectively, while in SMSS the CTOD values for initiation are 0.06 mm and 0.23 mm.

Toughness results from the *in situ* tests can now be compared to values obtained by using macro specimens, engineering standardized procedures having been applied in both cases. Table 3.2 shows that in the case of Zry-4 the 'engineering' J_{IC} value was higher for the small specimen (77.6 versus 69.0 kJ/m²). This can be explained by considering the differences in the dominant stress states due to the very different specimens' thickness.

3.2.3.2 Toughness/Micromechanisms Relationship

The observations in the SEM enable evidence to be obtained about the mechanisms that are acting in the processes of crack initiation and propagation. Figure 3.7 is an example in which the crack tip evolution can be nearly continuously observed and correlated with the P versus V record.

The presence of some toughening mechanisms in the materials studied in the present work was verified during the *in situ* SEM tests. As an example, crack deflection was present in most of the tested specimens as can be seen in Figure 3.7 corresponding to Zry-4 specimens. Crack branching can be also seen in Figure 3.8, in which the evolution of the blunting process followed by stable crack growth and branching can be followed in the series of images. Figure 3.9 shows, besides crack bridging, fiber fracture plus matrix/fiber debonding and pull-out in Arall. It is important to remark that it was verified that each load fall during the test was a consequence of fiber fracture. It was

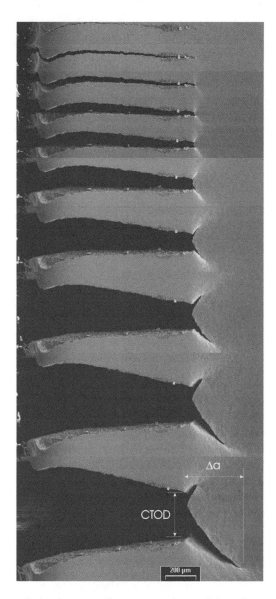

Figure 3.8 Crack growth development (from top to bottom) in soft martensitic stainless steel (SMSS). In the last image it is illustrated how $CTOD$ and Δa can be directly determined from the image for further $CTOD - \Delta a$ curve construction. In addition initiation values can be obtained directly at the moment where the initiation of slow stable crack growth is detected by observation

also attested that in notched specimens load drops were recorded before there was any evidence of stable crack growth initiation in the external aluminum alloy sheets [11].

Figure 3.10 shows a sequence of photographs where the crack grew through fractured hydrides in Zry-4. Also indicated is a fractured hydride inside the heavily damaged region that was avoided by the crack [9].

The presence of transformation-induced plasticity (TRIP) was verified in SMSS by means of other techniques [10], but no evidence of martensitic transformation was

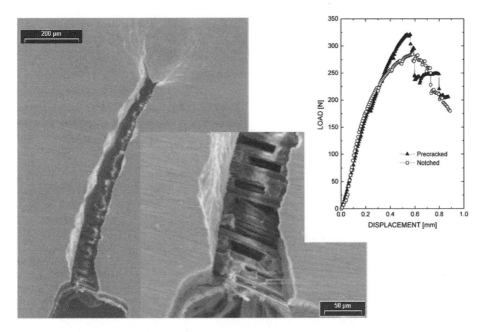

Figure 3.9   Mechanisms and *P–V* record in Arall 3 2/1: crack bridging and fiber fracture

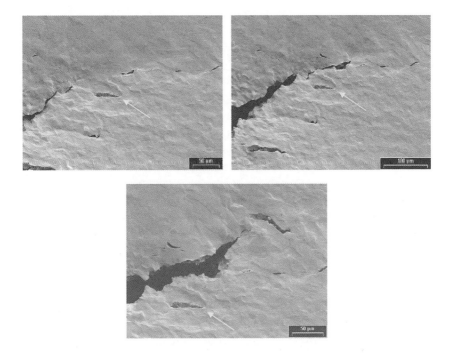

Figure 3.10   Crack growth through hydrides in Zry-4. The white arrow identifies a particular microcrack nucleated on a hydride and serves as a reference to follow the growth of the dominant macrocrack with further straining. The crack proceeds by the coalescence of some of the microcracks in the heavily deformed zone associated to the tip of the crack

observed during the tests, probably because of the small size of the retained austenite phase. However, stress-induced martensite was detected when loading notched SE(T) single crystalline specimens of Cu-Zn-Al shape-memory material(S-MM).

3.2.3.3 Limitations

The use of specimens of reduced size can result in some cases in difficulties in fulfilling the standardized mechanical testing procedures.

1. As is well known, plasticity, plane strain/plane stress conditions, and stable crack growth impose limitations on specimens size [8].

   (a) They are more restrictive for linear elastic fracture mechanics, where plastic zone sizes larger than 0.07 mm ($r_p < 0.02$ b, a, B) are not allowed for these specimens. It was not possible to obtain valid K_{IC} values for the materials tested.

   (b) As a consequence of the specimen size, the crack tip fields are closer to plane stress than to plane strain conditions. Therefore, in some cases the measured toughness could be larger than that obtained in larger specimens.

   (c) When R-curves are intended to be attained, only stable crack growth values of about 0.875 mm are valid in accordance to the standards ($\Delta a < 0.25$ b_0).

   (d) Where testing high toughness materials, the limits to J_{Pmax} established by the standard [8] are also transgressed.

2. Only phenomena that occur on the specimen surface can generally be observed. Most times the fracture is controlled by events occurring under the specimen surface. As an example, a certain amount of tunneling was verified after fracture in some of the SMSS specimens whilst there was no observable stable crack growth on their surface during the *in situ* test observations.

3. Spurious displacements: in spite of having been built in the stiffest mode possible, and as a consequence of the tiny displacements involved, there are some, other than the load point displacement, contributions to the total displacement, i.e. indentations, loose-fitting in the cinematic chain, clamp sliding, and clamp adjustment in tensile geometries. In order to eliminate these spurious displacements, the technique described in Section 3.2.2 and exemplified in Figure 3.6 was applied. It is similar to the one suggested by ASTM standards for J determination in SE(B) specimens. We understand that in this way the error sources are greatly reduced. The differences between the results corrected by indentations and the noncorrected ones were around 12% with a maximum discrepancy of 23% (Table 3.2).

One possible solution to drawbacks 1(b) and 2 could be the use of specimens with side grooves machined after the fatigue precracking procedure.

Despite the above limitations, *in situ* fracture toughness testing might become an unique tool when mechanisms and macroparameters are intended to be related in order to develop new materials or improve fracture properties. Elastic plastic fracture parameters (J and $CTOD$), including initiation and R-curves, can be determined in a simple way from precracked specimens.

3.3 CONCLUDING REMARKS

Several different situations can be found that can affect both the fracture toughness and the critical crack size, as well as the crack growth rate, especially in materials that are temperature and strain-rate sensitive.

An universal testing micromachine that allows the simultaneous observation of the specimen surface in an optical or electron microscope was employed to quantify fracture toughness properties (J_{IC}, $CTOD$, R-curves) and to correlate these values with the observed micromechanisms during the test. The applicability of the proposed *in situ* measurements was exemplified using a wide set of materials. It was found that:

- The initiation values of the different toughness parameters obtained by direct observation of the specimen surface were lower than those obtained by applying the standardized procedures to the P–V data recorded during the test.
- The engineering toughness values (J_{IC}, δ_i) obtained were slightly higher than the ones measured in macrospecimens.
- K_{JIC} obtained from J_{IC} resulted in higher values compared with K_{Pmax} obtained by using the maximum load. K_{Pmax} values determined from microspecimens were lower than the values corresponding to macrospecimens.
- It was also shown that the micromechanisms observed on the specimen surface during the test could be directly correlated with the corresponding point in the load-displacement record.

ACKNOWLEDGEMENTS

To CONICET (Consejo Nacional de Investigaciones Científicas y Técnicas, Argentina).

REFERENCES

[1] C.R. Farrar, N.A. Lieven and M.T. Bement (2005) 'An Introduction to Damage Prognosis', *Damage prognosis – For Aerospace, Civil and Mechanical Systems*, John Wiley & Sons, Ltd.

[2] T.L. Anderson (1994) *Fracture Mechanics. Fundamentals and Applications*, Second Edition, CRC Press, Boca Raton.

[3] A. Saxena (1998) *Nonlinear Fracture Mechanics for Engineers*, CRC Press, Boca Raton.

[4] M. Janssen, J. Zuidema and R.J.H. Wanhill (2002) *Fracture Mechanics*, Second Edition, Delft University Press, Delft.

[5] K. Wetzig and D. Schulze (Eds) (1995) *In Situ Scanning Electron Microscopy in Materials Research*, John Wiley & Sons, New York.

[6] K.H. Schwalbe (1995) 'Introduction to δ_5 as an Operational Definition of the CTOD and its Practical Use'. *Fracture Mechanics, 26. ASTM STP 1256*. American Society of Testing and Materials, pp. 763–778.

[7] A. Yawny, J. Malarria, E. Soukup and M. Sade (1997) 'A stage for *in situ* mechanical loading experiments in a scanning electron microscope (Philips 515) with a small chamber', *Rev. Sci. Instrum.* **68**, 150–154.

[8] ASTM E-1820-98 *Annual Book of ASTM Standards*, V. 03.01.

[9] G. Bertolino (2001) *Deterioro de las Propiedades Mecánicas de Aleaciones Base Circonio por Interacción con Hidrógeno*. PhD Dissertation, Instituto Balseiro, Bariloche, Argentina.

[10] P. Bilmes, C. Llorente and J.E. Perez Ipiña (2000) 'Toughness and microstructure of 13Cr4NiMo high strength steel welds'. *J. Mat. Engng and Performance*, **9**, 609–615.

[11] E.M. Castrodeza, A. Yawny, J.E. Perez Ipiña and F.L. Bastian (2002) 'Fracture micromechanisms in aramid fiber-metall laminates. In-situ SEM observations'. *J. Composite Mat.*, **36**, 387–400.

4

Predictive Modeling of Crack Propagation Using the Boundary Element Method

Paulo Sollero

Faculty of Mechanical Engineering, Universidade Estadual de Campinas, Brazil

4.1 INTRODUCTION

Materials science research for high strength, high stiffness and low weight materials imposed historical cycles of material usage for the aeronautic industry, from wood to aluminum and magnesium alloys and finally to composites. Fibre-matrix composites are increasingly replacing metals in the aerospace and automotive industries, and will tend to be the first material choice in many future designs. Their main advantages are their ability to be tailored for individual applications, their low weight and high stiffness, and their resistance to corrosion. The use of composites is often limited by the lack of data, and efficient methods are needed to evaluate the strength and lifetime of cracked composite structures.

Fracture mechanics methods have appeared as valuable tools for the design analysis and assessment of damage tolerance, but its development was directed to isotropic metals, with few applications to anisotropic composites. Moreover, many of the super-alloys, used for high temperature environments and high thermal conductivity requisites, exhibit anisotropic behaviour, as reported by Chan and Cruse [1]. The general equations for crack-tip stress and displacement fields in anisotropic bodies were firstly presented by Sih, Paris and Irwin [2], who observed that it was possible to extend fracture mechanics methods for the analysis of these bodies.

Damage Prognosis – For Aerospace, Civil and Mechanical Systems   Edited by D.J. Inman, C.R. Farrar, V. Lopes Junior and V. Steffen Junior   © 2005 John Wiley & Sons, Ltd

The finite element method has been, so far, the most popular technique for analysing fracture problems in composite materials, see for example Ref. [3]. Using this method, Chu and Hong [4] applied the J_k integral to mixed-mode crack problems for composite laminates. Crack growth, or the analysis of a problem for different crack sizes, is a very important topic in fracture mechanics analysis. However, finite element remeshing for each crack length tends to be time consuming, and the interior points of the domain are restricted to nodal points, or have to be interpolated from nodal values.

The boundary element method [5] is recommended for fracture mechanics analysis because of its ability to describe accurately both the stress and displacements fields, even near the crack tip where severe stress gradients occur. This is a very convenient feature for implementing path independent integrals, since various contours of integration can be generated easily and the path independence can be checked. Furthermore, the interior points have no mesh connectivity to the boundary points, and extensive remeshing is not required when the crack grows.

Applications of the boundary element method for stress distributions, mainly for circular and elliptical openings, in anisotropic plates and composite materials were presented in succession, see for example, Zhen and Brebbia [6] and Hwu and Yen [7].

Assembling the system of equations presents a difficulty in the application of the boundary element method to the fracture mechanics general problem, which is of a nonsymmetric mixed mode deformation. For a symmetric problem only one boundary of the crack needs to be modeled in a single region, corresponding to the symmetric part of the body. On the other hand, for a nonsymmetric problem the whole body needs to be modeled and the application of the conventional elastostatics formulation leads to an ill-conditioned problem, as the coincident crack boundaries are discretized. A pair of coincident elements on the crack faces give raise to identical boundary integral equations, as the source points are the same on both elements. The system of equations is underdetermined and a unique solution cannot be achieved.

Some special techniques have been introduced to overcome the modeling difficulty of applying the boundary element method to the general fracture mechanics problem. Among those applied to anisotropic materials, crack Green's function method [8] and the subregion method are the most preferred [9, 10].

The dual-boundary-element method is a more recent and accurate technique for overcoming this difficulty. The method is formulated in terms of two independent boundary integral equations: the displacement equation, which is the boundary integral equation of the boundary element method, and the traction equation. The displacement equation is applied on one crack face and the traction equation on the other. Using this method, general crack problems can be solved in a single-region analysis with increased accuracy and reduced mesh size. The method was introduced by Portela, Aliabadi and Rooke [11] and is suited for the analysis of crack-growth problems, when the discretization of a crack extension can be performed with new boundary elements, eliminating the remeshing problems found in other methods. Applications of the dual-boundary-element method for crack propagation problems were presented for isotropic materials by Portela [12].

Cracked structures submitted to fatigue loads or an aggressive environment are frequent in aeronautics, automotive, naval and oil industries. When the length of the crack is close to a critical length, the useful life of the structure or mechanical component is over. The replacement of this component by a new one is usually an expensive solution, even when it is possible to carry it out in the required time. A quicker alternative of lower cost is the repair of the damaged structure.

In this chapter, the dual-boundary-element method is applied for two-dimensional analysis of crack problems. A predictive technique for computing the crack growth path is presented and applied to cracked composite plates. The results are consistent with crack propagation theories and experiments.

4.2   DAMAGE AND FRACTURE MECHANICS THEORIES

4.2.1   Introduction

Material damage originates in the microscopic scale due to several mechanisms including continued stages of inhomogeneous slip and void growth in metals and fiber breakage or debonding in composites. The progressive damage may originate a macro-scopic crack and separation of crack faces due to loading and eventually leading to complete failure.

Lemaitre and Chaboche [13] claim that the domain of validity and use of the theory of damage mechanics is related to the evolution of the phenomena between the virgin state and macroscopic crack initiation, and that beyond this is the domain of fracture mechanics. Ladeveze and LeDantec [14] used damage mechanics to describe the matrix microcracking and fiber matrix debonding to model composite laminate damage at the elementary ply scale. Recently, Johnson and Simon [15] worked on the development and validity analysis of finite element codes that extend to macroscopic scale of the Ladeveze method. These codes were applied to simulate the collapse of aircraft and automotive composite structures under impact loads.

Salgado [16] points out that in the nineties the aircraft emphasis shifted from crack initiation to crack propagation life. Although the design criteria require severe control of notches and other stress concentrators, localized plasticity and fatigue crack occurrence cannot be eliminated. The current design methodology assumes that in the design stages, sensible areas will be set for frequent inspection in order to ensure that cracks are detected before they reach a critical size. This design methodology is called damage tolerance design [17].

Fracture mechanics is concerned with the analysis of materials and structures which contain detectable cracks, and provides the concepts and mathematical basis for the analysis of damage tolerance design. Linear elastic fracture mechanics is an engineering discipline that had it origins in the relationship between the energy release rate, the crack extension increment and the introduction of the stress intensity factor as a measure of the crack severity [18].

In this section, the crack deformation modes are reviewed. No mode III deformation is assumed in the present work and a general mixed mode (mode I + mode II) deformation is assumed, unless explicitly mentioned.

The stress intensity factors are introduced as a limiting condition of the stress fields near the crack tip in an anisotropic medium. This procedure is similar to the method frequently used to define the stress intensity factors for isotropic bodies, see for example Aliabadi and Rooke [19]. This similarity was observed in the work of Sih, Paris and

Irwin [2] and means that linear elastic fracture mechanics can be applied to generally anisotropic bodies with cracks in the same way as it is for isotropic bodies. An alternative mathematical formulation for the definition of the stress intensity factor for anisotropic bodies, based on the Rieman–Hilbert problem in complex function theory, is presented by Sih and Liebowitz [20].

The *J*-integral is an accurate technique for the evaluation of stress intensity factors. This path-independent contour integral was introduced by Rice [21] as the rate of change of potential energy for nonlinear constitutive behaviour for isotropic elastic–plastic materials. The relationship between the stress intensity factor and the *J*-integral for anisotropic materials are reviewed in this section.

4.2.2 Damage Mechanics

Damage models for laminated composites were recently reviewed by McCartney [22], who considered a physically based approach to damage development in laminates, where the initiation of ply crack is governed by energy balance principles. The aim was to address the industry demand for composite design codes that could predict damage from the microscopic scale to structures and avoid catastrophic failures.

The first step in a damage analysis of composites is to predict accurately the undamaged ply properties evaluated by analytical models as functions of the fiber and matrix properties. The next step should be the choice of the appropriate damage theory among stress transfer models considering thermal residual stresses and a damage evolution method. The analysis can be extended using a first ply failure theory and progressive ply cracking assessment.

However, extending a microscopic damage theory to the complete failure of the composite laminate is a research subject with few promising advances by numerical and probabilistic methods. Recent reviews on this subject consider that the failure criteria of fibrous composites cannot be considered a closed area of knowledge [23] and that the preliminary results of the homogenization method indicate that it will be successful for dealing with ply cracks in multiple plies [22].

The continuum damage mechanics model proposed by Ladeveze and LeDantec [14] was applied to predict the constitutive behavior of laminates by Johnson and Simon [15] and may be written as

$$\left\{ \begin{array}{c} \epsilon_{11} \\ \epsilon_{22} \\ \epsilon_{12} \end{array} \right\} = \left[\begin{array}{ccc} 1/E_1(1-d_1) & -\nu_{12}/E_1 & 0 \\ -\nu_{12}/E_1 & 1/E_2(1-d_2) & 0 \\ 0 & 0 & 1/G_{12}(1-d_3) \end{array} \right] \left\{ \begin{array}{c} \sigma_{11} \\ \sigma_{22} \\ \sigma_{12} \end{array} \right\} \tag{4.1}$$

where the four orthotropic elastic properties are supposed to be undamaged and d_k are the three components of the damage vector, such that $0 \le d_k \le 1$ and expressed as

$$d_k = f_k(Y_1, Y_2, Y_3) \tag{4.2}$$

where f_k are the evolution functions and the conjugate damage forces Y_l are defined from the strain energy function and for elastic damaging materials are written as

$$Y_1 = \sigma_{11}^2/(2E_1(1-d_1)^2),$$

$$Y_2 = \sigma_{22}^2/(2E_2(1-d_2)^2),\tag{4.3}$$

$$Y_3 = \sigma_{12}^2/(2G_{12}(1-d_3)^2).$$

This approach requires experimental evaluation of the damage evolution relations and as damage development does not necessarily lead to ultimate failure, a global failure criterion is also required.

As different failure modes are present in composite failure, and each of them is related to a failure probability, Mahadevan and Liu [24] recently proposed a probabilistic technique for predicting the ultimate strength of composite structures. The probabilistic progressive failure analysis was performed by evaluating a weakest link model and the correlation between different failures and sequences. The technique considered the ideas of stiffness reduction due to damage and the replacement of the damaged material, with microcracks, by an equivalent material of degraded elastic properties. This reliability based technique accounts for ply-level failure modes as matrix cracking, fiber failure and delamination. The component reliability is evaluated at each iterative step using ANSYS finite element code. The technique was applied to a composite aircraft wing structure to estimate the system failure probability.

4.2.3 Fracture Mechanics

Crack deformation modes, or failure modes, are characterized by the relative movement of upper and lower crack surfaces with respect to each other. The crack is assumed to lie in the x_1x_3 plane and the crack front to be parallel to the x_3 axis. The deformation due to the three basic modes are shown in Figure 4.1.

Mode I, or opening mode, is characterized by the symmetric separation of the crack faces with respect to the x_1x_2 and x_1x_3 planes. In mode II, or sliding mode, the crack

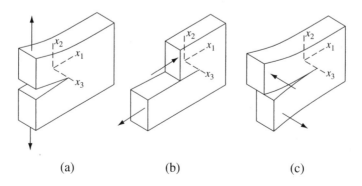

(a)   (b)   (c)

Figure 4.1   The three modes of cracking: (a) mode I, (b) mode II, (c) mode III

surfaces slide with respect to each other, symmetrically about the x_1x_2 plane but antisymmetrically about the x_1x_3 plane. Mode III, or tearing mode, is the one in which the crack faces also slide relative to each other, but antisymmetrically with respect to the x_1x_2 and x_1x_3 planes [18]. From these three basic modes any crack deformation can be represented by an appropriate linear superposition, as shown by Irwin [25].

A knowledge of the stress and displacement fields is essential in order to analyze the residual strength of anisotropic cracked bodies and predict the crack propagation path. The stresses and displacements in a small region surrounding the crack tip of an anisotropic plate in polar coordinates (r, θ) (see Figure 4.2) for loading modes I and II were presented in the works of Sih, Paris and Irwin [2] and Sih and Liebowitz [20].

Using superposition, the general state of stress and displacement near a crack tip can be considered as the sum of the components of the local modes of deformation in the opening and in the sliding modes, thus

$$\sigma_{ij} = \sigma_{ij}^I + \sigma_{ij}^{II}, \tag{4.4}$$

$$u_j = u_j^I + u_j^{II}, \tag{4.5}$$

where there is no implicit summation.

The stress intensity factors are the coefficients of the stress singularities at the crack tip, and can be defined for mode I as

$$K_I = \lim_{r \to 0}\{(2\pi r)^{\frac{1}{2}}\sigma_{22}(\theta = 0)\} \tag{4.6}$$

and similarly for mode II as

$$K_{II} = \lim_{r \to 0}\{(2\pi r)^{\frac{1}{2}}\sigma_{12}(\theta = 0)\}. \tag{4.7}$$

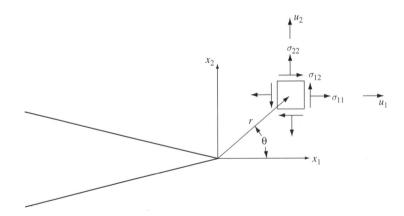

Figure 4.2   Displacements and stresses near the crack tip

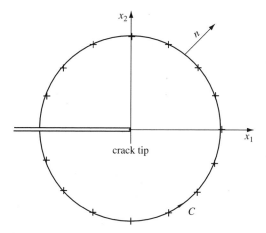

Figure 4.3   Circular contour for the evaluation of the *J*-integral

In order to define the *J*-integral, consider a generic contour *C* which is traversed in the counterclockwise sense between the two crack surfaces, as can be seen in Figure 4.3. The path independent *J*-integral is defined by Rice [21] as

$$J = \int_C (W \mathrm{d}x_2 - t_j u_{j,1}) \mathrm{d}C, \qquad (4.8)$$

where: $W = W(\epsilon_{ij})$ is the strain energy density; t_j are the traction components defined along the contour, given by $t_j = \sigma_{ij} n_i$, where σ_{ij} is the stress tensor and n_i are the components of the unit outward normal to the contour path. As so, the *J*-integral is the rate of energy released per unit of crack extension along the x_1 axis. It can be shown that the integral of equation (4.8) around the boundary of any simply connected region will vanish if W depends uniquely on ϵ_{ij} and no singularities occur within that region. No body forces are considered in this definition of the *J*-integral.

4.3   BOUNDARY ELEMENT FRACTURE MECHANICS

4.3.1   Introduction

As analytical solutions are only available for very limited cases, numerical methods tend to be the most frequent tools to solve crack problems. This section concentrates on boundary element solutions to anisotropic fracture mechanics problems.

The stress intensity factors were defined by equations (4.6) and (4.7) as the coefficients of stress singularities at the crack tip by a limiting condition of the stress field near the crack tip in an anisotropic medium. Hence, the stress intensity factors are a measure of the strength of the singularity and its evaluation is the essence of fracture mechanics analysis. A review of numerical methods for the evaluation of stress intensity factors can be found in Reference [26].

Techniques for computing stress intensity factors for orthotropic and anisotropic materials can be classified into three main groups: modified mapping collocation technique, crack Green's function method, quarter point elements and J_k-integral techniques.

Crack Green's functions were introduced in conjunction with the boundary element method by Snyder and Cruse [8]. The formulation used a special form of fundamental solution (Green's function) containing the exact form of the traction-free crack in an infinite medium, hence no modeling of the crack surface was required.

Traction singular quarter-point boundary elements were introduced by Martínez and Domínguez [27] for stress intensity factors computation for isotropic materials. The technique was used by Smith [28], who analyzed the influence of the length of the quarter-point element in the accuracy of the computed K_I and K_{II}.

The J-integral was extended to mixed mode loadings for isotropic materials by Hellen and Blackburn [29] who introduced the J_k-integral that enabled the separation of the coupled K_I and K_{II}. Chu and Hong [4] applied the J_k integral to mixed mode crack problems for anisotropic composite laminates using the finite element method.

This session presents the J-integral technique for the computation of stress intensity factors for anisotropic composite laminates by the boundary element method. A simple procedure was used for decoupling K_I and K_{II} based on the definition of the J-integral for anisotropic materials and the ratio of relative displacements at the crack tip [10]. This technique has been applied to several mode I and mixed mode numerical examples in glass, boron and graphite–epoxy composite laminates, in order to demonstrate its accuracy.

4.3.2 J_k-integral

The J_k-integral was introduced in order to extend the J-integral analysis to mixed-mode loading. The J_k-integral is defined as:

$$J_k = \int_C (Wn_k - t_j u_{j,k})\mathrm{d}C. \tag{4.9}$$

Notice that the J_1-integral is the Rice's path independent J-integral.

The J_k-integral can be shown [4] to be related to the stress intensity factors of a cracked homogeneous anisotropic plate by

$$J_1 = \alpha_{11}K_I^2 + \alpha_{12}K_I K_{II} + \alpha_{22}K_{II}^2, \tag{4.10}$$

$$J_2 = \beta_{11}K_I^2 + \beta_{12}K_I K_{II} + \beta_{22}K_{II}^2, \tag{4.11}$$

where the α_{ij} and β_{ij} are complex functions of the material elastic properties.

For traction free cracks the J_1-integral vanishes along the crack surfaces, whereas the J_2-integral would involve integration of highly singular integrands along each surface. In order to avoid this difficulty, an auxiliary relationship in terms of displacements ratios is developed here to be used together with J_1 for decoupling of the stress intensity factors K_I and K_{II}.

The coupling of the stress intensity factors has been a limiting factor in the analysis of cracked composite materials under mixed mode loading. However, a simple procedure can be introduced for the decoupling of mode I and mode II stress intensity factors, based on the ratio of relative displacements and equation (4.10), the relation of the J-integral for anisotropic materials, and K_I and K_{II}.

The relative sliding and opening displacements δ_n, for $\theta = \pm\pi$ are given by

$$\delta_1 = 2\sqrt{\frac{2r}{\pi}}(D_{11}K_I + D_{12}K_{II}) \tag{4.12}$$

and

$$\delta_2 = 2\sqrt{\frac{2r}{\pi}}(D_{21}K_I + D_{22}K_{II}), \tag{4.13}$$

where D_{ij} are functions of the complex parameters of the anisotropic material [10].

The ratio of relative displacements is

$$S = \frac{\delta_2}{\delta_1} = \frac{D_{21}K_I + D_{22}K_{II}}{D_{11}K_I + D_{12}K_{II}}, \tag{4.14}$$

and the ratio of stress intensity factors

$$F = \frac{K_I}{K_{II}} = \frac{SD_{12} - D_{22}}{D_{21} - SD_{11}}, \tag{4.15}$$

or

$$K_I = FK_{II}. \tag{4.16}$$

Substituting equation (4.16) into equation (10), and solving for K_{II} gives the following relationship:

$$K_{II} = \left(\frac{J_1}{\alpha_{11}F^2 + \alpha_{12}F + \alpha_{22}}\right)^{\frac{1}{2}} \tag{4.17}$$

Thus K_{II} and K_I are now decoupled and can be obtained from a knowledge of S, F, J_1 and α_{ij}.

The implementation of this procedure is straightforward. In order to evaluate the J_1-integral by equation (4.9), the integration path C is discretized into N line segments (see Figure 4.3) and the integration is carried out along each C^j segment as follows:

$$J_1 = \sum_{j=1}^{N}\left\{\int_{C^j}(W^j n_1^j - t_i^j u_{i,1}^j)\mathrm{d}C^j\right\}, \tag{4.18}$$

where N is the total number of segments. The internal strains are calculated using the usual definitions of strain in terms of displacement derivatives, and hence the internal stresses from Hooke's law.

4.4 PREDICTIVE MODELING OF CRACK PROPAGATION

4.4.1 Introduction

Crack growth analysis is a valuable technique in the assessment of damage tolerance and identification of critical areas for design and nondestructive inspection. The technique requires a crack growth criterion within the framework of fracture mechanics, which predicts the angle of crack initiation and propagation, and not only maximum strain or stress levels.

Failure theories for anisotropic materials are used for preliminary design purposes of composite structures, in order to impose acceptable stress and strain levels. An advantage of these theories, also known as strength theories, is their good correlation to experimental data.

Crack-growth theories are used to predict the direction and load at which a fracture propagates in an anisotropic material. The most simple and broadly used of these theories is the maximum circumferential stress theory, initially proposed by Erdogan and Sih [30] for the analysis of isotropic materials. The strain energy density theory proposed by Sih [31] has been proven to be very effective in solving mixed mode problems of crack growth in isotropic materials. The technique was extended to anisotropic materials by Sih [32].

The finite element has been a classical method for the analysis of fracture in composite materials. The method is suitable for a nonhomogeneous analysis and was applied by Eischen [33]. Crack initiation and propagation problems in orthotropic materials, using the maximum circumferential stress theory, was presented by Boone, Wawrzynek and Ingraffea [34].

The boundary element method has also been applied to crack propagation problems in orthotropic materials. Doblare *et al.* [35] applied the subregion technique and maximum circumferential stress theory to evaluate the ratio of mode I and mode II stress intensity factors and the angle of crack propagation. As the introduction of the artificial boundaries of the subregion technique is not unique it cannot be easily implemented as an automatic procedure in an incremental analysis of crack extension problems.

This section presents the dual boundary element method for crack propagation problems in anisotropic materials. An incremental crack extension analysis is performed to predict the crack path which is assumed piece-wise straight. After each increment a dual boundary element analysis is performed and the mixed mode stress intensity factors are computed by the *J*-integral technique.

4.4.2 Failure Theories

Failure modes in fiber-reinforced composite systems are very complex as they may involve fiber breakage, matrix cracking, fiber-matrix debonding and laminae delamination during a period of gradually increasing load [32]. The analysis of each of these failure modes, and the evaluation of its eventual dominance over the failure process, is beyond the scope of this work.

There are two main schools of thought for the analysis of cracked composite systems concerning the scaling of the continuum element relative to crack size and microstructure.

The first considers a global nonhomogeneous anisotropic medium composed by a stiff fiber isotropic medium embedded in an isotropic matrix. The crack tip can exist only in one phase of the composite system at a given instance, lying in the fiber, the matrix or at the interface between the two media. A similar approach was used by Selvadurai [36] for a boundary element analysis. The second assumes a global homogeneous anisotropic continuum such that crack extension takes place in an idealized material with the gross combined properties of the constituents. This approach excludes detailed consideration of local damage such as broken fibers, cracks in matrix, etc, and is the usual approach for boundary element applications (see for example Ref. [6]). The validity of these approaches depends on the relative size of crack and microstructure and the stress level. This work assumes the global homogeneous anisotropic continuum approach.

Failure theories for anisotropic materials were initially extensions of isotropic analysis to account for anisotropy. The diversity of mechanisms of failure of composites and the focus on the correlation to experimental data have resulted in a number of failure theories. Reviews of failure theories point out that there are no well-established fracture criteria used in the design of composite structures and that there is generally only a limited amount of dependable experimental data [18, 37].

Among the homogeneous anisotropic failure theories, the Tsai–Hill [38] criterion has been a classical approach. This theory states that in plane stress state the failure envelope is defined by

$$\left(\frac{\sigma_{11}}{X}\right)^2 - \frac{\sigma_{11}\sigma_{22}}{X^2} + \left(\frac{\sigma_{22}}{Y}\right)^2 + \left(\frac{\sigma_{12}}{S}\right)^2 = 1 \tag{4.19}$$

where X, Y and S are the uniaxial strengths parallel and perpendicular to fibers and in shear. Equation (4.19) was obtained by Tsai from a yield criterion for anisotropic materials proposed by Hill [39], by assuming the transversely isotropic condition for fiber-reinforced composites.

4.4.3 Crack Growth Theories

The theory of the maximum circumferential stress was derived by Erdogan and Sih [30] using a fracture mechanics approach to predict the maximum stress level causing the crack to propagate and the propagation direction. The criterion had different extensions to anisotropic materials and has been one of the most broadly used, see for example, Refs [34, 35]. The main advantages of the method are related to its simplicity and use of just the anisotropic elastic properties, without the need for any additional experimental data – usually not available to the analyst.

The maximum circumferential stress criterion states that the crack will propagate in the direction θ along with the circumferential stress $\sigma_{\theta\theta}$ reaches its maximum value. The combined stress field near a crack tip in an anisotropic material, for the superposition of modes I and II stresses was presented in equation (4.4). The circumferential stress is formulated in polar coordinates and can be obtained as in Ref. [40] by

$$\sigma_{\theta\theta} = \sigma_{11} \sin^2 \theta + \sigma_{22} \cos^2 \theta - 2\sigma_{12} \sin \theta \cos \theta. \tag{4.20}$$

Substituting the stress field equations in (4.4) and then in (4.20), the classical method for obtaining the maximum value of resulting equation (taking its first derivative to θ equal to zero and checking the sign of the second derivative) appears not to have a simple analytical solution. As a consequence, a numerical solution was implemented using bracketing and the golden-section search technique.

The strain-energy density criterion is a theory built on the framework of fracture mechanics that predicts the onset of failure as crack instability, or rapid crack propagation, through the critical density factor [32]:

$$S_c = r_c \left(\frac{\mathrm{d}W}{\mathrm{d}V} \right)_c \qquad (4.21)$$

with r_c being the radius of a core region surrounding the crack tip and $\mathrm{d}W/\mathrm{d}V$ is the strain energy density function. For a homogeneous anisotropic material this function is given by

$$\frac{\mathrm{d}W}{\mathrm{d}V} = \frac{1}{2}(\sigma_{11}\epsilon_{11} + \sigma_{22}\epsilon_{22} + 2\sigma_{12}\epsilon_{12}) \qquad (4.22)$$

where σ_{ij} are the stress field components for a crack in an anisotropic medium, as presented by Sih, Paris and Irwin [2]. Equation (4.22) yields an expression involving $\frac{1}{r}$, the coefficient of which is the strain energy density factor

$$S(\theta) = \frac{1}{\pi}(A_{11}K_I^2 + 2A_{12}K_I K_{II} + A_{22}K_{II}^2), \qquad (4.23)$$

where $A_{ij}(\theta)$ are functions of a complex variable μ_k, the characteristic property of the anisotropic material.

According to the strain energy density theory, unstable crack growth takes place in the radial direction θ_c along which S becomes minimum, or

$$\frac{\partial S}{\partial \theta} = 0 \text{ and } \frac{\partial^2 S}{\partial \theta^2} > 0. \qquad (4.24)$$

The maximum circumferential stress and the strain energy density criterion were applied in conjunction with the global failure theory of Tsai–Hill for predicting crack propagation paths in composites using the dual boundary element method [41] and some results will be shown in next section.

4.5   NUMERICAL RESULTS

4.5.1   Central Slant Crack in an Unidirectional Composite

Figure 4.4(a) shows the composite plate with a central slant crack used to compare the different propagation models. The slant crack forms an 'angle β to the tensile loading

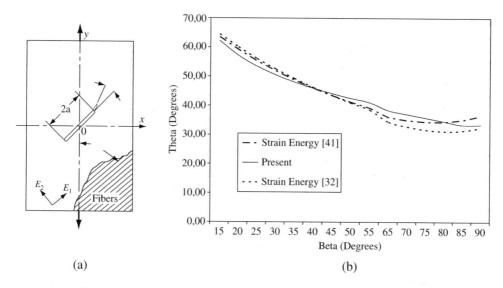

Figure 4.4 (a) Central slant crack in an unidirecional composite. (b) Comparison of the crack initiation angle θ

axis. The fibers of the unidirectional composite were aligned to the crack axis. The material is a Modulite II 5206 composite of which the gross mechanical properties are: $E_1 = 158.0\,\text{GPa}$, $E_2 = 15.3\,\text{GPa}$, $G_{12} = 5.52\,\text{GPa}$ and $\nu_{12} = 0.34$. Other material and model properties were: $X = 0.689\,\text{GPa}$, $Y = 0.028\,\text{GPa}$, $S = 0.062\,\text{GPa}$ and $r = 0.1\,\text{m}^{-3}$.

The results of the crack initiation angle θ obtained by the present implementation of the Tsai–Hill global failure theory, combined with the strain energy density criterion, are presented in Figure 4.4(b), compared to previous results obtained by implementations of the strain energy density criterion by Sollero, Mendonca and Aliabadi [41] and Sih [32].

Figure 4.4(b) shows a small difference (under 3 %) between the present values of the crack initiation angle θ and those found by Sih for values less than 60°. For larger values of the crack angle there is an increasing difference between these implementations, reaching 12 %.

4.5.2 Crack from a Hole Specimen

A crack from a hole specimen, shown in Figure 4.5, was designed in order to predict the crack propagation path. The material is a graphite–epoxy unidirectional laminate, with the fibers rotated at 60° and the orthotropic properties are: $E_1 = 144.8\,\text{GPa}$, $E_2 = 11.7\,\text{GPa}$, $G_{12} = 9.65\,\text{GPa}$ and $\nu_{12} = 0.21$.

The incremental crack growth analysis was performed with an initial crack length to width ratio $a/w = 0.15$. The initial and final steps of the crack path are shown in Figure 4.6. The crack propagates in the direction dictated by the minimum strain energy density criterion, in fiber breakage mode. As the crack approaches hole B, the path is affected by the stress field of this hole and tend to approach it.

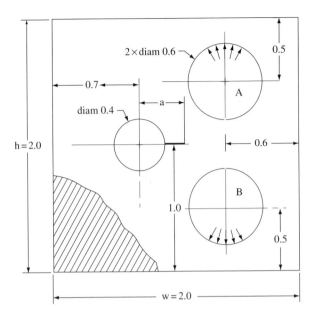

Figure 4.5   Crack from a hole specimen

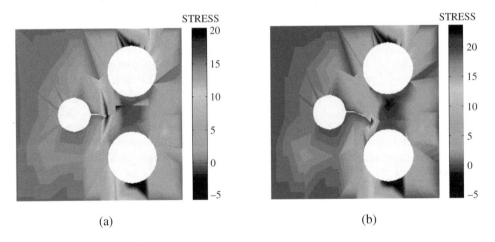

(a)   (b)

Figure 4.6   Crack growth path for the crack from a hole specimen: (a) initial step; (b) final step

4.6 CONCLUSIONS

A predictive modeling of crack propagation for anisotropic materials has been developed using the dual-boundary-element method. This analysis is performed in order to predict the crack propagation path. The stress intensity factors are evaluated after each increment of the crack propagation path by the J-integral technique.

The use of the dual-boundary-element method enabled the crack growth analysis to be performed in a single region with a substantial reduction in the remeshing efforts. The crack extension is achieved by introducing new boundary elements at the crack tip.

Failure and crack growth theories were reviewed and homogeneous and nonhomogeneous models were compared. The present approach was validated for the direction of crack propagation when compared with values from the literature.

A crack from a circular hole specimen was used to analyze the crack propagation path in an unidirectional laminate. The stress concentration field influenced the crack propagation path.

ACKNOWLEDGMENTS

I wish to thank Professor M.H. Aliabadi for his continuous advise and to my colleague Dr E.L. Albuquerque for his contribution. I also would like acknowledge the financial support of the Brazilian National Council for Scientific and Technological Development (CNPq).

REFERENCES

[1] Chan, K.S. and Cruse, T.A. Stress intensity factors for anisotropic compact-tension specimens with inclined cracks, *Engng. Fracture Mech.* **23**, 863–874 (1986).

[2] Sih, G.C., Paris, P.C. and Irwin, G.R. On cracks in linear anisotropic bodies, *Int. J. Fracture Mechanics* **3**, 189–203 (1965).

[3] Heppler, G.R. and Hansen, J.S. A high accuracy finite element analysis of cracked rectilinear anisotropic structures subjected to biaxial stress, in *Numerical Methods in Fracture Mechanics*, D.R.J. Owen and A.R. Luxmoore (Eds.), Proc. Sec. Int. Conference, Swansea, 223–237 (1980).

[4] Chu, S.J. and Hong, C.S. Application of the J_k integral to mixed mode crack problems for anisotropic composite laminates, *Engng. Fracture Mechanics* **35**, 1093–1103 (1990).

[5] Aliabadi, M.H. *The Boundary Element Method*, Volume 2, *Applications in Solids and Structures*, John Wiley & Sons, New York (2002).

[6] Zhen, J.Q. and Brebbia, C.A. Boundary element analysis of composite materials, in *Computer Aided Design in Composite Material Technology*, C.A. Brebbia, W.P. de Wilde and W.R. Blain (Eds.), Proc. Int. Conf., Southampton, Computational Mechanics, Southampton, 475–498 (1988).

[7] Hwu, C. and Yen, W.J. Green's functions of two-dimensional anisotropic plates containing an elliptic hole, *Int. J. Solids Structures* **27**, 1705–1719 (1991).

[8] Snyder, M.D. and Cruse, T.A. Boundary-integral equation analysis of cracked anisotropic plates, *Int. J. Fracture* **11**, 315–328 (1975).

[9] Tan, C.L. and Gao, Y.L. Boundary integral equation fracture mechanics analysis of plane orthotropic bodies, *Int. J. Fracture* **53**, 343–365 (1992).

[10] Sollero, P. and Aliabadi, M.H. Fracture mechanics analysis of anisotropic plates by the boundary element method, *Int. J. Fracture* **64**, 269–284 (1993).

[11] Portela, A., Aliabadi, M.H. and Rooke, D.P. The dual boundary element method: effective implementation for crack problems, *Int. J. Numerical Methods Engng.* **33**, 1269–1287 (1992).

[12] Portela, A. *Dual Boundary Element Incremental Analysis of Crack Growth*, PhD thesis, Wessex Institute of Technology (1992).

[13] Lemaitre, J. and Chaboche, J.L. *Mechanics of Solid Materials*, Cambridge University, Press, Cambridge (1990).

[14] Ladeveze, P. and LeDantec, E. Damage modelling of the elementary ply for laminated composites, *Composite Science and Technology* **43**, 257–267 (1992).

[15] Johnson, A.F. and Simon, J. Modelling fabric reinforced composites under impact loads, *Proc. EURO-MECH 400*: Impact and Damage Tolerance Modelling of Composite Materials and Structures, London, 1–8 (1999).

[16] Salgado, N. *Boundary Element Methods for Damage Tolerance Design or Aircraft Structures, Topics in Engineering*, Vol. 33, Computational Mechanics Publications, Southampton. 1998.

[17] *Federal Aviation Administration, Damage Tolerance Assessment Handbook*, Volume II: *Airframe Damage Tolerance Evaluation*, US Department of Transportation, DOT/FAA/CT-93/69.I, Springfield (1993).

[18] Kanninen, M.F. and Popelar, C.H. *Advanced Fracture Mechanics*, Oxford University Press, New York (1985).

[19] Aliabadi, M.H. and Rooke, D.P. *Numerical Fracture Mechanics*, Computational Mechanical Publications, Southampton, and Kluwer Academic Publishers, the Netherlands (1991).

[20] Sih, G.C. and Liebowitz, H. Mathematical theories of brittle fracture, in *Fracture: An Advanced Treatise*, H. Liebowitz (Ed.), Vol.II, Academic Press, New York (1968).

[21] Rice, J.R. A path-independent integral and the approximate analysis of strain concentration by notches and cracks, *J. Applied Mech.*, **35**, 379–386, (1968).

[22] McCartney, L.N. Physically based damage models for laminated composites, *Proc. Instn. Mech. Engrs. Part L: Materials: Design and Applications* **217**, 163–199 (2003).

[23] Paris, F. *A Study of Failure Criteria of Fibrous Composite Materials*, NASA Report CR-2001-210661, Langley Research Center, Hampton (2001).

[24] Mahadevan, S. and Liu, X. Probabilistic analysis of composite structure ultimate strength, *AIAA Journal* **40**, 1408–1414 (2002).

[25] Irwin, G.R. Analysis of stresses and strains near the end of a crack traversing a plate, *J. Applied Mech.*, **24**, 361–364, (1957).

[26] Aliabadi, M.H. Boundary element formulations in fracture mechanics, *ASME Appl. Mech. Rev.* **50**, 83–96 (1997).

[27] Martínez, J. and Domínguez, J. On the use of quarter point boundary elements for stress intensity factors computations, *Int. J. Numerical Methods Engng.* **20**, 1941–1950 (1984).

[28] Smith, R.N.L. The solution of mixed-mode fracture problems using the boundary element method, *Engng. Analysis* **5**, 75–80 (1988).

[29] Hellen, T.K. and Blackburn, W.S. The calculation of stress intensity factors for combined tensile and shear loading, *Int. J. Fracture*, **11**, 605–617, (1975).

[30] Erdogan, F. and Sih, G.C. On the crack extension in plates under plane loading and transverse shear, *J. Basic Engng.* **85**, 519–527 (1963).

[31] Sih, G.C. A special theory of crack propagation, in *Methods of Analysis and Solutions of Crack Problems, Mechanics of Fracture I*, G.C. Sih (Ed.), pp. XXI–XLV, Noordhoff (1973).

[32] Sih, G.C. *Mechanics of Fracture Initiation and Propagation*, Kluwer Academic, Dordrecht (1991).

[33] Eischen, J.W. *Fracture of Nonhomogeneous Materials*, PhD Thesis, University of Stanford, Stanford (1986).

[34] Boone, T.J., Wawrzynek, P.A. and Ingraffea, A.R. Finite element modelling of fracture propagation in orthotropic materials, *Engng. Fracture Mech.* **26**, 185–201 (1987).

[35] Doblare, M., Espiga, F., Gracia, L. and Alcantud, M. Study of crack propagation in orthotropic materials by using the boundary element method, *Engng. Fracture Mech.* **37**, 953–967 (1990).

[36] Selvadurai, A.P.S. Fracture mechanics of a reinforced solid: a boundary element approach, in *Advances in BEM in Japan and USA, Topics in Engineering* 7, M. Tanaka, C.A. Brebbia and R. Shaw (Eds.), Proc. Int. Conf., Computational Mechanics, Southampton, 107–117 (1990).

[37] Rowlands, R.E. Strength (failure) theories and their experimental correlation, in *Handbook of Composites, Failure Mechanics of Composites* 3, Chapter II, G.C. Sih and A.M. Skudra (Eds.), Elsevier, Amsterdam (1985).

[38] Tsai, S.W. Strength theories of filamentary structures, in *Fundamental Aspects of Fiber Reinforced Plastic Composites*, Chapter 1, R.T. Schwartz and H.S. Schwartz (Eds.), Interscience, New York (1968).

[39] Hill, R. *The Mathematical Theory of Plasticity*, Oxford University Press, Oxford (1950).

[40] Timoshenko, S.P. and Goodier, J.N. *Theory of Elasticity*, Third Edition, McGraw-Hill, New York (1970).

[41] Sollero, P., Mendonca, S.G. and Aliabadi, M.H. Crack growth prediction in composites using the dual boundary element method, in *Progress in Durability Analysis of Composite Systems*, K.L. Reifsnider and D.A. Dillard (Eds.), Proc. Int. Conf., Blacksburg, pp. 7.62–7.65 (1997).

5

On Friction Induced Nonideal Vibrations: A Source of Fatigue

José M. Balthazar[1] and Bento R. Pontes[2]

[1] State University of São Paulo at Rio Claro (UNESP-Rio Claro), Rio Claro, Brazil
[2] State University of São Paulo at Bauru (UNESP-Bauru), Bauru, Brazil

5.1 PRELIMINARY REMARKS

Fatigue is common problem considered in the design of all structural and machine components that are subjected to repeated or fluctuation loads. We mention that cyclic loading can alter the stress – strain response of a material and we note that fatigue is an important aspect of material cycle behavior and, on the other hand, fatigue damage is caused by cyclic plasticity, i.e. cyclic stresses cause damage only to the extent that they plastically deform the material. The subject of fatigue is complex and can be studied in many different ways. Although vibration-based methods are generally accepted as tools for examining structural integrity, they still pose a lot of questions and challenges. Besides the major questions of vibration-based procedures related to the use of modal characteristics, frequency or time-domain data, the selection of measurement points, the frequency selection, the problem with nonlinear behavior of the system becomes a major difficulty in a lot of cases. In general, any structure contains nonlinearities stemming from nonlinear material behavior, geometric nonlinearities, nonlinearities in the supports or connections, and so on. We assumed that one could determine, either theoretically or experimentally, all the variables that describe the behavior of a vibrating system and the introduction of a damage (defect) in a structure can be regarded as an

Damage Prognosis – For Aerospace, Civil and Mechanical Systems   Edited by D.J. Inman, C.R. Farrar, V. Lopes Junior and
V. Steffen Junior   © 2005 John Wiley & Sons, Ltd

additional nonlinearity. The introduction of a new nonlinearity, as well as its growth, changes the vibrational response of the structure and is expected to influence its nonlinear dynamic behavior. The idea of employing some nonlinear dynamics characteristics, and especially the state–space representation of the structural vibration response for damage diagnosis purposes, is a relatively new one. This approach applies nonlinear dynamics tools for the purposes of reconstruction of a model space and employs some nonlinear dynamics invariants for the purpose of condition monitoring instead of traditional methods. It has also been observed that some characteristics related to state–space representation of the structural vibration time signals change with the initiation and the accumulation of cracks in structures (Trendafilova and Brussels, 2001).

5.1.1 *Some Observations on Modeling and Friction*

We know that model validation procedures and numerical predictions should theoretically be revisited as soon as new measurements and new data from the damaged system become available. These iterations should not be restricted to parametric calibration experiments. When new measurements become available, the values of calibrated parameters can surely be confirmed or recalibrated. In addition, the model form should also be allowed to evolve. For example, a material model that represents the mechanism by which energy is dissipated through friction or impact between two components should be augmented with a crack propagation model once it is assessed that a crack has formed and is growing (Farrar *et al.*, 2001). Friction plays an important role, in a wide variety of engineering problems. Note that the literature is rich in reviews of this subject (see for instance, Oden and Martins, 1985; Armstrong *et al.*, 1994; Fenny *et al.*, 1998; Ibrahim and Evin, 1994a, b; Berger, 2002). We know from the current literature that examples of friction vibrators, which appear in everyday engineering or life problems, can be easily found. We can mention some of them, such as squealing railway wheels on narrow curves, grating brakes of trains or automobiles, chattering machine tools, rating joints of robotic manipulators, drill strings and so on (see for instance Hinrichs *et al.*, 1997).

We also remarked that classes of vibrating systems that present an interaction phenomenon due to dry friction are those machines that use flexible elements such as wakes or belts. Note that the belt transmits motion between pulleys or interacts with the other mechanisms through surface contact. An analysis of the nonlinear dynamics of such a system became fundamental for the solution of engineering problems concerning the control of vibrations for functionality purposes. Therefore, several scientific investigations of interacting phenomena have been carried out with the objective of developing new design and control techniques (for example, Pontes, 2003; Thomsen, 1999). It is agreed in the literature that in all of these above examples, self-sustained (excited) vibrations occur due to dry friction (as an example, see Popp and Stelter, 1990). The characteristics of dry friction in mechanical systems produce two main effects: energy dissipation and self-excitation effects (Hagedorn, 1988). Self-excitation occurs in engineering systems where dry-friction forces have significant influence on the system operation. Because of this, dry friction has been the object of many experimental investigations. Mathematical models have been proposed to study dry friction static and dynamic properties (Canudas de Wit *et al.*, 1995; McMillan, 1997; Liang and Fenny, 1998; Hinrichs *et al.*, 1998). However, systems with dry friction, in general, are nonsmooth and may cause difficulties in both theoretical and numerical analyses. The direct calculations of Lyapunov exponents cannot be used in nonsmooth systems and a special approach to the analysis of stick–slip, slip–slip and slip–stick motion transitions should be developed (Hinrichs *et al.*, 1997). The explanation of the so-called stick–slip phenomenon can be found in a lot of the literature and here we mention only some of them (Magnus, 1986; McMillan, 1997; Hinrichs *et al.*, 1998; Liang and Fenny, 1998; Thomsen, 1999).

There are in the literature a number of works on the theory of oscillations with self-excitation. Self-excitation occurs in cases where the oscillator possesses a nonlinear behavior dependent on damping forces, that is, the damping force tends to increase the amplitude of the oscillations when they present small amplitude, and tends to decrease them when they present large amplitude. A possible steady-state motion is found when the system gains energy during part of the cycle and loses energy during the remaining part of the cycle, in such a way that at the end of each cycle the net energy is null. Several authors studied this type of problem. Nayfeh and Mook (1979) studied self-excited oscillations using perturbation methods. We also know that in many engineering problems, stick–slip vibrations are not desired and should to be avoided due the data base precision of motion, safety of operation, noise and wear (see, for instance, Andronov *et al.*, 1996; Hagedorn, 1988; Fenny *et al.*, 1998; Serrarens *et al.*, 1998; Van de Grande *et al.*, 1999).

Next we will present a short comment on fretting fatigue as relating to the goals of this chapter.

5.1.2 *Fretting Fatigue*

We may examine the friction interface problem, with various approaches. Among many of them, we know that fretting fatigue refers to the degradation of mechanical properties of a material in a region of contact where cyclically tangential loads occur in the presence of normal or clamping loads. We note that such processes involve cyclic tangential load motion only near the edges of contact (slip region) while the central region remains in full contact, normally referred as stick region (Naboulsi and Nicholas, 2003). We also know that in addition to limiting cycle oscillations, it is possible to have unstable vibrations. These may be produced in the presence of high coefficients of friction corresponding to frictional seizure. These vibrations are much more chaotic than the limit cycle cases, yet they can still predict by a deterministic model. We remarked that a better representation of the coefficient of friction and possibly its dependence on pressure and sliding velocity is needed to more precisely model friction-induced instability in the presence of high local pressures and surface damages (Tworzydlo *et al.*, 1999). In order to develop a study of friction and to predict dynamical systems response and performance, a robust friction model must be employed.

5.1.2.1 *Cumulative Fatigue*

Predicting the life of parts stressed above the endurance limit is an empirical procedure. For the large percentage of mechanical and structural parts subjected to randomly varying stress cycle intensity (e.g., automotive suspension and aircraft structural components), the prediction of fatigue life is further complicated. An analytical procedure, that is used by prediction of fatigue life, often called the 'linear cumulative-damage rule' (Juvinall and Marshek, 1991; Shigley *et al.*, 2003). The simple concept proposed is: if a part is cyclically loaded at a stress level causing failure in 10^5 cycles, each cycle of this loading consumes one part in 10^5 of the life of the part; if other stress cycles are interposed corresponding to a life of 10^4 cycles, each of these consumes one part in 10^4 of the life, and so on. When, on this basis, 100 % of the life has been consumed, fatigue is predicted. The Palmgren–Miner rule may be expressed by following the equation in which $n_1, n_2, \ldots n_k$ represent the number of cycles at specific overstress levels, and $N_1, N_2, \ldots, N_k$ represent the life (in cycles) at these overstress levels, as taken from the appropriate S–N curve. The S–N curves are constructed from the results of the fatigue tests using values defining: the **fatigue strength** S_f (that is, the intensity of reversed stress causing failure after a given number of cycles) and when there is the **endurance limit** S_n (that is, the highest level of alternating stress that can be withstood indefinitely without failure. Numerous tests have established that ferrous materials have an endurance limit.). Fatigue failure is predicted when (Juvinall and Marshek, 1991; Shigley *et al.*, 2003):

$$\frac{n_1}{N_1} + \frac{n_2}{N_2} + \cdots + \frac{n_k}{N_k} = 1 \quad \text{or} \quad \sum_{j=1}^{j=k} \frac{n_j}{N_j} = 1$$

5.1.3 *Friction-Induced Self-Excited Vibrations with Limited Power Supply*

We taken into account those self-excited vibrations that are attributed to friction-induced dynamic coupling between certain degrees of freedom of the system. Numerical results included a limited power supply in order to excite the structure to take into account the interactions between them and a limited power supply, that is nonideal vibrations (see, for example, Kononenko, 1969; Nayfeh and Mook, 1979; Balthazar *et al.*, 2003). An ideal energy source is one that acts on the oscillating system, but does not experience any reciprocal influence from the system. Such an ideal source ensures a specified law of action of the oscillating system independently of the conditions of motion of such a system, like an external force of given frequency and amplitude, constant velocity, or voltage in an electric circuit, etc. The action of the ideal energy source on an oscillating system can be conveniently represented by means of an explicit function of time. A nonideal energy source is one that acts on an oscillating system and at the same time experiences a reciprocal action from the system. Alteration in the parameters of the system may be accompanied by alterations in the working conditions of the energy source. This interaction may become especially active when the energy source has very limited power. In practice almost all energy sources have a limited number of imperfections for particular conditions. Note that, since the influence of a nonideal energy source of an oscillating system depends on the state of its motion, it is impossible to express this action as an explicit function of time. Thus an oscillating system with a nonideal source must be considered as autonomous. In summary, as simple example, an ideal (linear) system, excited by a harmonic force (well defined) obeys an equation

$$\ddot{X} + \mu\dot{X} + \omega^2 X = f\cos(\Omega t). \tag{5.1}$$

where both amplitude (f) and frequency (Ω) are assumed to be known beforehand, in other words there is a certain ideal energy source that produces the exciting force. This force has constant parameters (amplitude and frequency) that are independent of both motion and of the vibrating system. The parameters of the vibrating system are damping coefficient μ and natural frequency ω^2. Note that the dynamic equation of only one DC motor is

$$I\ddot{\varphi} = L_m(\varphi, \dot{\varphi}) - H(\varphi, \dot{\varphi}) \tag{5.2}$$

where I is the moment of inertia of the rotating mass; L_m is the driving torque of the source of energy (DC motor) and H is the resisting torque applied to the rotor. We remarked that instead of L_m in equation (5.2) we should introduce an additional equation describing the internal dynamical process of the energy source. For example, we mention that for a DC motor this will describe the processes of the electrical circuits. We also remarked that the characteristics of the mechanical energy source specify the relationship between the torque (L_m) and the velocity of rotation ($\dot{\varphi}$) which is kept constant while the relationship is found. Usually the characteristic curve for a DC motor is taken as a linear function of $\dot{\varphi}$. Usually the functions L_m and H are assumed to be known, and to be continuous for values of $\dot{\varphi}$ in the range of interest. A nonideal system, is described by the equations (taken only one DC motor)

$$\ddot{X} + \mu\dot{X} + \omega^2 X = Q(X, \dot{X}, \varphi, \dot{\varphi}) \tag{5.3a}$$
$$I\ddot{\varphi} = L_m(\varphi, \dot{\varphi}) - H(\varphi, \dot{\varphi}) + R(\varphi, \dot{\varphi}, X, \dot{X}) \tag{5.3b}$$

where Q and R are the terms expressing the coupling of the vibrating system with the nonideal source of energy. It is important to note that for the majority of motions (Kononenko, 1969), equation (5.3 b) will given in the following form

$$I\ddot{\varphi} = \varepsilon\{L_m(\varphi, \dot{\varphi}) - H(\varphi, \dot{\varphi}) + R(\varphi, \dot{\varphi}, X, \dot{X})\} \tag{5.3c}$$

where ε is a small parameter of the problem. Usually the problem is solved by using perturbation methods. It is known that systems excited by ideal sources are typically predictable near resonance, i.e. frequency response curves for the system closely approximate the system behavior over the entire frequency domain. In contrast, a system driven by a nonideal energy source may exhibit some peculiar deviations from this well-defined behavior. In addition we also know that, in the ideal system a DC motor operating on a structure requires a certain input (power) to produce a certain output (RPM) regardless of the motion of the structure. For a nonideal system this may be not the case; hence we are interested in what happens to the DC motor, input and output, as the response of the system changes. If we consider the region before resonance on a typical frequency response curve, we note that as the power supplied to the source increases, the RPM of the motor increases accordingly. However, this behavior doesn't continue indefinitely. That is, the closer the motor speed moves toward the resonant frequency the more power the source requires increasing the motor speed, that is, a large change in the power supplied to the motor results in a small change in the frequency, but a large increase in the amplitude of the resulting vibrations. Thus, near resonance it appears that additional power supplied to the DC motor increases the amplitude of the response while having little effect on the RPM of the motor. This phenomenon is referred to as the Sommerfeld effect, that is, the structural response or vibrations provide a kind of energy sink. One of the problems faced by designers is how to drive a system through resonance and avoid this kind of energy sink. In this study, we will clearly illustrate the problem as well the benefits associated with driving a system through resonance. Note that the behavior of friction and systems with limited power supply (nonideal) was analyzed by few authors, in different aspects (Alifov and Frolov, 1977). Some authors took into account the chaotic behavior of the nonideal system (Pontes *et al.*, 2000; Pontes *et al.*, 2003a, b).

5.1.4 Control of Nonideal Vibrating Problems

In this chapter, we address the problem of stick–slip compensation of a 2-DOF mass-spring-belt system that interacts with an energy source of limited power supply. As the equilibrium of the system is not the origin, a conventional feedback controller does not remove the steady-state error. Typically, adding an integral action to the feedback law could eliminate the steady-state error. However, undesirable limit cycling can occur because of the slipping friction force. The alternative proposed here to regulate the position of the mass is the use of a switching control law combining a state feedback term and a discontinuous term. The discontinuous term is made active only in the region near the desired reference (Southward *et al.*, 1991). Then, here, we use a kind of control, studied before by Tataryn *et al.*, (1996). We also mention that:

(i) They developed an experimental study on the industrial manipulator (robot) to evaluate the performances of four selected friction compensation techniques, two modified integral actions and two robust nonlinear compensators.

(ii) The main element of the controller was a simple PD control.

(iii) The friction compensation techniques used and performances obtained: (1) rate-varying integral (RVI) – eliminated steady-state errors originating from stick friction but did not eliminate tracking errors from slip friction; (2) reset–offset integral (ROI) – a better performance reducing the startup errors and oscillations caused by the interaction of friction and integral gain; (3) discontinuous nonlinear proportional feedback (DNPF) (similar to this work) – eliminated the steady-state errors due to static friction but had no slip friction compensation. Also, its discontinuous performance about desired set points produced control signal oscillations; (4) smooth robust nonlinear feedback (SRNF) – eliminated the majority of the positional errors caused by stick–slip friction in both regulating and tracking tasks, with the least amount to unwanted control signal oscillation.

(iv) The compensation of viscous friction on the actual responses was not directly addressed. The effect of viscous friction was, however, observed and addressed.

(v) The nonlinear robust compensators utilized position measurements and knowledge of the maximum estimate for stick friction.

(vi) The compensation gains have to be readjusted to achieve optimal results. We remarked that an extension of this work to a two-degrees of freedom system can be seen in (Pontes *et al.* 2003a).

Finally, in order to conclude these preliminary remarks, we mention that this chapter is organized as follows: in Section 5.2 we analyze the nonlinear dynamics of ideal and nonideal stick–slip vibrations; in Section 5.3 we discuss a switching control for stick–slip vibrations; and in Section 5.4 we present the conclusions of the this chapter.

5.2 NONLINEAR DYNAMICS OF IDEAL AND NONIDEAL STICK–SLIP VIBRATIONS

In the present section an oscillating block-belt-motor system as in (Pontes *et al.*, 2000) and (Pontes *et al.*, 2003a, b) is analyzed. The analyzed nonideal system is described by the oscillating block-belt dynamical system described by Figure 5.1.

The motion equations can be written as follows (Pontes *et al.*, 2000, 2003a, b)

$$m\ddot{x} + c\dot{x} + kx = F_f(x, \dot{x}, v_{\text{Rel}}) + F_0 \cos(\omega_E t) \tag{5.4a}$$

$$I\ddot{\varphi} = T_{Motor}(\dot{\varphi}) - rF_f \tag{5.4b}$$

where $x, \dot{x}$ and $\ddot{x}$ are the oscillating block displacement, velocity and acceleration, respectively; m the block mass; c the viscous damping coefficient; k the elastic constant; F_0 and ω_E the amplitude and the frequency of the external excitation force, respectively; $v_{\text{Rel}} = v_B - \dot{x} = r\dot{\varphi} - \dot{x}$ the relative velocity between block and belt; v_B the belt velocity; r the radius of the belt pulley or transmission rate; $\varphi, \dot{\varphi}$ and $\ddot{\varphi}$ are the angular displacement, velocity and acceleration of the DC motor shaft (power source), respectively; I the inertia moment of the system rotate part; $T_{Motor}(\cdot)$ the mechanical torque, described by

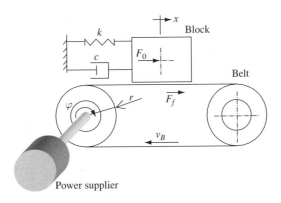

Figure 5.1 Mechanical system with energy source: mass block-belt-motor system

the DC motor dynamic characteristic; $F_f(\cdot)$ the friction force interaction function which represents the static and dynamic friction effects (Andronov *et al.*, 1996; Hagedorn, 1988). Also are defined F_D the dynamic friction force amplitude; F_S the static friction force amplitude; $\omega_N = \sqrt{k/m}$ the natural frequency of the mass-spring system; $f_0 = F_0/m$ the amplitude of the external excitation force normalized by the block mass. The rotational motion equation (5.4 b) describes the interaction between the power source and the friction force. The mechanical torque T_{Motor}, can be described by the stationary characteristic curves given in Figure 5.2.

In an ideal system, a system of equations is obtained considering the equilibrium between the motor torque and the required torque for the oscillating block-belt system. However, the angular acceleration is null and the angular velocity is constant. Then, the ideal system model may be represented by the oscillating system equation only, the first equation of the system (5.4 a). The model for the friction force $F_f(\cdot)$ used in this work is based on the model introduced by Karnopp (1985) and Southward *et al.* (1991). The model for the friction force F_f combines the stick and slip forces as follows:

$$F_f = F_{slip}(\nu_{Rel}[\lambda(\nu_{Rel})] + F_{stick}(F_R)[1 - \lambda(\nu_{Rel})] \tag{5.5}$$

with

$$\lambda(\nu_{Rel}) = \begin{cases} 1 & |\nu_{Rel}| > \eta \\ 0 & |\nu_{Rel}| \le \eta \end{cases}, \quad \eta > 0 \tag{5.6}$$

where F_{slip} is the dynamic friction force representing the friction force during the slipping, F_{stick} is the static friction force for the sticking mode, F_R is the resultant force on the block mass and η defines a region near-zero velocity as $|\nu_{Rel}| < \eta$ where $\eta \ll \nu_{Rel}$. The friction force term $F_{stick}(\cdot)$, also referred to as static friction force, represents the friction at zero velocity. This force exists when there is no relative motion

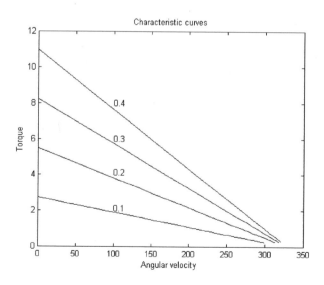

Figure 5.2 Torque stationary characteristic curves of the DC motor for a given applied voltage, for torque constant $K_T = 0.1; 0.2; 0.3; 0.4$ without current limitation

between the mass and the belt. With relative motion, the mass will break free and enter into a slipping mode. The model adopted for the static friction $F_{stick}(\cdot)$ is of the form:

$$F_{stick}(F_R) = \begin{cases} F_s^+ & F_R \geq F_s^+ > 0 \\ F_R & F_s^- < F_R < F_s^+ \\ F_s^- & F_R < F_s^- < 0 \end{cases} \tag{5.7}$$

where F_s^+ and F_s^- denote the upper and lower bounds of the friction respectively. These bounds may have different values. When the block is not moving relative to the driving belt, that is, when $v_{Rel} = 0$, there is a transition from stick to slip when F_R exceeds the stick friction force limit levels. The slipping friction force $F_{slip}(\cdot)$ provides the friction force at nonzero relative velocity and is represented as:

$$F_{slip}(v_{Rel}) = F_d^+(v_{Rel} - \eta)\mu(v_{Rel}) + F_d^-(v_{Rel} + \eta)\mu(-v_{Rel}) \tag{5.8}$$

where $\mu(\cdot)$ is the right-continuous Heavyside step function, $F_d^+(\cdot)$ and $F_d^-(\cdot)$ define the slipping force for positive and negative velocity, respectively. Upper and lower limits for $F_d^+(\cdot)$ and $F_d^-(\cdot)$ are $F_s^+(\cdot)$ and $F_s^-(\cdot)$, respectively, as the slipping force limits near zero velocity do not in general exceed the respective static friction force limits. The slipping friction force is also referenced as dynamic friction force. Figure 5.3 illustrates the slipping friction model adopted in this work. After some manipulation, note that the mass block-belt dynamical system and the rotational motion equations describe the analyzed nonideal problem (see equations 5.4 a and 5.4 b). Defining the state variables $x_1 = x$, $x_2 = \dot{x}$, $x_3 = \varphi$, $x_4 = \dot{\varphi}$, the nonideal system model may be represented by the state space equation system thus:

$$\frac{dx_1}{d\tau} = \dot{x} = x_2; \quad \frac{dx_2}{d\tau} = \ddot{x} = -(2\xi\omega_N)x_2 - (\omega_N)^2 x_1 - \frac{F_f}{m} + f_0\cos(\omega_E\tau)$$

$$\frac{dx_3}{d\tau} = \dot{\varphi} = x_4; \quad \frac{dx_4}{d\tau} = \ddot{\varphi} = \frac{(T_{motor} - rF_f)}{I} \tag{5.9}$$

where $\xi = \frac{c}{C_{CR}} = \frac{c}{2m\omega_N}$ is the damping ratio; $\omega_N = \sqrt{\frac{k}{m}}$ the natural frequency of the mass-spring system. The ideal system model may be represented by the state space equation system

$$\frac{dx_1}{d\tau} = \dot{x} = x_2; \quad \frac{dx_2}{d\tau} = \ddot{x} = -(2\xi\omega_N)x_2 - (\omega_N)^2 x_1 - \frac{F_f}{m} + f_0\cos(\omega_E\tau). \tag{5.10}$$

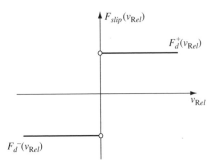

Figure 5.3   The slipping force model

Next we will do some numerical simulations on ideal and nonideal problems, in order to understand the behavior of these vibrating systems. A bifurcation diagram is obtained as a set of Poincaré maps when varying chosen control parameters. The system control parameter considered here is the dynamic friction force amplitude F_D and the Poincaré map points are obtained when the oscillating system trajectory crosses the zero relative velocity, that is, the motion transition points of stick–slip oscillations. By this bifurcation diagram, as shown in Figure 5.4, the influence of the friction force amplitudes on the slip–stick, slip–slip and slip–stick transition points of the ideal system are investigated. The motion transition points as a function of the dynamic friction force amplitude are characterized by bifurcational-like behavior. For a large difference between the friction force values during stick–slip oscillations, the motion presented three transition points.

Two sets of results were obtained on the dependence of the nonideal system on the motor characteristic. The torque constant values assumed are $K_T = 0.5$ and $K_T = 0.05$. In both cases, the same applied voltage values were used. Each set of results show the block and motor angular velocity, the phase portrait velocity displacement, the Poincaré map and the frequency spectrum. The results for the torque constant $K_T = 0.5$ and $K_T = 0.05$ are presented in Figures 5.5 and 5.6 respectively. A torque constant $K_T = 0.5$ represents an energy source with a high power supply. The results showed that for a torque constant $K_T = 0.05$, the oscillating system response suffers more influence from the energy source. The interaction between the motor and the oscillating system is evidenced in the different phase portraits and Poincaré maps also showed in these figures. Note that the sequence of results presented in Figures 5.5 and 5.6 show the occurrence of quasiperiodic (regular) dynamics and nonperiodic (irregular) dynamics. For instance, an irregular motion with chaotic characteristic is detected in Figure 5.6 for a torque constant $K_T = 0.05$.

Next we will discuss a kind of control for the above ideal and nonideal problems.

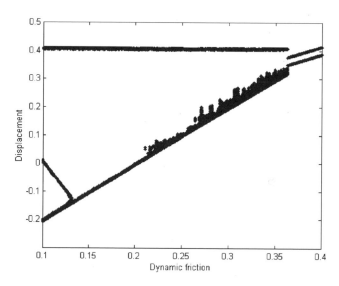

Figure 5.4   Bifurcation diagram depending on the dynamic friction force amplitude F_D of the ideal system ($\nu_B = 0.14$, $F_s = 0.4$)

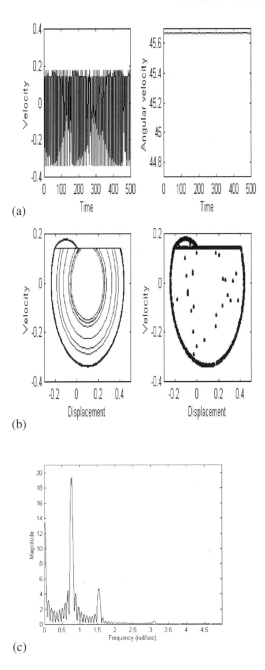

Figure 5.5 Nonideal system behavior for $K_T = 0.5$, $\nu_B \approx 0.14$, $F_D = 0.1$ and $F_S = 0.4$: (a) velocity and angular velocity time responses; (b) phase portrait velocity-displacement and Poincaré section (6400 points); (c) frequency spectrum

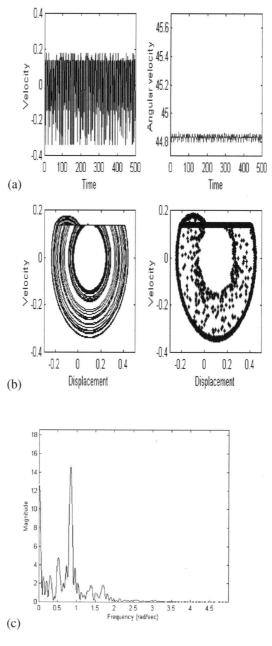

Figure 5.6 Nonideal system behavior for $K_T = 0.05$, $\nu_B \approx 0.14$, $F_D = 0.1$ and $F_S = 0.4$: (a) velocity and angular velocity time responses; (b) phase portrait velocity-displacement and Poincaré section (6400 points); (c) frequency spectrum

5.3 SWITCHING CONTROL FOR IDEAL AND NONIDEAL STICK–SLIP VIBRATIONS

The equation of motion for the oscillating block-belt system is (Pontes *et al.*, 2003a):

$$m\ddot{x} + c\dot{x} + kx = F_f(x, \dot{x}, \nu_{Rel}) + F \tag{5.11}$$

where x and $\dot{x}$ are the oscillating block displacement and velocity, respectively; m the block mass, c *the* viscous damping coefficient, k the elastic constant; F the control force; $\nu_{Rel} = \nu_E - \dot{x} = r\dot{\varphi} - \dot{x}$ the relative velocity between block and belt; ν_B the belt velocity; r the radius of the belt pulley or transmission rate, $\dot{\varphi}$ the angular velocity of the DC motor shaft; F_f the friction force interaction function which represents the static and dynamic friction effects. The model for the friction force F_f used in this work is based on the model introduced by Karnopp (1985) and Southward *et al.* (1991), presented in Section 5.2.

5.3.1 The Switching Control Strategy

An alternative to the PID stabilization of the position of the mass is a switching control law combining a state feedback term and a discontinuous term dependent on the mass position. The discontinuous term is made active only in the region near the desired reference. The same strategy can be applied to the problem of tracking a periodic orbit. The proposed control law follows the development found in Southward *et al.* (1991) for a 1-DOF system. In the state-space form (5.11) takes the form:

$$\dot{x}_1 = x_2$$
$$\dot{x}_2 = -\frac{c}{m}x_2 - \frac{k}{m}x_1 + \frac{1}{m}(F_f(x_1, x_2, \nu_{Rel}) + \frac{1}{m}F \tag{5.12}$$

with $x_0 = x(0)$.

5.3.2 Asymptotic Stabilization Problem

We consider initially the state feedback controller law:

$$F = -K_1 x_1 - K_2 x_2 \tag{5.13}$$

The equilibrium point for system (5.12) using (5.5) with control (5.13) is obtained and using F_{slip} (5.8) for $\eta = 0$ one gets the equilibrium point:

$$\overline{x}_1 = \frac{F_{slip}(\nu_{Rel})}{(k + K_1)}, \quad \overline{x}_2 = 0 \tag{5.14}$$

Assuming $\nu_B > 0$, we have $\nu_{Rel} = \nu_B - x_2 > 0$ which implies that the equilibrium point is in the positive part of the x_1 axis in the phase plane. For the friction model adopted $F_{slip}(\nu_{Rel} > 0) = F_d^+$ which yields $\overline{x}_1 = \frac{F_d^+}{(k+K_1)}$. The equilibrium point $\overline{x}_1$ is not the origin

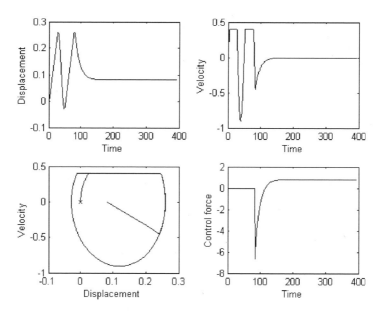

Figure 5.7 Ideal case: system behavior for a feedback controller with $K_1 = 10$ and $K_2 = 20$. The initial condition is marked as *

and we have a steady-state error. At this equilibrium point, the trajectory is in the slipping mode and the feedback control force does not compensate the dynamic friction force. The steady-state error obtained for this situation is illustrated in Figure 5.7 for $\nu_B = 0.4\,\text{m/s}$, $|F_s^+| = |F_s^-| = 10\,\text{N}, |F_d^+| = |F_d^-| = 4\,\text{N}, \xi = 0.05$ and $\omega_N = \sqrt{k/m} = 2\pi\,\text{rad/s}$.

5.3.3 The Switching Term

To compensate the stick-slip friction we add to (5.13) a switching term of the form:

$$F_s = K_1 x_c(x_1) \tag{5.15}$$

where

$$x_c(x_1) = \begin{cases} 0 & \gamma_u < x_1 \\ (\gamma_u - x_1) & 0 < x_1 \le \gamma_u \\ 0 & x_1 = 0 \\ (\gamma_l - x_1) & \gamma_l \le x_1 < 0 \\ 0 & x_1 < \gamma_l \end{cases} \tag{5.16}$$

with $\gamma_u = \bar{x}_1 + \varepsilon$, $\gamma_l = -\bar{x}_1 - \varepsilon$, $\varepsilon > 0$. The choice of the function $x_c(x_1)$ as in (5.16) guarantees that $-\bar{x}_1 \le (x_1 + x_c(x_1)) \le \bar{x}_1$ is only satisfied for $x_1 = 0$, where the compensation force limits are $\tilde{F}_s^+ = F_s^+ + K_1\varepsilon$, $\tilde{F}_s^- = F_s^- - K_1\varepsilon$. When the mass block is in the interval $-\bar{x}_1 \le x_1 \le \bar{x}_1$, any positive value for ε is sufficient to have the feedback force greater than the static friction force limits. This guarantees that the solution of (5.12) moves towards the origin (Pontes, 2003a). The stability proof was presented in Pontes (2003).

5.3.4 *Simulation Results*

In this section, we present numerical results for both position regulation and tracking problems of the system (5.12) for different conditions considering the switching feedback control force, where $\omega_N := \sqrt{k/m} = 2\pi \, \text{rad/s}$. The numerical results presented show the occurrence of different types of motion. In what follows, the results (Figures 5.8 to 5.10) show the responses to initial conditions with and without the switching term.

5.3.4.1 *Position Regulation Problem*

Figure 5.8 shows the responses to initial conditions with the switching term for the ideal case. The solution moving to the origin, removing the steady-state error observed in Figure 5.7 due to the stick-slip friction.

5.3.4.2 *Periodic Orbits Tracking Problem*

In this section we present results for the tracking of a sinusoidal trajectory. We consider here the design of the control input F of (5.12) such that the solution of the closed-loop system follows a prespecified periodic orbit. The stability of the feedback system can be analyzed using the same procedure as for the position regulation problem (Pontes, 2003). In Figures 5.9 and 5.10, the responses were obtained with the proposed controller for the nonideal case. In Figures 5.9(a) and 5.9(b), the stick–slip compensation forces $|\tilde{F}_s^+| = |\tilde{F}_s^-|$ are lower and higher than the static friction bounds, $|F_s^+| = |F_s^-| = 10 \, \text{N}$.

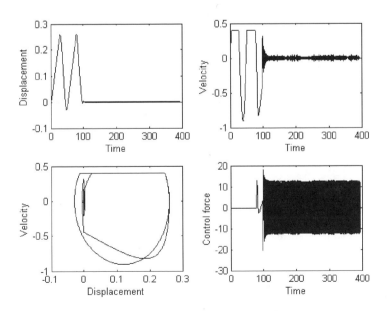

Figure 5.8 Ideal case: responses with the switching term for $x_0 = 0, K_1 = 10, K_2 = 20, \gamma_u = \gamma_l = 0.001, |\tilde{F}_s^+| = |\tilde{F}_s^-| = 12$

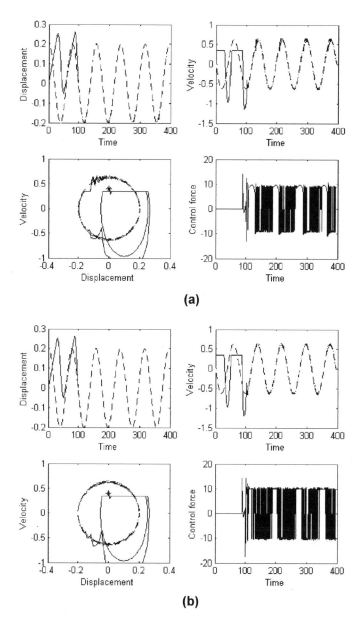

Figure 5.9 Nonideal case: responses with the switching term for $K_1 = 10, K_2 = 20$, $\gamma_u = \gamma_l = 0.01$; $x_0 = (0,0.4)$. (a) $|\tilde{F}_s^+| = |\tilde{F}_s^-| = 9.0$ N and (b) $|\tilde{F}_s^+| = |\tilde{F}_s^-| = 10.1$ N. The dashed line indicates the desired periodic orbit $x_1 = 0.2\sin\pi t$. The initial condition is marked as *

The results in Figure 5.9 (a) show a steady-state error greater around the stick mode when the influence of the static friction is dominant. These results show the efficiency of the switching term, mainly in the stick mode.

Figure 5.10 showed the chaotic motion behavior of nonideal case before control action and the tracking of periodic orbits with the switching term. The chaotic motion

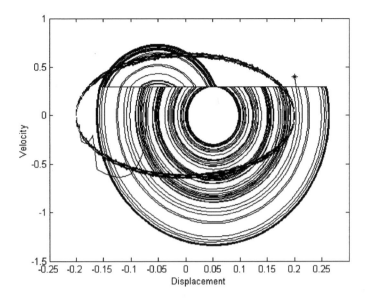

Figure 5.10 Nonideal case: chaotic motion before control action and tracking of periodic orbits with the switching term for $|F_s^+| = |F_s^-| = 10\,\mathrm{N}, |F_d^+| = |F_d^-| = 2\,\mathrm{N}, \nu_B = 0.3\,\mathrm{m/s}, K_1 = 10,$ $K_2 = 20, \gamma_u = \gamma_l = 0.01$ and $|\tilde{F}_s^+| = |\tilde{F}_s^-| = 10.1\,\mathrm{N}$. The dashed line indicates the desired periodic orbit $x_1 = 0.2sin\pi t$. The initial condition is marked as *

appearance was attributed to interaction between the friction force and the energy source (limited power supply). In this work a switching feedback controller for stick–slip compensation of a system, which interacts, with an energy source of limited power supply was developed. A switching-control law combining a state feedback term and a discontinuous term was proposed to regulate the position of the mass, and the problem of tracking a desired periodic trajectory was also considered. The results demonstrated that the feedback system is robust with respect to the friction force, which is assumed to be within known upper and lower bounds. An illustrative example of the control of chaotic motion was presented to the tracking problem of periodic trajectory (Pontes *et al.*, 2003a).

5.4 SOME CONCLUDING REMARKS

Through vibrations monitoring of a machine are detected their conditions of operation. By using identification of the damage we may identify the spoil of the components (fatigue or wear) due to the interactions of the surfaces with stick–slip damping (associated with fretting fatigue) and/or dynamics loads (oscillations or impacts).

The knowledge of the reason and behavior of stick–slip vibrations is a fundamental part of the interpretation of machine operation conditions, for we may distinguish the different signals and identify the occurrence of the partial cracks surfaces (on propagation and with damping) and/or complete fracture (impact and damping) in complex structural elements, for example in aircraft (wings).

The present work complements a preliminary investigation concerning stick–slip vibrations such as those related to damping models and Nonideal excitation energy sources. We also present a strategy of control, in order to suppress these vibrations, which may be a source of fatigue on machine components.

ACKNOWLEDGMENTS

The authors acknowledge support by FAPESP, Fundação de Apoio à Pesquisa do Estado de São Paulo; CNPq, Conselho Nacional de Desenvolvimento Científico e Tecnológico, both are Brazilian Research Funding Agencies, and Pan American Advance Study Institute on Damage Prognosis (October 19–30, Florianópolis, Sta. Catarina, Brasil, 2003).

REFERENCES

Alifov, A. and Frolov K.V. (1977) Oscillations in a system with dry friction and restrict excitations, *Mekanika Twerdo Tela*, **12**(4), 57–65.

Andronov A.A., Vitt, A.A. and Khaikin, S.E. (1996) *Theory of Oscillations*, Pergamon Press, London.

Armstrong-Helouvry, B., Dupont, P. and Canudas de Wit, C. (1994) A survey of models, analysis tools and compensation methods for the control of machines with friction, *Automatica*, **30**, 1083–1138.

Balthazar, J.M., Weber, H.I., Brasil, R.M.L.R.F., Fenili, A., Belato, D., Felix, J.L.P. and Garzeri, F.J. (2003) An overview on non-ideal vibrations, *Meccanica*, **38**, 613–624.

Berger E.J (2002) Friction modeling for dynamic system simulation, *Applied Mechanics Reviews – ASME*, **55**, 535–577.

Canudas De Wit, C., Olsson, H., Astrôm, K.J. and Lischinsky, P. (1995) A new model for control of systems with friction, *IEEE Trans. Automatic Control*, **40**, 419–425.

Farrar, C.R., Sohn, H., Hemez, F.M., Anderson, M.C., *et al.* (2001) Damage Prognosis: Current Status and Future Needs, *Los Alamos National Laboratory, Copyrighted Material, Report LA-14051-MS*.

Fenny, B., Guran, A., Hinrichs, N. and Popp, K. (1998) A historical review on dry friction and stick–slip phenomenon, *Applied Mechanics Reviews, Trans. ASME*, **51**, 321–341.

Hagedorn, P. (1988) *Nonlinear Oscillations*, Oxford Science Publications, New York.

Hinrichs, N., Oestreich, M. and Popp, K. (1997) Dynamics of oscillators with impact and friction, *Journal of Chaos, Solitons and Fractals*, **8**, 535–558.

Hinrichs, N., Oestreich, M. and Popp, K. (1998) On modeling of friction oscillators, *Journal of Sound and Vibration*, **216**, 435–459.

Ibrahim, R.A. (1994a) Friction induced vibration, chatter, squeal, and chaos – Part 1: Mechanics of contact and friction, *Applied Mechanics Reviews*, **47**, 209–226.

Ibrahim, R.A. (1994b) Friction induced vibration, chatter, squeal, and chaos – Part 2: Dynamics and modeling, *Applied Mechanics Reviews*, **47**, 227–253.

Juvinall, R.C. and Marsheck, K.M. (1991) *Fundamentals of Machine Component Design, Second Edition*, John Wiley & Sons, New York.

Karnopp, D. (1985) Computer simulation of stick–slip friction in mechanical dynamic systems, *Journal of Dynamic Systems, Measurement, and Control, Trans. ASME*, **107**, 100–103.

Kononenko, V.O. (1969) *Vibrating Systems with Limited Power Supply*, Iliffe Books, London.

Liang, J.W. and Fenny, B.F. (1998) Dynamical friction behavior in the forced oscillator with a compliant contact, *ASME J. Appl. Mech.*, **65**, 250–257.

McMillan, A.J. (1997) A nonlinear friction model for self-excited vibrations, *Journal of Sound and Vibration*, **205**, 323–335.

Magnus, K. (1986) *Schwingungen*, Teubner, Stuttgart.

Naboulsi S. and Nicholas, T. (2003) Limitations of the Coulomb friction assumption in fretting fatigue analysis, *International Journal of Solids and Structures*, **40**, 6497–6512.

Nayfeh, A.H. and Mook, D.T. (1979) *Nonlinear Oscillations*, Wiley Interscience, NY.

Oden, J.T. and Martins, J.A.C. (1985) Models and computational methods for dynamic friction phenomena, *Computing Methods in Applied Mechanics and Engineering*, **52**, 527–634.

Pontes, B.R. (2003) *On Dynamics and Control of Nonlinear Systems, Excited by Ideal and Non-Ideal Energy Sources*, PhD Dissertation, Department of Electrical Engineering, School of University of São Paulo at São Carlos, São Carlos, SP, Brazil.

Pontes, B.R., Oliveira, V.A. and Balthazar, J.M. (2000) On friction driven vibrations in a mass block-belt-motor system with a limited power supply, *Journal of Sound and Vibration*, **234**, 713–723.

Pontes, B.R., Oliveira, V.A. and Balthazar, J.M. (2003a) On the control of a non-ideal engineering system: A friction-driven oscillating system with limited power supply, *Materials Science Forum*, **440–441**, 353–362.

Pontes, B.R., Balthazar, J.M. and Oliveira, V.A. (2003b) Stick–slip chaos in a non-ideal self-excited system, In: *ASME International Design Engineering Technical Conference, September 2–6, 2003, Chicago, Illinois*, pp.1–6, CD-ROM (publication DETC 2003-VIB 48628).

Popp, K. and Stelter, P. (1990) Stick-slip vibrations and chaos, *Philosophical Transactions of the Royal Society London*, Ser. A, **332**, 89–195.

Serrarens, A.F.A., Molengraft, M.J.G., Kok, J.J. and Steen, L. (1998) H-Infinite control for suppressing stick–slip in oil well drillstrings, *IEEE Control Systems Magazine*, **18**, 305–325.

Shigley, J.E., Budynas, R.G. and Mischke, C.R. (2003) *Mechanical Engineering Design*, McGraw-Hill, New York.

Southward, S.C., Radcliffe, C.J. and MacCluer, C.R. (1991) Robust nonlinear stick–slip friction compensation, *Journal of Dynamics Systems, Measurements and Control*, **113**, 639–645.

Tataryn, P.D., Sepehri, N. and Strong, D. (1996) Experimental comparison of some compensation techniques for the control of manipulators with stick–slip friction. *Control Engineering Practice*, **4**, 1209–1219.

Thomsen, J.J. (1999) Using fast vibrations to quench friction-induced oscillations, *Journal of Sound and Vibration*, **228**, 1079–1102.

Trendafilova, I. and Brussels, H.V. (2001) Nonlinear dynamics tools for the motion analysis and condition monitoring of robot joints, *Mechanical Systems and Signal Processing*, **15**, 1141–1164.

Tworzydlo, W.W., Hamzeh, O.N. and Zaton, W. (1999) Friction-induced oscillations of a pin-on-disk slider: Analytical and experimental studies, *Wear*, **236**, 9–23.

Van de Grande, B.L., Van Campen, D.H. and De Kraker, A. (1999) An approximate analysis of dry-friction-induced stick–slip vibrations by a smoothing procedure, *Nonlinear Mechanics*, **19**, 157–169.

6

Incorporating and Updating of Damping in Finite Element Modeling

J.A. Pereira and R.M. Doi

Unesp – Ilha Solteira, Department of Mechanical Engineering, Ilha Solteira – SP, Brazil

6.1 INTRODUCTION

The entire life-cycle development of a product involves an iterative engineering effort, including different analyses, eventually some redesign and, mainly, engineering judgment. In this context, numerical models play an important part in obtaining a reliable product, as they can be used for simulation and prediction of the conditions of operation of the model and their respective limitations (Garcia, 2001; Oberkampf, 2001). In the case of structural dynamic analysis, finite elements can be a valuable tool, mainly due to its great versatility and ease of simulation of an endless number of modeling conditions rapidly and with relatively low cost. However, finite-element models are idealized models, based on simplifications that the analyst presupposes to be reasonable in representing the real physical system. In general, properties of components like joints and connections are difficult to obtain accurately (Ren and Beards, 1995; Ratcliffe and Lieven, 2000; Mottershead and Friswell, 2002), which may lead to unreliable results.

Another difficulty with finite element modeling is the incorporating of damping (Adhikari, 2000; Doi and Pereira, 2002) in the model. The inclusion and the understanding of the mechanism of damping into structural modeling are very difficult. It is common to face different damping mechanisms, like viscous, hysteretic and Coulomb damping in a model. Structures containing joints and connections are even more difficult to evaluate since these components cause concentrated damping in certain

Damage Prognosis – For Aerospace, Civil and Mechanical Systems   Edited by D.J. Inman, C.R. Farrar, V. Lopes Junior and V. Steffen Junior   © 2005 John Wiley & Sons, Ltd

regions, making damping to be nonproportional along the structure. Although, it is also possible to obtain an equivalent viscous damping through the equivalence of the dissipated damping energy (Richardson and Potter, 1975) of the model, and its inclusion in finite element modeling it is not common.

In the structural dynamics area, more representative models have been obtained by comparing the results of finite element analysis with measured experimental data and, if necessary, updating the parameters of the finite element model based on those measured data (Friswell and Mottershead, 1995; Pereira, 1996; Jones and Turcotte, 2002; Link, 2001). However, for reasons of mathematical simplification, the damping is not usually considered in finite element modeling (Mottershead *et al.*, 2000; Modak *et al.*, 2002). Consequently, no matter how close the experimental and analytical models, in terms of natural frequencies, there will probably be some discrepancy between the behavior of the two models due to damping.

Attempts to include damping in finite element model updating have been reported (Adhikari, 2000; Doi and Pereira, 2002; Göge and Link, 2002), but the many unknowns and limited information about damping in structural modeling suggest that the theme is still not sufficiently known scientifically, and therefore demands much study and development for real applications. This is still a subject that requires more research and development in order to benefit substantially the structural area (Cudney, 2001).

This chapter discusses the modeling of damping of a reasonable complex structure made up of bars and joints. A finite-element model updating procedure incorporating the effect of damping in that model has been used to improve the model of the structure. The updating approach is based on the measured FRF(s) and it seeks to obtain a more representative finite-element model, using the measured FRF(s) as reference. Correlation techniques are used to evaluate the representation of the finite-element model. The updating process, initially, is based on the correction of physical and geometric parameters of the model. In the second stage, a model of nonproportional damping formulated from the previous updated matrices is included in the finite-element model. The updating of the damping, in this case, is made by adjusting the damping coefficients of each distinct region of the model.

6.2 THEORETICAL FUNDAMENTALS

6.2.1 Model Updating

From a mathematical point of view, finite element model updating involves a process of minimization of a residue defined by the difference between the analytical and experimental data. Various strategies have been proposed to express this difference between the models; it could be a residue vector defined in terms of the modal parameters (Link, 2001), in terms of the frequency response function (Zang *et al.*, 2001), in terms of the spatial properties (mass/stiffness orthogonality, total mass) (Heylen, 1987) and others (Müller-Slany *et al.*, 1999). Sensitivity-based finite element model updating is a very promising technique, as it can improve the representativeness of the model by minimizing these residues in relation to a set of selected updating parameters. In a general way, it seeks changes in the finite element parameters that minimize the difference between the analytical and experimental data. By using the FRF(s) based formulation (Larson and Sas, 1992; Pereira, 1996), the residue can be written in terms of the spatial parameters of the model and the measured response properties. It leads to an expression that combines the experimental and analytical information of the model.

$$\{\varepsilon\} = \left[f^{A}\left(p_1 + \Delta p_1, p_2 + \Delta p_2, \ldots, p_{np} + \Delta p_{np}\right) \right] \{\Delta^{(E-A)}\} \tag{6.1}$$

where np is number of updating parameters; $\{\Delta^{(E-A)}\}$ is the known difference between experimental and analytical data; and f^{A} the function of the spatial parameters of the analytical model.

By minimizing a norm of the residue vector, $\{\varepsilon\}$, for each selected updating parameter, one may find changes of some parameters that improve the model. An objective function J for the solution of the problem can be defined by taking the sum of the square of the residue vector:

$$J = \{\varepsilon\}^{T}\{\varepsilon\} \tag{6.2}$$

The proposed sensitivity-based model updating method uses the frequency response function, FRF(s), instead of modal parameters to define the residue (Lammens, 1995; Pereira, 1996). This could present some advantages, since the FRF(s) are directly measured quantities, which can avoid errors during the extraction of the modal parameters. Furthermore, a huge amount of data is available to formulate the problem, which may improve the stability of the updating set of equations. However, it should be pointed out that some authors argue that the direct use of FRF(s) instead of the modal parameters in the updating process, does not bring great benefits because, in spite of the reduced number of points of data, the quality and the amount of information of the modal model is the same (Friswell and Penny, 1997).

The proposed method describes the discrepancy between the analytical and experimental models as a force residue. The input residue expresses the difference between the experimental force $\{F^{E}\}$ and the analytical force $\{F^{A}\}$ necessary to give an analytical displacement $\{X^{A}\}$ equal to the experimental displacement $\{X^{E}\}$.

$$\{\varepsilon\} = \{F^{A}\} - \{F^{E}\} \tag{6.3}$$

Rewriting equation (6.3) in terms of the analytical dynamic stiffness matrix $[Z^{A}]$ of the structural model, we get expression (6.4).

$$\{\varepsilon(\{p\})\} = [Z^{A}(\{p\})]\{X^{A}\} - \{F^{E}\} \tag{6.4}$$

where: $[Z^{A}] = [K] + i\omega[C] - \omega^{2}[M] = [H(\omega)]^{-1}$; $[K]$, $[M]$ and $[C]$ are stiffness, mass and damping matrices respectively; $[H]$ is the receptance matrix; and $\{p\}$ parameters of the model.

For an unit loading applied at the jth degree of freedom and taking into account that $\{X^{A}\} = \{X^{E}\}$, the residue can be expressed in terms of the dynamic stiffness matrix and the measured FRF(s).

$$\{\varepsilon\} = [Z^{A}(\{p\})]\{H^{E}\}_j - \{1\}_j \tag{6.5}$$

Equation (6.5) represents a set of nonlinear equations, for which the solution for the p-parameters of interest could exist or not. Assuming that the solution exists and that the components of the matrix $[Z^A]$ behave in an almost linear fashion, this matrix can be linearised by a truncated Taylor series (equation 6.6).

$$[Z^A(\{p\})] = [Z^A(\{p_0\})] + \sum_{i=1}^{np} \frac{\partial [Z^A(\{p\})]}{\partial p_i} \Delta p_i + \dots \tag{6.6}$$

Substituting equation (6.6) into equation (6.5), one could define the residue in terms of the first order sensitivity of the dynamic stiffness matrix.

$$\{\varepsilon^{lin}(\{p\})\} = [Z^A(\{p_0\})]\{H^E\}_j + \sum_{i=1}^{np} \frac{\partial [Z^A(\{p\})]}{\partial p_i} \Delta p_i \{H^E\}_j - \{1\}_j \tag{6.7}$$

Equation (6.7) requires analytical and experimental model compatibles (size/topology), and its solution gives the updated parameters of the model. In this chapter, the compatibility of the model size is discussed in terms of the reduction of the analytical model (Kidder, 1972). The relationship between $[Z]$ and $[H]$ first derivatives is used to calculate the derivatives.

$$\frac{\partial [Z^A]}{\partial p_i} = [Z^A] \frac{\partial [H^A]}{\partial p_i} [Z^A] \tag{6.8a}$$

$$\frac{\partial [H^A]}{\partial p_i} = [Z^A] \frac{\partial [Z^A]}{\partial p_i} [Z^A] \tag{6.8b}$$

Once the analytical model is larger than the experimental one, the above equations permit the definition of an exact expression for evaluating the derivatives of the reduced dynamic stiffness matrix $[Z^{Red}]$ (Larsson and Sas, 1992) in terms of the active degrees of freedom for a given frequency ω_k.

$$\{\varepsilon^{lin}(p)\} = [Z_k^{Red}(\{p_0\})]\{H_k^E\}_j + \sum_{j=1}^{np} \frac{\partial [Z_k^{Red}(\{p\})]}{\partial p_i} \Delta p_i \{H_k^E\}_j - \{1\}_j \tag{6.9}$$

A better conditioned system can be obtained by multiplying expression (6.9) by the inverse of the analytical dynamic stiffness matrix, making it less sensitive to noise on the measurement.

$$\{\tilde{\varepsilon}_k(p)\} = \sum_{i=1}^{np} [H_k^{Red}]^0 \frac{\partial [Z_k^{Red}(\{p\})]}{\partial p_i} \{H_k^E\}_j \Delta p_i + \{H_k^E\}_j - \{H_k^A\}_j \tag{6.10}$$

The solution of the linearized problem for a given frequency ω_k, may be calculated by least-squares approach. The resulting matrix of simultaneous updating equations, in a compact form, is given by the expression (6.11).

$$[S_k]\{\Delta p\} = \{\Delta H_k\}_j \tag{6.11}$$

where: $[S_k]$ is known sensitivity matrix; $\{\Delta p\}$ the unknown parameter changes; and $\{\Delta H_k\}$ the known vector difference.

Equation (6.11) is valid at the updating frequency point's ω_k, and it defines a set of m-equations in the updating parameters. The elements of the sensitivity matrix, $[S_k]$, are obtained from the analytical dynamic stiffness matrix and the difference vector, $\{\Delta H_k\}$, from the analytical and measured frequency response functions, at each selected frequency point. The elements of the vector $\{\Delta p\}$ are the unknown parameter corrections to be estimated.

The updating set of equation is solved by using the singular value decomposition (SVD) method (Golub and Loan, 1985) and the calculated changes of the parameters, Δp_i, are used to obtain, in an iterative way, an estimate of the updated mass, stiffness or damping matrices. If the discrepancy between the models is not yet sufficiently small, the procedure is repeated. An important feature of this procedure is that it operates at an element or region level and the changes of the parameters are directly related to this element or region, thus keeping the physical meaning of the updating of the model. The p-parameters can represent any physical or geometrical properties of the model and they can be adjusted independently or proportionally to each other.

6.2.2   Model Correlation and Parameters Sensitivity

The representativeness of the finite-element model can be evaluated through correlation techniques. In this chapter, the correlation of the analytical and experimental models will be evaluated through the relative difference of the natural frequencies, comparison of the mode shapes and the comparison of the FRF(s).

The frequency difference is described by direct comparison and the mode shape correlation is expressed by using the well-known modal assurance criterion (MAC) values.

$$\text{MAC} = \frac{\left|\{\psi^E\}^T\{\psi^A\}\right|^2}{\left(\{\psi^A\}^T\{\psi^A\}\right)\left(\{\psi^E\}^T\{\psi^E\}\right)} \tag{6.12}$$

MAC-value of 1 indicates a perfect correlation while MAC-value of 0 indicates absence of correlation between the experimental, $\{\psi^E\}$, and analytical, $\{\psi^A\}$, mode shapes.

The correlation of the FRF(s) is evaluated qualitatively through a graphic visualization of the superimposed analytical and experimental FRF(s) for equivalent measured points. It is also evaluated quantitatively by using the frequency response assurance criterion (FRAC) (Zang et al., 2001), which is defined, for a jth degree of freedom, by equation (6.13):

$$\text{FRAC}_j = \frac{\left|\{H^E(\omega_k)\}_j^T\{H^A(\omega_k)\}_j\right|^2}{\left(\{H^E(\omega_k)\}_j^T\{H^E(\omega_k)\}_j\right)\left(\{H^A(\omega_k)\}_j^T\{H^A(\omega_k)\}_j\right)} \tag{6.13}$$

Values of FRAC of 1 indicate a perfect correlation between the analytical and experimental FRF(s), while values of 0 indicate absence of correlation. It must be observed that finite element modeling without damping can present FRF(s) with differences of magnitude caused by absence of damping, resulting in low FRAC values even they visually present an acceptable correlation.

Although the analysis of the sensitivity of the parameters is not a direct method of location of error in modeling, it plays an important role in finite element model updating. It can give some information concerning the influence of each parameter or set of parameters on the behavior of the model. Calculation of the sensitivity of the parameters allows the identification of which components or parameters of the structure are more sensitive to small variations, indicating those components where updating is likely to work or not. However, the most sensitive regions are not necessarily the regions of the model that are inadequately represented in the analytical modeling, although they deserve special attention. The knowledge of the most sensitive parameters is quite relevant information for the selection of the updating parameters since it can avoid the selection of insensitive parameters, thus improving the performance of the updating process (Pereira, 1996).

The eigenvalues sensitivity of the model for a p-parameter is obtained by writing the motion equation of the model for the rth mode:

$$[K]\{\psi\}_r + \lambda_r[M]\{\psi\}_r = 0 \tag{6.14}$$

and taking its derivatives for a p parameter (Fox and Kapoor, 1968; Vanhonacker, 1985).

$$\frac{\partial[K]}{\partial p}\{\psi\}_r + [K]\frac{\partial\{\psi\}_r}{\partial p} - \frac{\partial\lambda_r}{\partial p}[M]\{\psi\}_i - \lambda_r\frac{\partial[M]}{\partial p}\{\psi\}_r - \lambda_r[M]\frac{\partial\{\psi\}_r}{\partial p} = 0 \tag{6.15}$$

Premultiplying equation (6.15) by the sth mode and taking advantage of the orthogonality properties of mass and stiffness matrices, the resulting equation can be rearranged and if one assumes that the mode shapes $r = s$, the first order sensitivity for the rth eigenvalue can be calculated by expression (6.16).

$$\frac{\partial\lambda_r}{\partial p} = \{\psi\}_r^T\frac{\partial[K]}{\partial p}\{\psi\}_r - \lambda_r\{\psi\}_r^T\frac{\partial[M]}{\partial p}\{\psi\}_r \tag{6.16}$$

For some specific elements the derivatives of mass and stiffness matrices can be easily obtained by analytical derivation of the matrices in relation to the updating parameters. However, for a more complex element, it is not feasible and the expression is evaluated numerically.

6.2.3 Incorporating of Damping in the Finite Element Model

In mechanical structures the effect of damping could come from a simple mechanism of damping or even from a combination. Different types of damping mechanisms have been found like viscous damping, hysteretic damping, Coulomb damping and others. Often they are responsible for the dissipation of the energy of the system (Richardson and Potter, 1975). Different mathematical models have been proposed to represent damping (Nashif et al., 1985; Tomlinson, 2001; Adhikari, 2000; Clough and Penzien, 1993) and the viscous damping model is one of the most used, mainly because of its simplicity and practicality for modal analysis purposes. This chapter describes the effect of damping in the finite-element model in terms of proportional and nonproportional damping matrices.

The proportional damping mechanism describes damping as being distributed uniformly along the whole structure and the nonproportional model describes damping as being distributed discretely in each region of the structure. In the first case, the damping could be defined as proportional to the mass matrix by a coefficient α, proportional to the stiffness matrix by a coefficient β, or proportional to the mass and stiffness matrix.

$$[C] = \alpha[M] + \beta[K] \tag{6.17}$$

From the orthogonality properties of the mass and stiffness matrices, equation (6.17) permits the definition of the modal damping ratio in terms of the natural frequency and the proportionality coefficients (Clough and Penzien, 1993). In this case, the damping ratio is inversely proportional to the frequency by the coefficient α and directly proportional to the frequency by the coefficient β equation (6.18).

$$\zeta_r = \frac{\alpha}{2\omega_r} + \frac{\beta\omega_r}{2} \tag{6.18}$$

In the case of nonproportional damping, i.e. the model presents different values of damping for each region, the damping matrix is defined from the damping properties of each distinct region of the model. The damping matrix of each region is calculated from expression (6.19).

$$[C_R] = \alpha_R[M_R] + \beta_R[K_R] \tag{6.19}$$

where: $R = 1, 2, \ldots, NuR$; and NuR is number of different regions or materials.

The distribution of the damping, in this case, is not uniform and the damping matrix of the whole model is defined by a combination of those matrices of each distinct region. Figure 6.1 shows schematically the mass, stiffness and damping matrices of a structure that presents proportionality damping coefficients different for region a and region b. The calculated damping matrix is included in the modeling and in this case, the FRF(s) calculated from the finite element model will contain the damping.

6.2.4 Strategy of the Updating

The finite-element model updating approach involves the effect of damping on the formulation of the set of updating equations. The sensitivity of the dynamic stiffness is defined in terms of the partial derivative of the dynamic stiffness matrix in relation to the set of updating parameters including geometrical, physical and damping coefficients.

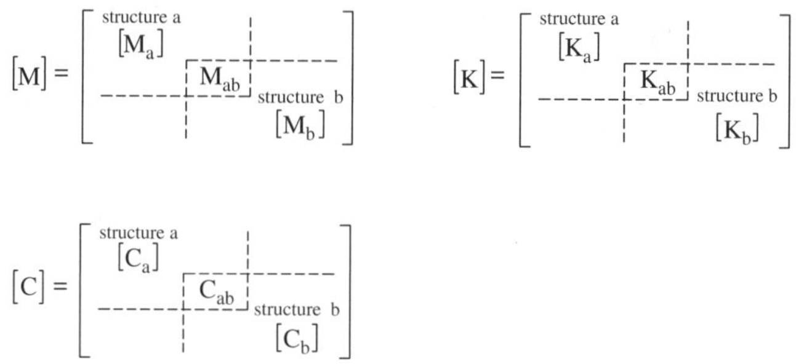

Figure 6.1 Matrices of the system with combined properties

The updating strategies to correct the parameters is accomplished in two very defined consecutive stages.

Stage 1
The model is updated in terms of the mass and stiffness matrices, without considering the damping. The updating is accomplished by correcting directly the selected physical and/or geometric parameters of the model, aiming at minimizing the discrepancy between analytical and experimental data. The procedure operates at an element or macro-region level of the finite element model and the changes to the model, during the updating process, could directly be related to the physical or geometric parameters of each element or macro-region. As the procedure makes the updating at local level, the symmetry, positive definiteness and connectivity characteristics of the FEM remain uncorrupted and the updated model can easily be compared with the original one. This feature facilitates the physical understanding of the changes of the model. The updating of the matrices are calculated by the expressions (6.20) and (6.21):

$$[K(\{p + \Delta p\})] = \left(\sum_{R=1}^{NuR} [K(\{p\})]_R \right) - [K(\{p\})]_{R^{th}} + [K(\{p + \Delta p\})]_{R^{th}} \tag{6.20}$$

$$[M(\{p + \Delta p\})] = \left(\sum_{R=1}^{NuR} [M(\{p\})]_R \right) - [M(\{p\})]_{R^{th}} + [M(\{p + \Delta p\})]_{R^{th}} \tag{6.21}$$

where $\overline{\sum}$ indicates a sum in the finite element sense; and R^{th} a selected element or region.

Stage 2
The updating process is accomplished by updating only the damping matrix. The previous updated matrix is not altered and the new adjusted model is obtained by updating the proportionality coefficients α_R and β_R, Eq. (6.22) or updating the modal damping ratio (equation 6.23).

$$[C(\{\alpha + \Delta \alpha, \beta + \Delta \beta\})] = \left(\sum_{R=1}^{NuR} [\alpha_R [M^U]_R + \beta_R [K^U]_R] \right) - [\alpha_R [M^U]_{R^{th}}$$
$$+ \beta_R [K^U]_{R^{th}}] + [(\alpha_R + \Delta \alpha_R)[M^U]_{R^{th}} + (\beta_R + \Delta \beta_R)[K^U]_{R^{th}}] \tag{6.22}$$

$$[C(\{\zeta + \Delta \zeta\})] = [C(\{\zeta\})] - [C(\zeta_r)] + [C(\zeta_r + \Delta \zeta_r)] \tag{6.23}$$

where $\{\zeta\}^T = \{\zeta_1 \zeta_2 \cdots \zeta_m\}$; and m is the number of active modes.

Since the previous updated parameters are unaffected during the updating of the damping, the coefficients α and β, can be updated at a global level, region level or even element level. This feature of the proposal can accommodate a nonuniform distribution of damping in the model.

6.3 APPLICATION

This section discusses the application of finite element model updating for a frame space structure. The test aims to evaluate the potential of the proposal to analyze a reasonably complex structure, involving the modeling of joints and connections. Joints and connections are structural components that present an additional difficulty to be represented in finite element modeling, and also, they can present an effect of concentrated damping of the model. The damping, in this case, will not be distributed uniformly in the model, thus characterizing a nonproportional damping. This effect becomes an additional obstacle for obtaining a more representative finite element model. Therefore the study of a structure presenting these characteristics is relevant to evaluate the potentiality of the proposed updating approach.

The structure used for this study is a space frame structure 3.0 m long, 0.707 m wide and 0.5 m high (Figure 6.2). It consists of bars connected by joints made of steel spheres, the bars are steel tubes with a conical part at each extremity. The bars are screwed into the spheres by using steel screw bolts. The bolts are tightened similarly in order to minimize the effect of nonlinearity under the loading conditions. The material of the components of the structure is steel, except the conical part of the bar, which is made of cast iron.

6.3.1 Structure Modeling

The analytical model of the structure was generated by finite elements, using beam element and concentrated mass. The modeling of the structure was performed by separating the structure into three distinct regions denominated *cylindrical components, conical components* and *screw components*, as detailed in Figure 6.3. Element of mass were added in the connection points among the screws and the conical components to take into consideration some effects of concentrated mass due to the spherical joints and also to the effect of the screws head.

Figure 6.4 illustrates details of the discretization of the mesh of the finite element of the structure. The model contains 174 beam elements and 66 mass elements, totaling 159 nodes, which correspond to 954 degrees of freedom. The mesh was generated by taking

Figure 6.2 Spatial frame structure test

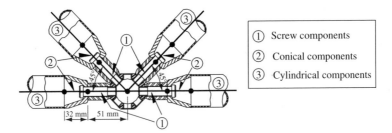

Figure 6.3   Details of the finite elements modeling of the joints

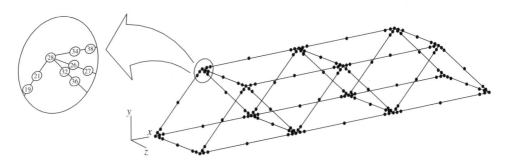

Figure 6.4   Finite elements model mesh

into account the topology of the measuring points of the experimental model, in such a way that the analytical model presents nodes corresponding to all of the measuring points. The structure was analyzed in a free–free condition.

The experimental test was realized taking into account the finite-element model aiming to obtain compatible geometry for the models. The excitation was of the impulsive type, applied at point 6 in the z-direction, using an instrument hammer. The responses were measured in a frequency range of 0–80 Hz, with three-axial accelerometer positioned at 43 points. The excitation and measuring points coincide with the joints and the medium position of each bar, see Figure 6.5.

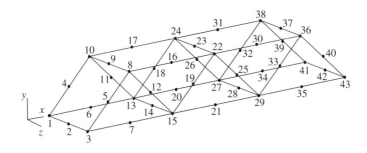

Figure 6.5   Model experimental mesh

6.3.2 Initial Correlation of the Models

The comparison of the models initially demands compatibility between the mesh of the FE model and the mesh of the experimental model. In this case the meshes of the analytical and experimental models present a geometric equivalence among the degrees of freedom, i.e. they were defined in a common system of reference. The difference among the number of degrees of freedom of the models is solved by using dynamic reduction.

The initial correlation of the models for the first seven mode shapes shows a satisfactory correlation of the mode shapes, MAC-values above 90%, exceptional for the seventh mode shapes. The differences between the natural frequencies are around 6%. Figure 6.6 shows the superposition of the FRF(s) of the models. The sixth and seventh modes are not sufficiently separated, which caused difficulties in the extraction of the seventh mode shape by taking the available FRF(s). Probably it will demand a multiexcitation test. In spite of the discrepancy of the frequencies, the FRF(s) present the same tendency, confirming that the initial finite-element model represents the same structure and that an appropriate updating of the finite element parameters could lead to a more representative finite element model.

6.3.3 Selection and Updating Parameters

The structure presents a considerable complexity and has a wide range of updating parameters, which makes the choice of the updating set of parameters not so obvious. Eigenvalue sensitivity analysis was used to identify the parameters most susceptible to changes and to define the more appropriate parameters for updating. The sensitivity is calculated in accordance with the different components of the finite element model. The model is separated into three distinct regions and the eigenvalue sensitivity of the parameters of each region is evaluated. The parameters are physical and geometric properties of the model, totaling 15 parameters as specified in Table 6.1.

Figure 6.7 shows the average sensitivity of the components for the first seven eigenvalues of the model. In the range of analysis, the most sensitive parameters are

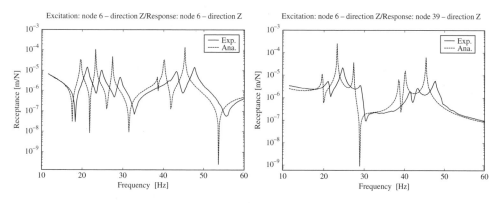

Figure 6.6 Superposition of FRF(s) experimental and analytical before the updating

Table 6.1 Numbering of the parameters for the sensibility analysis

Components	Physical parameters	Parameters p_i
Cylindrical and screws	Specific mass of the steel	1
	Poisson's coefficient of the steel	2
	Module of elasticity of the steel	3
Conical	Specific mass of the cast iron	4
	Poisson's coefficient of the cast iron	5
	Module of elasticity of the cast iron	6

Components	Geometric parameters	Parameters p_i
Cylindrical	Cross-section area	7
	Inertia for bending around axis z	8
	Inertia for bending around axis y	9
Conical	Cross-section area	10
	Inertia for bending around axis z	11
	Inertia for bending around axis y	12
Screws	Cross-section area	13
	Inertia for bending around axis z	14
	Inertia for bending around axis y	15

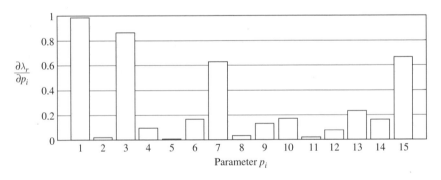

Figure 6.7 Average sensitivity of the structure

parameters 1, 3, 7 and 15, which correspond respectively to the modulus of elasticity of the steel components, specific mass of the steel components, the cross-section area of the cylindrical components and the momentum of inertia in y-direction of the screws components.

The selection of the set of updating parameters was carried out taking into account the average sensitivity of the parameters, as well as some engineering judgment. The values of the physical properties of the elements of the model, in this case, are more reliable when compared with their geometric properties, due to inaccuracy in the estimation of these properties. The screws components and mechanism of tightening the connections (see detail in Figure 6.3) present a quite complex geometry that takes the calculation of their properties and estimating less precise. Therefore, the selection of the updating parameters involves the geometric parameters, more specifically the inertia of the screw components. The parameters of inertia I_Y are selected and, due to the symmetry of the screws, the parameters of inertia, I_Z, are also updated in the same proportion of the inertia I_Y, aiming at obtaining physical consistence of the model.

6.3.4 Updating of the Physical and/or Geometric Parameters

In this stage, the effect of damping is still not considered in the updating of the finite-element model, only the geometric parameters were updated. Figures 6.8 and 6.9 show the evolution of the updated parameters and the relative difference of frequencies of the models.

Table 6.2 shows the correlation of the models after updating. The differences between the natural frequencies of the models were reduced, as compared with the initial correlation, showing a sensible improvement in the correlation of the models for the

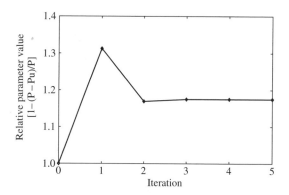

Figure 6.8   Updating parameters evolution

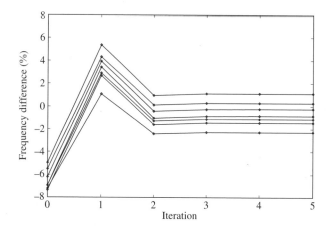

Figure 6.9   Evolution of the relative differences of the natural frequencies

Table 6.2   Initial and final correlation of the models

Experimental model		Analytical model			Updated model		
Mode	ω^E [Hz]	ω^A [Hz]	$\frac{\omega^A-\omega^E}{\omega^E}100\%$	MAC (%)	ω^U [Hz]	$\frac{\omega^A-\omega^U}{\omega^E}100\%$	MAC (%)
1	21.15	19.62	−7.23	98.5	20.67	−2.27	98.5
2	24.62	23.26	−5.52	99.6	24.69	0.28	99.6
3	29.12	27.31	−6.22	98.8	29.05	−0.24	98.8
4	41.44	38.57	−6.76	94.2	40.97	−1.13	94.0
5	43.13	40.14	−6.93	97.2	42.77	−0.83	96.9
6	47.63	45.27	−4.96	98.4	48.09	0.97	92.0
7	48.87	45.28	−7.35	19.8	48.16	−1.45	23.3

first seven natural frequencies. The MAC-values continue to indicate a good correlation among the analytical and experimental mode shapes, with the exception of the seventh mode, which still presents low MAC-values.

Figure 6.10 presents the superposition of the FRF(s) of the updated model and the experimental one. The figure shows that the behavior of the updated model is very close to the experimental one. A discrepancy in the amplitude of the FRF(s) at the resonance is expected because, in the first stage, the procedure did not incorporate the damping in the finite-element model.

Table 6.2 and Figure 6.9 showed a sensible improvement in the correlation of the models after the updating of the parameters, I_Y and I_Z of the screws. However, the quantitative correlation of FRF(s) is not satisfactory, calculation of the FRAC-values shows very low values, which come mainly from absence of the damping in the finite-element model, as will be discussed in the next section.

6.3.5   Incorporation and Updating of the Damping in the Modeling

The incorporation of damping in the modeling is intended to obtain a more realistic finite-element model. However, the inclusion of damping in the analytic model is not a

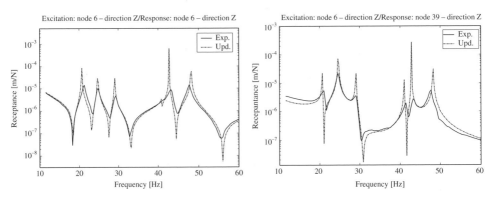

Figure 6.10   Superposition of FRF(s) experimental and updated

simple task, since the most of damping mechanisms are not well known and a previous evaluation of the amount of damping that should be introduced into the finite element model is not straightforward. That value should be arbitrated by engineer judgment or estimated from experimental ratio damping. At this point, it is important to emphasize that the estimate of damping can present errors of the order of 40 % to 200 % (Doebling, 2001), mainly for reasonably complex structures.

In this chapter, the values of the damping parameters of the experimental model were initially used to estimate the damping matrix used in the finite-element model. The damped FRF(s) of the finite-element model were calculated and then compared with measured ones to verify the correlation of the models. In the case of poor correlation, the damping parameters were updated.

The estimating of the damping firstly involved an analysis to verify some tendency in the distribution of the modal damping ratio of the experimental model aiming to define the type of damping to be used in the model. Figure 6.11 shows that the distribution presents a tendency to be inversely proportional to the frequency, for the first six mode shapes.

Additionally, an analysis of the physical configuration of the structure shows that some regions of the structure contain joints and connections that present different intensity of damping, i.e. the damping is not uniformly distributed in the whole structure. In this case, it is difficult to verify a tendency in the behavior of the damping, making the choice of a more appropriate damping model quite difficult. So, some nonproportional models were tried. The model of damping formulated from the mass and stiffness matrices diverged and the model formulated from the stiffness was not realistic. A model formulated from the mass matrix was more appropriate model in the active range of frequency as discussed in the following.

The nonproportional damping formulated from the mass matrix of the model was defined by taking a coefficient of proportion α_R for each region of the model. It was defined as coefficient of proportionality α_1 for the cylindrical and conical components and coefficient α_2 for the screw components region of the connections. The coefficients α_1 and α_2 were initially assumed to have the same values. They were estimated from the

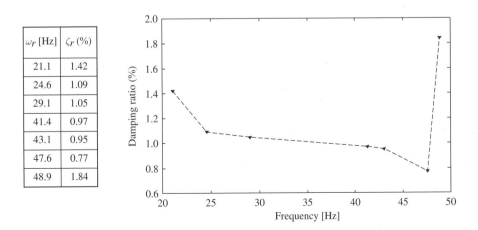

ω_r [Hz]	ζ_r (%)
21.1	1.42
24.6	1.09
29.1	1.05
41.4	0.97
43.1	0.95
47.6	0.77
48.9	1.84

Figure 6.11   Distribution of the modal damping ratio of the structure

average values, of the damping ratio. The updating process consisted of the updating of the damping coefficients α_1 and α_2 simultaneously and in an independent way, in the belief that the components present different levels of damping as compared to each other.

Figure 6.12 shows the evolution of the updating parameters. The updated coefficients reach the stability from the fourth iteration and it is possible to verify that the coefficient of region 2 suffered an increase and the coefficient of region 1 a reduction in relation to their initial values. It is worth verifying that the value of the updated coefficient in region 2 is much bigger than the coefficient of region 1, which means that the updated damping is physically consistent, and is in agreement with the supposition that the region of the joints presents a larger amount of damping. Figure 6.13 shows the super-position of FRF(s) from the experimental, analytical and updated models. It shows that

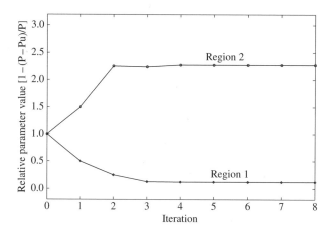

Figure 6.12 Evolution of the updating parameters

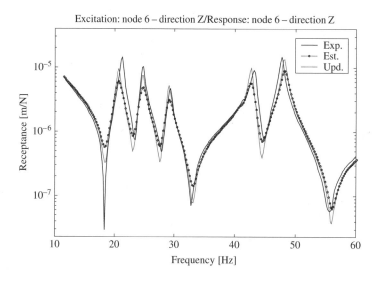

Figure 6.13 Superposition of FRF(s)

the values of the damping coefficients initially estimated, resulted in a high damping if compared with the experimental model. After the updating of those coefficients, the new updated model presented better agreement.

Figures 6.14 and 6.15 show a comparison of the mode shapes and the FRF(s) of the experimental and updated models. After the incorporation and updating of the damping, the correlation of the FRF(s) improves considerably, as can seen through the FRAC-values in the bar graphs.

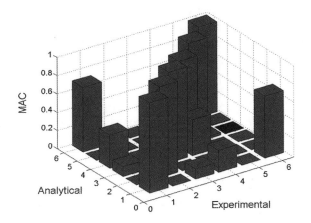

Figure 6.14   MAC-values after updating

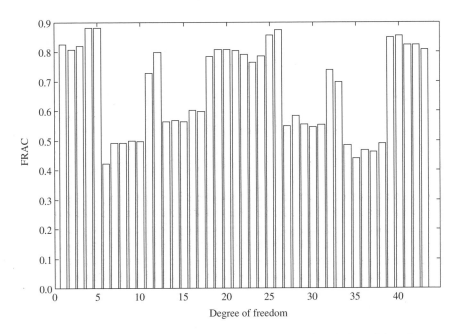

Figure 6.15   FRAC-values after the incorporation and updating of damping

6.4 CONCLUSION

The chapter discusses a finite-element-model updating procedure incorporating the effects of damping in the finite-element model. The updating process first makes an update of the physical and geometric parameters of the model, and later an update of the damping. The updating approach is based on the measured FRF(s) and seeks to obtain a more representative finite element taking into account the damping. It uses a nonproportional damping. Correlation techniques are used to evaluate the representativness of the finite-element model. It was shown that sensibility analysis is very useful for selection of the updating parameters, mainly when the structure presents considerable complexity and offers a wide range of updating parameters. The model of nonproportional damping formulated from the mass matrix is shown to be appropriate to represent the damping of the structure for the used range of active frequency, presenting good stability in the process of convergence of the proportionality coefficients α. The obtained results showed that the methodology was able also to update damping parameters of the finite element model, which makes it a promising tool for practical applications.

REFERENCES

Adhikari, S. (2000) *Damping Models for Structural Vibration*, PhD Thesis, Engineering Department, Trinity College, Cambridge University, Cambridge, UK.

Clough, R.W. and Penzien, J. (1993) *Dynamics of Structures*, McGraw-Hill, New York, USA

Cudney, H. (2001) Discussions and Debates at SD2000, in Ewins, D.J. and Inman, D.J. (Eds), *Structural Dynamics 2000: Current Status and Future Directions*, Research Studies Press, Baldock, pp. 413–427.

Doebling, S.W. (2001) The strategic questions posed at SD2000, in Ewins, D.J. and Inman, D.J. (Eds.), *Structural Dynamics 2000: Current Status and Future Directions*, Research Studies Press, Baldock, pp. 389–411.

Doi, R.M. and Pereira, J.A. (2002) Analysis of the Effects of Damping in the Models Validation, *Proceedings of the 26th International Seminar on Modal Analysis*, Leuven, Belgium (CD-ROM).

Fox, R.L. and Kapoor, M.P. (1968) Rates of changes of eigenvalues and eigenvectors, *AIAA Journal*, **6**, 2426–2429.

Friswell, M.J. and Mottershead, J.E. (1995) *Finite Element Model Updating in Structural Dynamics*, Kluwer Academic Publishers, The Netherlands.

Friswell, M.I. and Penny, J.E. (1997) The Practical Limits of Damage Detection and Location Using Vibration Data, *Proceedings of the 11th VPI&SU Symposium on Structural Dynamics and Control*, Blacksburg.

Garcia, J. (2001) The need for computational model validation, *Experimental Techniques*, **25**(2), 31–33.

Göge, D. (2003) Automatic updating of large aircraft models using experimental data from ground vibration testing, *Aerospace Science and Technology*, **7**(1), 33–45.

Göge, D. and Link, M. (2002) Assessment of computational model updating procedures with regard to model validation, *Aerospace Science and Technology*, **7**, 47–61.

Golub, G.H. and van Loan, C.F. (1985) *Measuring Vectors, Matrices, Subspaces, and Linear System Sensitivity*, The Johns Hopkins University Press, Baltimore.

Heylen, W. (1987) *Optimization of Model Matrices of Mechanical Structures Using Experimental Modal Data*, PhD Thesis, Katholieke Universiteit Leuven, Belgium.

Jones, K. and Turcotte, J. (2002) Finite element model updating using antiresonant frequencies, *Journal of Sound and Vibration*, **252**, 717–727.

Kidder, R.L. (1972) Reduction of structure frequency equations, *AIAA Journal*, **11**, 892.

Lammens, S. (1995) *Frequency Response Based Validation of Dynamic Structural Finite Element Model*, Leuven, PhD Thesis, Katholieke Universiteit Leuven, Faculteit Der Toegepaste Wetenchappen, Leuven, Belgium.

Larsson, P.O. and Sas, P. (1992) Model Updating Based on Forced Vibration Testing Using Numerically Stable Formulations, *Proceedings of the 10th International Modal Analysis Conference* (CD-ROM).

Link, M. (2001) Updating of Analytical Models Review of Numerical Procedures and Application Aspects, in Ewins, D.J., Inman, D.J. (Eds), *Structural Dynamics 2000: Current Status and Future Directions*, Research Studies Press, Baldock, pp. 193–223.

Modak, S.V., Kundra, T.K. and Nakra, B.C. (2002) Use of an updated finite element model for dynamic design, *Mechanical System and Signal Processing*, **16**, 303–322.

Mottershead, J.E., Mares, C., Friswell, M. and James, S. (2000) Selection and updating of parameters for an aluminum space-frame model, *Mechanical Systems and Signal Processing*, **14**, 923–944.

Mottershead, J.E. and Friswell, M. (2002) Model Updating of Joints and Connections, *Proceedings of the International Conference on Structural Dynamics Modelling: Test, Analysis, Correlation and Validation*, Madeira Island.

Müller-Slany, H.H., Pereira, J.A. and Weber, H.I. (1999) *Schadensdiagnose für elastomechanicsche Strukturen auf der Basis adaptierter Diagnosemodelle und FRF-Daten*, in Di-Schwingungstafung'99, Berichte, Al.. Vdi-Schwingungstafung, 1999, pp. 323–340.

Nashif, A.D., Jones, D.I.G. and Henderson, J.P. (1985) *Vibration Damping*, John Wiley & Sons, New York.

Oberkampf, W.L. (2001) What are Validation Experiments?, *Experimental Techniques*, **25**, 35–40.

Pereira, J.A. (1996) *Structural Damage Detection Methodology using a Model Updating Procedure Based on Frequency Response Functions – FRF(s)*, PhD Thesis, Universidade Estadual de Campinas, Faculdade de Engenharia Mecânica, Campinas-SP, Brazil.

Ratcliffe, M.J. and Lieven, N.A.J. (2000) A generic element based method for joint identification, *Mechanical System and Signal Processing*, **14**, 3–28.

Ren, Y. and Beards, C.F. (1995) Identification of joint properties of a structure using FRF data, *Journal of Sound and Vibration*, **186**, 567–587.

Richardson, M. and Potter, R. (1975) Viscous vs. Structural Damping in Modal Analysis, *Proceedings of the 46th Shock and Vibration Symposium*.

Tomlinson, G.R. (2001) State of the Art Review: Damping, in Ewins, D.J., Inman, D.J. (Eds), *Structural Dynamics 2000: Current Status and Future Directions*, Research Studies Press, Baldock, pp. 369–386.

Vanhonacker, P. (1985) Sensitivity Analysis of Mechanical Structures; *Proceedings of 10th International Seminar on Modal Analysis*, Leuven, Belgium.

Zang, C., Grafe, H. and Imregun, M. (2001) Frequency-domain criteria for correlating and updating dynamic finite element models, *Mechanical Systems and Signal Processing*, **15**, 139–155.

Part II
Monitoring Algorithms

7

Model-Based Inverse Problems in Structural Dynamics

Valder Steffen Jr and Domingos A. Rade

Federal University of Uberlândia, Campus Santa Monica, Uberlândia, Brazil

7.1 INTRODUCTION

Since the early 1960s, optimization techniques have been developed and extensively used aiming at the improvement of mechanical, aeronautical and spatial structures. These days, due to the development of computer sciences and sophisticated hardware, we use the so-called automated design synthesis, which embraces a number of computer codes involving CAD, finite elements, optimization techniques, graphical facilities, parallel computing, etc. Typical applications are mostly devoted to direct problems in engineering, in which a mathematical model is used to predict the response of the system.

Other types of problem frequently encountered in engineering analysis are the *inverse problems*, in which a system model must be constructed or refined, given a set of measured inputs and/or outputs. This means that the dynamic characteristics of the model are unknown *a priori* and must be estimated.

A number of problems in engineering can be considered as inverse problems, such as finite-element model updating, experimental modal analysis, parameter estimation, damage identification, and input force identification. In general, the solution of inverse problems involves a curve-fitting procedure in which a set of model parameters has to be determined so as to provide an optimal correlation with the experimental data.

Damage Prognosis – For Aerospace, Civil and Mechanical Systems   Edited by D.J. Inman, C.R. Farrar, V. Lopes Junior and V. Steffen Junior   © 2005 John Wiley & Sons, Ltd

Consequently, optimization techniques must be employed to minimize an error function that is written to represent the difference between the model-predicted and experimental data.

In the context of structural health monitoring and damage prognosis, two important inverse problems are model updating and damage identification (localization and extent evaluation). Concerning the first problem, it is of paramount importance to have a reliable model of the system, in such a way that the damaged structure can have its dynamical behavior predicted. The updating problem can be formulated as a parameter identification problem in which the parameters correspond to correction factors to be applied to an initial model. The second problem can be formulated in a similar way and the unknown parameters are understood as damage indicators. Consequently, design paradigm can evolve from the conservative 'safe life' to the more scientific 'damage tolerant' status.

This chapter includes a review of the theory about vibration analysis of discrete systems and the formulation underlying model updating, parameter identification and damage detection. Classical optimization techniques, as well as pseudorandom optimization methods, are briefly presented. A number of applications illustrate the theoretical aspects presented.

7.2   THEORY OF DISCRETE VIBRATING SYSTEMS

Finite-element modeling of continuous vibrating systems provides equivalent discrete multi-degree-of-freedom systems upon which numerical analyses must be performed in order to predict the dynamic behavior of the modeled structures. In this section the basic theory underlying vibration analysis of linear, time invariant discrete structural systems is reviewed.

7.2.1   Eigenvalue Problem

One way to characterize the dynamic behavior of a vibrating structure is to evaluate its modal properties (eigenvalues and eigenvectors). Given a mathematical model of the structure, this can be done by solving the so-called *eigenvalue problem*.

7.2.1.1   Undamped Systems

For an undamped, nongyroscopic system the eigenvalue problem is given by:

$$[\mathbf{K} - \lambda_r \mathbf{M}]\mathbf{x}_r = \mathbf{0} \quad r = 1, 2, \ldots, N, \quad \text{with } \lambda_r = \omega_r^2 \tag{7.1}$$

In the equation above, λ_r designates the rth eigenvalue and $\mathbf{x}_r$ the corresponding eigenvector. Physically, ω_r represents the rth natural frequency of the structure and $\mathbf{x}_r$ is the corresponding natural vibration mode. Given the properties of the mass and stiffness matrices, it can be shown that all the eigenvalues and eigenvectors are real quantities and $\lambda_r \geq 0$ (the null eigenvalues corresponding to rigid body modes that an insufficiently restrained structure can exhibit).

Assuming that all the eigenvalues are distinct, it can be shown that the following *orthogonality relations* are satisfied by the eigenvectors:

$$\mathbf{x}_r^T \mathbf{M} \mathbf{x}_s = \eta_r \delta_{rs} \qquad \mathbf{x}_r^T \mathbf{K} \mathbf{x}_s = \lambda_r \eta_r \delta_{rs} \quad r, s = 1, 2, \ldots, N \tag{7.2}$$

where η_r designates the generalized mass (or modal mass) associated to the rth eigenpair and δ_{rs} is the Kronecker's delta ($\delta_{rs} = 0$ for $r \neq s$; $\delta_{rs} = 1$ for $r = s$). Since eigenvectors can be arbitrarily scaled, it is common practice to normalize them so as to have unit modal masses. In this case, we have:

$$\mathbf{x}_r^T \mathbf{M} \mathbf{x}_s = \delta_{rs} \qquad \mathbf{x}_r^T \mathbf{K} \mathbf{x}_s = \lambda_r \delta_{rs} \quad r, s = 1, 2, \ldots, N \tag{7.3}$$

7.2.1.2 Viscously Damped Systems

For the more general case of viscously damped systems the eigenvalue problem is written:

$$[\mathbf{A} - s_r \mathbf{U}] \mathbf{z}_r = \mathbf{0}, r = 1, 2, \ldots, 2N \tag{7.4}$$

The solutions of (7.4) are constituted by the pairs $(s_r, \mathbf{z}_r)$, which can be grouped in the following matrices:

- $\mathbf{Z} = [\mathbf{z}_1 \ \mathbf{z}_2 \ \ldots \ \mathbf{z}_{2N}] \in \mathbb{C}^{2N,2N}$: modal matrix   (7.5)
- $\mathbf{S} = \text{diag}\{s_1 \ s_2 \ldots s_{2N}\} \in \mathbb{C}^{2N}$: spectral matrix   (7.6)
- $\mathbf{N} = \text{diag}\{n_1 \ n_2 \ \ldots \ n_{2N}\} \in \mathbb{C}^{2N}$: generalized mass matrix   (7.7)

As opposed to the eigenvalues and eigenvectors of undamped systems, which are real quantities, the eigensolutions of systems with general viscous damping can be either real or complex. For lightly damped structures (which is the most frequent situation), the eigenvalues and eigenvectors are complex and appear in complex-conjugated pairs. In this case, the complex eigenvalues can be expressed in the form:

$$s_r = -\zeta_r \omega_r + j\omega_r \sqrt{1 - \zeta_r^2} \qquad \bar{s}_r = -\zeta_r \omega_r - j\omega_r \sqrt{1 - \zeta_r^2} \tag{7.8}$$

where ω_r and ζ_r designate, respectively, the undamped natural frequency and modal damping factor corresponding to the rth vibration mode.

The eigensolutions satisfy the following orthogonality relations with respect to the state matrices:

$$\mathbf{z}_r^T \mathbf{U} \mathbf{z}_s = n_r \delta_{rs} \qquad \mathbf{z}_r^T \mathbf{A} \mathbf{z}_s = s_r n_r \delta_{rs} \quad r, s = 1, 2, \ldots, 2N \tag{7.9}$$

7.2.1.3 Gyroscopic Systems

For the case of viscously damped gyroscopic systems, the eigenvalue problem is written:

$$[\mathbf{A} - s_r \mathbf{U}] \mathbf{z}_r = \mathbf{0}, \ r = 1, 2, \ldots, 2N \tag{7.10a}$$

where $\mathbf{U}$ is a nonsymmetric matrix.

In this case, it is convenient to define the *adjoint problem* as follows:

$$[\mathbf{A} - s_r^* \mathbf{U}^T] \mathbf{z}_r^* = \mathbf{0}, \quad r = 1, 2, \ldots, 2N \tag{7.10b}$$

The complete set of eigensolutions is then formed by the complex quantities $(s_r, s_r^*, \mathbf{z}_r, \mathbf{z}_r^*)$, $r = 1, 2, \ldots, 2N$, where s_r and s_r^* are the *left and right eigenvalues*, respectively, and $\mathbf{z}_r^*$ and $\mathbf{z}_r$ are the *right and left eigenvectors*, respectively. It can be verified that: $s_r^* = \bar{s}_r$, where the bar indicates complex conjugation.

It can be demonstrated that the following biorthogonality relations are valid:

$$\left(\bar{\mathbf{z}}_r^*\right)^T \mathbf{U} \mathbf{z}_s = n_r \delta_{rs} \qquad \left(\bar{\mathbf{z}}_r^*\right)^T \mathbf{A} \mathbf{x}_s = s_r n_r \delta_{rs} \quad r, s = 1, 2, \ldots, 2N \tag{7.11}$$

7.2.2 Frequency Response Functions

Another important response model of vibrating structures comprises the so-called frequency response functions (FRFs), which characterize the dynamic behavior of the structure in the frequency domain.

7.2.2.1 Undamped Systems

For an undamped system, the equations of motion in the frequency domain can be written in the form: $\mathbf{X}(\omega) = \mathbf{H}(\omega)\mathbf{F}(\omega)$, where

$$\mathbf{H}(\omega) = [\mathbf{K} - \mathbf{M}\omega^2]^{-1} \tag{7.12}$$

is the receptance, or, more generally, *frequency response function* (FRF) matrix.

The general term $H_{ij}(\omega)$ of the frequency response matrix can be interpreted as the frequency-dependent ratio of the amplitude of the harmonic response at coordinate i, to the amplitude of the harmonic excitation force applied at coordinate j, i.e.,

$$H_{ij}(\omega) = \frac{X_j(\omega)}{F_j} \tag{7.13}$$

By manipulating equation (7.12) and using orthogonality relations (7.3), the FRF matrix can be alternatively expressed in terms of eigenvalues and eigenvectors, as follows:

$$\mathbf{H}(\omega) = \mathbf{X}\boldsymbol{\eta}^{-1}[\boldsymbol{\Lambda} - \omega^2 \mathbf{I}]^{-1}\mathbf{X}^T \quad \text{or} \quad \mathbf{H}(\omega) = \sum_{r=1}^{N} \frac{\mathbf{x}_r \mathbf{x}_r^T}{\eta_r(\omega_r^2 - \omega^2)} \tag{7.14}$$

$$H_{ij}(\omega) = \sum_{r=1}^{N} \frac{x_{ir} x_{jr}}{\eta_r(\omega_r^2 - \omega^2)} \tag{7.15}$$

where x_{ir}, x_{jr} designate, respectively, the ith and jth components of the rth eigenvector.

7.2.2.2   FRF Poles and Zeros

By making use of the definition of the inverse of a matrix, equation (7.12) can be rewritten as:

$$\mathbf{H}(\omega) = \frac{\operatorname{adj}(\mathbf{K} - \mathbf{M}\omega^2)}{\det(\mathbf{K} - \mathbf{M}\omega^2)} \qquad (7.16)$$

The general term $H_{ij}(\omega)$ can be expressed as the ratio of two polynomials, as follows:

$$H_{ij}(\omega) = (-1)^{i+j}\left(K_{ij} - M_{ij}\omega^2\right) \frac{\det\left(^{[ji]}\mathbf{K} - ^{[ji]}\mathbf{M}\omega^2\right)}{\det(\mathbf{K} - \mathbf{M}\omega^2)} \qquad (7.17)$$

where $^{[ji]}\mathbf{M} \in \mathbf{R}^{N-1,N-1}$, $^{[ji]}\mathbf{K} \in \mathbf{R}^{N-1,N-1}$ denote, respectively, the matrices obtained by deleting, from matrices $\mathbf{M}$ and $\mathbf{K}$, lines j and columns i. The real zeros of the denominator correspond to the natural (resonance) frequencies of the system, being also known as systems poles, whilst the zeros of the numerator correspond to the values of the forcing frequency for which the amplitude of the harmonic response vanishes. They are known as the *antiresonance frequencies*. The antiresonance frequencies of the FRF $H_{ij}(\omega)$ are denoted herein as $^{[ij]}\omega_r$, $r = 1, 2, \ldots$

It can be demonstrated (Rade, 1994) that the antiresonance frequencies of a driving point FRF $H_{ij}(\omega)$ correspond to the natural frequencies of a structural configuration with modified boundary conditions, characterized by the grounding of coordinate i. On the other hand, the antiresonance frequencies of a transfer FRF $H_{ij}(\omega)$ can be interpreted as the natural frequencies of the structure subjected to a noncollocated feedback control loop $f_j = -k_{ij}X_i$, with $k_{ij} \to \infty$. Such interpretations are illustrated in Figure 7.1.

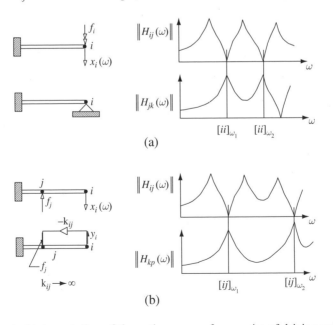

Figure 7.1   Physical interpretation of the antiresonance frequencies of driving point FRFs (a) and transfer FRFs (b)

It should be noted that, while the resonance frequencies are global properties, which are independent of the particular FRF considered, the antiresonance frequencies are local properties, which means that their number and values depend on the particular FRF considered, as indicated by indices i and j in equation (7.17). As demonstrated by Rade and Silva (1999), the antiresonance frequencies play a fundamental role in a number of situations, such as in the design of dynamic vibration absorbers, and can be used in the same way as the system poles for FE model updating and structural failure detection. These authors also demonstrate that the antiresonance frequencies of a given FRF $H_{ij}(\omega)$ are the real eigenvalues of the following nonsymmetric eigenvalue problem:

$$\left({}^{[ji]}\mathbf{K} - {}^{[ij]}\lambda_r \, {}^{[ji]}\mathbf{M} \right)^{[ij]}\mathbf{x}_r = 0 \tag{7.18}$$

For a transfer FRF ($i \neq j$) as opposed to a driving point FRF ($i = j$), matrices ${}^{[ij]}\mathbf{K}$ and ${}^{[ij]}\mathbf{M}$ are not symmetric, so that one has a non self-adjoint system which requires the following adjoint eigenvalue problem to be defined:

$$\left({}^{[ji]}\mathbf{K}^T - {}^{[ij]}\lambda_r^* \, {}^{[ji]}\mathbf{M}^T \right)^{[ij]}\mathbf{x}_r^* = 0 \tag{7.19}$$

The right and left antiresonance eigenvectors satisfy the following biorthogonality relations:

$$\left({}^{[ij]}\mathbf{x}_r^* \right)^T {}^{[ji]}\mathbf{M} \, {}^{[ij]}\mathbf{x}_s = \delta_{rs} \, {}^{[ij]}\eta_s \tag{7.20a}$$

$$\left({}^{[ij]}\mathbf{x}_r^* \right)^T {}^{[ji]}\mathbf{K} \, {}^{[ij]}\mathbf{x}_s = \delta_{rs} \, {}^{[ij]}\lambda_r \, {}^{[ij]}\eta_s \tag{7.20b}$$

where ${}^{[ij]}\eta_r$ are generalized masses.

7.2.2.3 Viscously Damped Systems

For the case of a general damped system, neglecting initial conditions, it is possible to write:

$$\mathbf{H}(\omega) = (j\omega\mathbf{U} - \mathbf{A})^{-1} \tag{7.21}$$

By manipulating equation (7.21) and using orthogonality relations (7.9), the FRF matrix for a system with general viscous damping can be expressed in the following forms, in terms of eigenvalues and eigenvectors:

$$\mathbf{H}(\omega) = \mathbf{Z}\mathbf{N}^{-1}[s\mathbf{I} - \mathbf{S}]^{-1}\mathbf{Z}^T = \sum_{r=1}^{2N} \frac{\mathbf{z}_r\mathbf{z}_r^T}{n_r(j\omega - s_r)} \tag{7.22}$$

or:

$$H_{ij}(\omega) = \sum_{r=1}^{2N} \frac{z_{ir}z_{jr}}{n_r(j\omega - s_r)} \tag{7.23}$$

where z_{ir}, z_{jr} designate, respectively, the ith and jth component of the rth complex eigenvector.

7.3 RESPONSE SENSITIVITY

The finite element matrices **M**, **C** and **K** contain all the information regarding the physical and geometrical characteristics of the modeled structure. An important issue one has frequently to deal with concerns the use of a finite-element model to predict *how* and *how much* the dynamic behavior of the modeled structure varies as the result of variations introduced (voluntarily or not) into the values of a given set of physical and/ or geometrical parameters. This problem is placed within the scope of the so-called analysis of modified structures (Arora, 1976). Specifically, in the context of damage prognosis, it is of paramount importance to be able to assess the structure's behavior after some damage occurs. Such assessment can be made by using a validated finite-element model in combination with some kind of numerical analysis.

In general terms, sensitivity analysis concerns the establishment of relationships between the structural response and a set of structural parameters. A comprehensive presentation of the subject can be found in the book by Haug *et al.* (1986). From a mathematical/numerical point of view, sensitivity analysis can be performed by computing the partial derivatives of the structure response with respect to each parameter of interest. Although higher-order derivatives can be calculated, first-order derivatives are sufficient in most cases.

Assuming that the parameters of interest, forming a vector **p**, appear in the finite element matrices, the dependency between the response and such parameters can be expressed symbolically as follows:

$$\mathbf{r} = \mathbf{r}(\mathbf{M}(\mathbf{p}), \mathbf{C}(\mathbf{p}), \mathbf{K}(\mathbf{p})) \tag{7.24}$$

where vector **r** represents the structure's responses (static responses, time-domain dynamic responses, frequency responses or eigensolutions). The sensitivity of the structural response with respect to a given parameter p_i, evaluated for $p_i = p_i^0$, is defined as:

$$\left.\frac{\partial \mathbf{r}}{\partial p_i}\right|_{p_i^0} = \lim_{\Delta p_i \to 0} \frac{\mathbf{r}\left(\mathbf{M}(p_i^0 + \Delta p_i), \mathbf{C}(p_i^0 + \Delta p_i), \mathbf{K}(p_i^0 + \Delta p_i)\right) - \mathbf{r}\left(\mathbf{M}(p_i^0), \mathbf{C}(p_i^0), \mathbf{K}(p_i^0)\right)}{\Delta p_i}$$

$$\tag{7.25}$$

In the equation above it should be understood that the variation Δp_i is to be added to the value p_i^0, which can be either the value of the parameter of interest in the baseline model or the current value of this parameter in a iterative procedure, such as that frequently employed in nonlinear structural optimization.

The sensitivities can be calculated from finite element models by two main different means:

(a) by using finite differences. According to this method, the sensitivity of the response with respect to a given parameter is calculated by solving successively the model equations for two values of parameter p_i: p_i^0 and $p_i^0 + \Delta p_i$. From these responses, the sensitivity given by (7.25) is numerically estimated as follows:

$$\left.\frac{\partial \mathbf{r}}{\partial p_i}\right|_{p_i^0} \approx \frac{\mathbf{r}\left(\mathbf{M}(p_i^0 + \Delta p_i), \mathbf{C}(p_i^0 + \Delta p_i), \mathbf{K}(p_i^0 + \Delta p_i)\right) - \mathbf{r}\left(\mathbf{M}(p_i^0), \mathbf{C}(p_i^0), \mathbf{K}(p_i^0)\right)}{\Delta p_i} \tag{7.26}$$

(b) by computing the analytical derivatives of the model matrices with respect to the parameters. This method, which is presented in detail in the following, applies when the parameters of interest appear explicitly in the finite element matrices.

7.3.1 Sensitivity of Eigenvalues and Eigenvectors

Let us consider the eigenvalue problem for a N degree-of-freedom finite element model of a non-self-adjoint system featuring general viscous damping, represented by the equations below. It is assumed that all the eigenvalues are distinct.

$$[\mathbf{A} - s_r\mathbf{U}]\mathbf{z}_r = \mathbf{0}, \ r = 1, 2, \ldots, 2N \tag{7.27a}$$

$$[\mathbf{A} - s_r^*\mathbf{U}^T]\mathbf{z}_r^* = \mathbf{0}, \ r = 1, 2, \ldots, 2N \tag{7.27b}$$

It is assumed that the right and left eigenvectors are scaled so as to satisfy:

$$\left(\bar{\mathbf{z}}_r^*\right)^T\mathbf{U}\mathbf{z}_s = \delta_{rs}, \qquad \left(\bar{\mathbf{z}}_r^*\right)^T\mathbf{A}\mathbf{z}_s = \delta_{rs}s_r \qquad r, s = 1, 2, \ldots, 2N \tag{7.28}$$

with the complementary condition:

$$\mathbf{z}_r^T\mathbf{W}\mathbf{z}_r = 1 \tag{7.29}$$

where $\mathbf{W}$ is a positive-definite matrix which possibly depends on the parameter p_i.

Upon derivation of equations (7.28) with respect to a given parameter p_i, we get:

$$\frac{\partial\left(\bar{\mathbf{z}}_r^*\right)^T}{\partial p_i}\mathbf{U}\mathbf{z}_s + \left(\bar{\mathbf{z}}_r^*\right)^T\frac{\partial\mathbf{U}}{\partial p_i}\mathbf{z}_s + \left(\bar{\mathbf{z}}_r^*\right)^T\mathbf{U}\frac{\partial(\mathbf{z}_s)}{\partial p_i} = 0 \tag{7.30a}$$

$$\frac{\partial\left(\bar{\mathbf{z}}_r^*\right)^T}{\partial p_i}\mathbf{A}\mathbf{z}_s + \left(\bar{\mathbf{z}}_r^*\right)^T\frac{\partial\mathbf{A}}{\partial p_i}\mathbf{z}_s + \left(\bar{\mathbf{z}}_r^*\right)^T\mathbf{A}\frac{\partial(\mathbf{z}_s)}{\partial p_i} = \delta_{rs}\frac{\partial s_r}{\partial p_i} \tag{7.30b}$$

The derivatives of the left and right eigenvectors are then expressed as linear combinations of the corresponding modal bases, i.e.:

$$\frac{\partial\mathbf{z}_r}{\partial p_i} = \sum_{s=1}^{2N}\mathbf{z}_p q_{rp}^i \quad \text{and} \quad \frac{\partial\mathbf{z}_r^*}{\partial p_i} = \sum_{s=1}^{2N}\mathbf{z}_p q_{rs}^{i\ *} \tag{7.31}$$

where q_{rs}^i and $q_{rs}^{i\ *} \in C$ are the coefficients of linear combination.

By manipulating the equations above, one obtains:

$$\frac{\partial s_r}{\partial p_i} = \left(\bar{\mathbf{z}}_r^*\right)^T\left[\frac{\partial\mathbf{A}}{\partial p_i} - s_r\frac{\partial\mathbf{U}}{\partial p_i}\right]\mathbf{z}_r \quad r = 1, 2, \ldots, 2N \tag{7.32a}$$

$$\frac{\partial s_r^*}{\partial p_i} = \text{conj}\left(\frac{\partial s_r}{\partial p_i}\right) \quad r = 1, 2, \ldots, 2N \tag{7.32b}$$

$$q_{rs}^i = \frac{(\bar{\mathbf{z}}_r^*)^T \left[\frac{\partial \mathbf{A}}{\partial p_i} - s_s \frac{\partial \mathbf{U}}{\partial p_i}\right] \mathbf{z}_s}{s_s - s_r} \qquad r, s = 1, 2, \ldots, 2N, r \neq s \qquad (7.33a)$$

$$q_{rs}^{i\;*} = \frac{(\bar{\mathbf{z}}_r^*)^T \left[\frac{\partial \mathbf{A}}{\partial p_i} - s_r \frac{\partial \mathbf{U}}{\partial p_i}\right] \mathbf{z}_s}{s_r - s_s} \qquad (7.33b)$$

$$q_{rr}^{-i\;*} + q_{rs}^i = -(\bar{\mathbf{z}}_r^*)^T \frac{\partial \mathbf{U}}{\partial p_i} \mathbf{z}_r \qquad (7.33c)$$

$$q_{rr}^i = -\sum_{\substack{s=1 \\ s \neq r}}^{2N} q_{sr}^i \mathbf{z}_r^T \mathbf{W} \mathbf{z}_r - \frac{1}{2} \mathbf{z}_r^T \frac{\partial \mathbf{W}}{\partial p_i} \mathbf{z}_r \qquad (7.33d)$$

Finally, $q_{rr}^{i\;*}$ are obtained from (7.33c).

7.3.2   Antiresonance Sensitivities

Similarly to the system global eigenvalues, analytical expressions can be derived for the sensitivities of the antiresonance eigenvalues with respect to model parameters. By applying equations (7.18) and (7.19) to the eigenvalue problem, a similar procedure to that used to develop the expressions of the sensitivities of the global eigenvalues, the following expression is obtained for the sensitivity of the antiresonance eigenvalues of a FRF $H_{ij}(\omega)$ (Flanelly, 1971; Rade, 1994):

$$\frac{\partial^{[ij]}\lambda_r}{\partial p_i} = \frac{1}{[ij]\eta_r} \left(^{[ij]}\mathbf{x}_r^*\right)^T \left[\frac{\partial \mathbf{K}}{\partial p_i} - ^{[ij]}\lambda_r \frac{\partial \mathbf{M}}{\partial p_i}\right] {}^{[ij]}\mathbf{x}_r \qquad r, s = 1, 2, \ldots \qquad (7.34)$$

7.3.3   Sensitivity of Frequency Response Functions

For a mechanical system featuring general viscous damping, the FRF matrix is given by: $\mathbf{H}(\omega) = \mathbf{Z}^{-1}(\omega)$
where:

$$\mathbf{Z}(\omega) = -\mathbf{M}\omega^2 + i\omega\mathbf{C} + \mathbf{K} \qquad (7.35)$$

is the dynamic stiffness matrix. By deriving the relation: $\mathbf{H}(\omega) \cdot \mathbf{Z}(\omega) = \mathbf{I}$ with respect to a given design parameter p_i, one obtains:

$$\frac{\partial \mathbf{H}(\omega)}{\partial p_i} = -\mathbf{H}(\omega) \frac{\partial \mathbf{Z}(\omega)}{\partial p_i} \mathbf{H}(\omega) \qquad (7.36)$$

By combining equations (7.35) and (7.36), the following expression is obtained for the sensitivity of the FRFs with respect to parameter p_i:

$$\frac{\partial \mathbf{H}(\omega)}{\partial p_i} = -\mathbf{H}(\omega)\left(-\omega^2 \frac{\partial \mathbf{M}}{\partial p_i} + i\omega \frac{\partial \mathbf{C}}{\partial p_i} + \frac{\partial \mathbf{K}}{\partial p_i}\right)\mathbf{H}(\omega) \tag{7.37}$$

7.4 FINITE-ELEMENT MODEL UPDATING

The comparison of the dynamic responses measured from vibration tests performed on a given structural system to the corresponding responses predicted by finite-element modeling reveals, in most cases, discrepancies, whose magnitude depends mainly on the degree of complexity of the structure. Assuming that the experimental errors are small, the observed differences are then attributed to imperfections of the finite-element model, which can be originated from various sources, such as:

(a) simplification of the theory adopted to derive the model (disregard of nonlinear phenomena, warping of cross sections, shear deformations, etc);
(b) difficulty in modeling of dissipative effects;
(c) discretization errors (insufficient mesh refinement, simplification of geometry);
(d) numerical errors occurring during the resolution of the equations of motion;
(e) uncertainties of the values of some parameters of the system used as inputs to generate the finite-element model, resulting from the difficulty in estimating accurately the values of physical and geometrical properties such as material properties, cross section dimensions and localized stiffness in welded, bolted or riveted joints.

If the differences observed between the experimental and FE-predicted responses are considered to be unacceptable, one is then led to search for corrections of the finite-element model so as to reduce, as much as possible, the discrepancies between the two sets of responses. Focusing on the last source of error listed above (parameter uncertainty), the finite element adjustment procedure can be formulated as an optimization problem, where one tries to find optimal mass, stiffness and damping matrices, which lead to minimal differences between experimental and analytical responses.

Several methods of finite-element-model updating exploring the dynamic responses have been developed during the last 20 years. The reader should refer to the book by Friswell and Mottershead (1995) for a description of a number of such methods, as well as the discussion of various theoretical and practical aspects related to the subject.

It is recognized that, regardless of the particular procedure adopted, the success of the finite-element-model correction is limited by the fact that, in practical situations, one has to use incomplete experimental data (both in frequency and spatial domains), since:

(a) the number of sensors used in the experimental tests is usually much smaller then the number of degrees-of-freedom of the finite-element model, mainly due to the cost of the sensors and multichannel acquisition systems. Moreover, some coordinates of the test structure are inaccessible to instrumentation. The measurement of rotational quantities (angular displacements, velocities and accelerations) is still difficult and costly. However, such difficulties related to limited number of instrumented coordinates tend to be minimized (or, at least, reduced), by the increasing popularization of laser-based techniques for vibration measurement, which are capable of providing high spatial density measurements.

(b) vibration measurements are always performed in a limited frequency band, generally encompassing only a few vibration modes of the system.

The main consequences of the limitations mentioned above are:

(a) there is, in most cases, a mismatch of the dimensions of the vectors containing the measured and analytical responses. This leads to the necessity to use some kind of data adaptation in order to compare both sets of data and to use them in numerical procedures of model updating. Two main types of data adaptation are used: the *reduction* (or *condensation*) of the finite-element matrices and vectors to the dimensions of the vectors of experimentally measured responses, or the *expansion* of the response vectors to the dimension of the finite-element model.

(b) the solution of the finite-element updating problem is not unique, in the sense that the effectiveness of the adjusted model can only be ensured in the range of experimental data used for updating.

Moreover, from the numerical standpoint, data incompleteness is frequently associated with poor numerical conditioning of the equations to be solved. Numerical ill conditioning, combined with the unavoidable contamination of the experimental responses by measurement noise, can have a dramatic influence on the obtained solution.

In the following, a method of finite element model adjustment based on eigensolution sensitivities is reviewed.

7.4.1 Model Updating Method Based on the Minimization of a Residual Formed from the Differences between Eigenvalues and Eigenvectors of the Associated Undamped System

The basic idea is to search for corrections to be applied to the physical and/or geometrical properties (material modulus, mass density, cross-section dimensions, etc.) of some regions of the original finite-element model. In this case, a parameter identification problem can be formulated as an optimization problem, in which the cost function represents the differences between the measured data and the corresponding predictions of the finite-element model, and the design variables are correction parameters to be applied to the values of a previously selected set of physical and/or geometrical properties of the original model.

The local methods set out to produce the so-called *knowledge-based models*, which preserve the connectivity and topology of the original finite-element model, while providing clear physical meaning for the correction parameters.

To enable the formulation of a parameter identification problem, local methods require an adequate parameterization of the modeling errors. This can be done by regarding the finite-element-model matrices as the result of assembling the matrices corresponding to a given number of substructures (groups of finite elements, also named macroelements), to each of which one correction parameter is ascribed. According to this strategy, for a viscously damped finite element model, the corrected matrices are expressed as follows:

$$\mathbf{M} = \mathbf{M}^0 + \Delta\mathbf{M} = \mathbf{M}^0 + \sum_{i=1}^{n_m} m_i \mathbf{M}_i^0 \tag{7.38}$$

$$\mathbf{K} = \mathbf{K}^0 + \Delta\mathbf{K} = \mathbf{K}^0 + \sum_{i=1}^{n_k} k_i \mathbf{K}_i^0 \tag{7.39}$$

$$\mathbf{C} = \mathbf{C}^0 + \Delta\mathbf{C} = \mathbf{C}^0 + \sum_{i=1}^{n_c} c_i \mathbf{C}_i^0 \tag{7.40}$$

where $\mathbf{M}^0$, $\mathbf{K}^0$ and $\mathbf{C}^0$ denote the matrices of the original finite element model and m_i, c_i are the unknown dimensionless parameters to be applied to the inertia, stiffness and damping matrices of the ith substructure, respectively, with the understanding that if one of such parameters is zero, no modeling error is attributed to the corresponding substructure.

The model-updating problem is then reduced to finding an optimal set of values for the correction parameters, which leads to the smallest difference between the measured dynamic responses and the model-predicted counterparts. Based on this principle, various updating methods have been developed (Friswell and Mottershead, 1995) using different types of experimental data and cost functions. Some of those methods are reviewed in what follows.

By denoting by $\mathbf{p}$ the vector containing the whole set of inertia and stiffness correction parameters m_i, k_i (damping is neglected), in its simplest form, the method consists in the iterative resolution of the following optimization problem:

$$\min_{p} J(\mathbf{p}) = \mathbf{r}_\lambda^T(\mathbf{p}) \mathbf{r}_\lambda(\mathbf{p}) + \mathbf{r}_x^T(\mathbf{p}) \mathbf{r}_x(\mathbf{p}) \tag{7.41}$$

with side constraints: $p_i^L \le p_i \le p_i^U$.

In equation (7.41), the so-called *residual vectors* represent the dimensionless differences between model-predicted and experimental eigensolutions, as follows:

- $$\mathbf{r}_\lambda^T(\mathbf{p}) = [\Delta\lambda_1(\mathbf{p})\ \Delta\lambda_2(\mathbf{p})\ \ldots\ \Delta\lambda_m(\mathbf{p})]^T \in R^m \tag{7.42}$$

where:

$$\Delta\lambda_i(\mathbf{p}) = 2\frac{\lambda_i^{(ex)} - \lambda_i^{(fe)}(\mathbf{p})}{\lambda_i^{(ex)} + \lambda_i^{(fe)}(\mathbf{p})} \quad i = 1, 2, \ldots, m \tag{7.43}$$

- $$\mathbf{r}_X^T(\mathbf{p}) \left[\Delta\mathbf{x}_1^T(\mathbf{p})\ \Delta\mathbf{x}_2^T(\mathbf{p})\ \ldots\ \Delta\mathbf{x}_m^T(\mathbf{p})\right]^T \in R^{c.m} \tag{7.44}$$

where:

$$\Delta\mathbf{x}_i(\mathbf{p}) = 2\frac{\mathbf{x}_i^{(ex)} - \mathbf{L}\mathbf{x}_i^{(fe)}(\mathbf{p})}{\left\|\mathbf{x}_i^{(ex)}\right\| + \left\|\mathbf{x}_i^{(fe)}(\mathbf{p})\right\|} i = 1, 2, \ldots, m \tag{7.45}$$

In equation (7.45), matrix $\mathbf{L} \in \mathbf{R}^{c,N}$, is a Boolean matrix (whose elements are either 0 or 1), which enables the localization of the c coordinates of the experimental eigenvectors on the larger set of N coordinates of the finite element model.

According to the Gauss–Newton method, the experimental eigensolutions are related to the corresponding finite element undamped predictions through linearized Taylor series, as follows:

$$\lambda_r^{(ex)} = \lambda_r^{(fe)} + \sum_{i=1}^{n_k} \frac{\partial \lambda_r^{(fe)}}{\partial k_i} \Delta k_i + \sum_{i=1}^{n_m} \frac{\partial \lambda_r^{(fe)}}{\partial m_i} \Delta m_j \tag{7.46}$$

$$\mathbf{x}_r^{(ex)} = \mathbf{L}\mathbf{x}_r^{(fe)} + \sum_{i=1}^{n_k} \frac{\partial \left(\mathbf{L}\mathbf{x}_r^{(fe)}\right)}{\partial k_i} \Delta k_i + \sum_{i=1}^{n_m} \frac{\partial \left(\mathbf{L}\mathbf{x}_r^{(fe)}\right)}{\partial m_i} \Delta m_i \quad r = 1, 2, \ldots, m \tag{7.47}$$

In equations (7.46) and (7.47), one recognizes the sensitivities of the eigensolutions with respect to stiffness and inertia parameters. For the case of undamped models, taking into account the error parameterization scheme expressed in equations (7.38)–(7.40), and assuming that the finite element eigenvectors are normalized according to:

$$\mathbf{x}_r^{(fe)^T} \mathbf{M} \mathbf{x}_r^{(fe)} = 1 \quad \text{and} \quad \mathbf{x}_r^{(fe)^T} \mathbf{K} \mathbf{x}_r^{(fe)} = \lambda_r, \quad r = 1, 2, \ldots, m$$

the eigensolution sensitivities can be computed using the following equations:

- Sensitivity of the eigenvalues w.r.t. stiffness correction parameters:

$$\frac{\partial \lambda_r^{(fe)}}{\partial k_i} = \mathbf{x}_r^{(fe)^T} \mathbf{K}_i \mathbf{x}_r^{(fe)} \tag{7.48}$$

- Sensitivity of the eigenvalues w.r.t. inertia correction parameters:

$$\bullet \frac{\partial \lambda_r^{(fe)}}{\partial m_i} = -\lambda_r^{(fe)} \mathbf{x}_r^T \mathbf{M}_i \mathbf{x}_r^{(fe)} \tag{7.49}$$

- Sensitivity of the eigenvectors w.r.t. stiffness correction parameters:

$$\bullet \frac{\partial \mathbf{x}_r^{(a)}}{\partial k_i} = \sum_{s=1}^{p} \mathbf{x}_s c_{sr} \tag{7.50}$$

where:

$$c_{sr} = -\frac{1}{\left(\lambda_s^{(fe)} - \lambda_r^{(fe)}\right)} \mathbf{x}_s^{(fe)^T} \mathbf{K}_i \mathbf{x}_r^{(fe)}, s \neq r \quad \text{and} \quad c_{rr} = 0 \tag{7.51}$$

- Sensitivity of the eigenvectors w.r.t. inertia correction parameters:

$$\bullet \frac{\partial \mathbf{x}_r^{(fe)}}{\partial m_i} = \sum_{s=1}^{p} \mathbf{x}_s^{(fe)} d_{sr}, \tag{7.52}$$

where:

$$d_{sr} = -\frac{\lambda_r^{(\text{fe})}}{\left(\lambda_s^{(\text{fe})} - \lambda_r^{(\text{fe})}\right)} \mathbf{x}_s^{(\text{fe})^T} \mathbf{M}_i \mathbf{x}_r^{(\text{fe})}, s \neq r \tag{7.53a}$$

$$d_{rr} = -\frac{1}{2} \mathbf{x}_s^{(\text{fe})^T} \mathbf{M}_i \mathbf{x}_r^{(\text{fe})} \tag{7.53b}$$

Given the derivatives presented above, equations (7.46) and (7.47) can be solved iteratively for the inertia and stiffness correction parameters. Such iterative scheme is necessary since the linearized Taylor series is accurate only for small variations in the correction parameters. Starting from an initial finite-element model, in each iteration (k), the following set of linear estimation equations has to be solved:

$$\begin{bmatrix} \mathbf{S}_\lambda^K & \mathbf{S}_\lambda^M \\ \mathbf{S}_x^K & \mathbf{S}_x^M \end{bmatrix}^{(k)} \begin{bmatrix} \Delta \mathbf{k} \\ \Delta \mathbf{m} \end{bmatrix}^{(k)} = \begin{bmatrix} \Delta \lambda \\ \Delta \mathbf{x} \end{bmatrix}^{(k)} \tag{7.54}$$

System (7.54) can be cast into the following compact form:

$$\mathbf{S}^{(k)} \Delta \mathbf{p}^{(k)} = \Delta \mathbf{z}^{(k)} \tag{7.55}$$

where:

- $\mathbf{S}^{(k)} \in \mathbf{R}^{m(2c+1), n_k+n_m}$ is the sensitivity matrix or Jacobian matrix of the eigensolutions of the finite element model;
- $\Delta \mathbf{p}^{(k)} \in R^{n_k+n_m}$ is the vector formed by the increments of the correction parameters;
- $\Delta \mathbf{z}^{(k)} \in R^{m(2c+1)}$ is the vector containing the differences between the finite element and experimental eigensolutions.

At each iteration, the vector of parameter increments can be calculated as the normal least square solution:

$$\Delta \mathbf{p}^{(k)} = \left(\mathbf{S}^{(k)^T} \mathbf{S}^{(k)} \right)^{-1} \mathbf{S}^{(k)^T} \Delta \mathbf{z}^{(k)} \tag{7.56}$$

provided that $m(2c+1) \geq n_k + n_m$ (number of equations greater than the number of correction parameters) and the rank of the sensitivity is $n_k + n_m$. Iterations are carried out until a convergence criterion is satisfied. Typical criteria are:

$$\|\Delta \mathbf{p}^{(k)}\| \leq \text{TOL}_\mathbf{p} \qquad \|\Delta \mathbf{z}^{(k)}\| \leq \text{TOL}_\mathbf{z}$$

Some improvements have been proposed in the basic updating procedure outlined above. The first one consists in weighting differently the estimation equations, according to the confidence one ascribes to the corresponding measured eigenvalues and eigenvectors. From the practical point of view, this possibility is of paramount importance since it is recognized that eigenvalues (natural frequencies) are usually identified, from experimental modal analysis, much more accurately than the eigenvectors (mode shapes). As a result, the mode-shape data is less reliable than the natural frequency data so that different weighting of the corresponding estimation equations is justified. The relative accuracy can be incorporated in the updating procedure by altering the cost function (7.41) as follows:

$$J(\mathbf{p}) = \mathbf{r}_\lambda^T(\mathbf{p})\mathbf{W}_\lambda\mathbf{r}_\lambda(\mathbf{p}) + \mathbf{r}_x^T(\mathbf{p})\mathbf{W}_x\mathbf{r}_x(\mathbf{p}) \tag{7.57a}$$

or:

$$J(\mathbf{p}) = [\mathbf{r}_\lambda^T(\mathbf{p})\ \mathbf{r}_x^T(\mathbf{p})]\mathbf{W}\left\{ \begin{array}{c} \mathbf{r}_\lambda(\mathbf{p}) \\ \mathbf{r}_x(\mathbf{p}) \end{array} \right\} \tag{7.57b}$$

where $\mathbf{W} = \begin{bmatrix} \mathbf{W}_\lambda & \mathbf{0} \\ \mathbf{0} & \mathbf{W}_x \end{bmatrix}$ is a positive definite weighting matrix, which is a diagonal matrix whose elements are usually given by the reciprocals of the variance of the corresponding quantities identified from experiments. Accounting for this weighting, the so-called *weighted least square solution* of the system of estimation equations is obtained as:

$$\Delta(\mathbf{p})^{(k)} = \left(\mathbf{S}^{(k)^T}\mathbf{W}\mathbf{S}^{(k)}\right)^{-1}\mathbf{S}^{(k)^T}\mathbf{W}\Delta\mathbf{z}^{(k)} \tag{7.58}$$

One difficulty frequently encountered in model updating computations is that the sensitivity matrix $\mathbf{S}^{(k)}$ in equation (7.54) can be rank deficient. This is likely to occur when linear dependency exists among some of its columns.

When the columns of $\mathbf{S}^{(k)}$ are close to being linearly related, then the parameter estimation problem is said to be *ill conditioned*, being characterized by a strong sensitivity of the estimated parameters with respect to disturbances (noise), which inescapably contaminates the experimental data. In general, ill conditioning arises when insufficient experimental data is used to enable the correction parameters to be estimated uniquely. To avoid such a situation, one alternative is to increase, as much as possible, the amount of experimental data available for model updating. Such a strategy is called *enlargement of the knowledge space* of the structure (Lallement and Cogan, 1992) One possibility is to exploit simultaneously the dynamic responses of various structural configurations obtained by introducing concentrated masses or springs to the original structures. Such a procedure has been named *perturbed boundary condition (PCB) testing* (Nalitolela *et al.*, 1990; Chen *et al.*, 1993; Lammens *et al.*, 1993). In the limiting case when the mass and/or stiffness additions assume very large values, the structural modifications actually represent grounding of the coordinates where they are applied. Rade (1994) and Rade and Lallement (1996) developed methods enabling access to the eigensolutions of the modified structures by performing computations on the experimental data of the original configuration, without the need to perform actual complementary tests. By exploiting the interpretation of the antiresonance frequencies in terms of boundary condition modifications, the same authors (Rade and Lallement, 1998) suggested exploring the antiresonance eigenvalues and eigenvectors of both driving point and transfer FRFs in combination with resonance eigensolutions for updating finite-element models, verifying that such strategy can lead to effective enrichment of the experimental data and, as a result, improvement of numerical conditioning of system of estimation equations.

7.4.2 Localization of Dominant Modeling Errors

When dealing with finite element models of complex structural systems, it is not feasible to work with a large number of correction parameters, since this would result in a very large number of unknowns to be estimated in the model updating computations. As noted by Friswell and Mottershead (1995), such situation may be characterized by ill-conditioned systems of estimation equations, which can be the result of:

(a) the tendency to have linear combinations between the columns of the observation matrix, which expresses the fact that the corresponding correction parameters cannot be estimated independently from the experimental data available;
(b) the fact that some correction parameters may have little influence on the dynamic responses used in the model updating procedure. As a result, the columns of the observation matrix corresponding to these parameters tend to have terms whose magnitude is very small as compared with the others. Such differences in scaling can be a source of ill conditioning.

Most frequently, modeling errors are concentrated in a certain regions of the structure, in which difficulty in estimating the values of some physical and geometrical parameters arises. Such is the case, for example, with regions containing spot-weldments, bolted or riveted joints, whose actual stiffness is not easy to estimate.

Thus, from the practical standpoint, the resolution of the problem of model adjustment must start by localizing dominant errors and searching for corrections for those parameters only, which generally implies a significant reduction of the number of unknowns.

Lallement and Piranda (1990) suggested an error localization method, which is known as the best subspace technique, which is described in the following.

Given a system of estimation equations such as (7.55), written in the form:

$$\mathbf{S}\Delta\mathbf{p} = \Delta\mathbf{z} + \boldsymbol{\epsilon} \qquad (7.59)$$

one searches to find, among the n_{col} columns of $\mathbf{S}$, a reduced number of columns which forms a vector basis spanning the right subspace vector $\Delta\mathbf{z}$, in such a way that the representation error expressed by $\boldsymbol{\epsilon}$ is minimal.

The selection of columns is made one by one, according to the following procedure: To select the first column, n_{col} systems of equations are solved:

$$\mathbf{S}_{\cdot j}\mathbf{x}^{(1)} = \Delta\mathbf{z} + \boldsymbol{\epsilon}_j^{(1)}, j = 1 \text{ to } n_{col} \qquad (7.60)$$

where $\mathbf{S}_{\cdot j}$ indicates the jth column of $\mathbf{S}$.

To each normal least-square solution of (7.60), the corresponding error vector is given by:

$$\boldsymbol{\epsilon}_j^{(1)} = \mathbf{S}_{\cdot j}\hat{\mathbf{x}}^{(1)} - \Delta\mathbf{z} = \left[\mathbf{S}_{\cdot j}(\mathbf{S}_{\cdot j}^T\mathbf{S}_{\cdot j})^{-1}\mathbf{S}_{\cdot j}^T - \mathbf{I}\right]\Delta\mathbf{z} \qquad (7.61)$$

The first selected column $\mathbf{S}_{\cdot\bar{j}}$ is that corresponding to the minimum valued of the Euclidian norm of the error vector $\varepsilon_j^{(1)}$. The corresponding representation error is defined as follows:

$$e^{(1)} = \min_j \frac{\left\|\varepsilon_j^{(1)}\right\|}{\|\Delta\mathbf{z}\|} \times 100 = \frac{\left\|\varepsilon_{\bar{j}}^{(1)}\right\|}{\|\Delta\mathbf{z}\|} \times 100$$

For the selection of the second column, $2n_{\text{col}} - 1$ systems of the type:

$$[\mathbf{S}_{\cdot\bar{j}}\,\mathbf{S}_{\cdot k}]\mathbf{x}^{(2)} = \Delta\mathbf{z} + \varepsilon_{\bar{j}k}{}^{(2)}, k = 1 \text{ to } 2n_{\text{col}}, k \neq \bar{j}$$

are solved by least-squares. The column selected is the one leading to the minimal representation error:

$$e^{(2)} = \min_k \frac{\left\|\mathbf{e}_{\bar{j}k}\right\|}{\|\Delta\mathbf{z}\|} \times 100$$

The column selection process is iterated until no significant reduction in the value of the representation error is obtained by including further columns.

7.5 REVIEW OF CLASSICAL OPTIMIZATION TECHNIQUES

We can define the general problem of nonlinear optimal design as the determination of the values of design variables x_i $(i = 1, \ldots, n)$ such that the objective function attains an extreme value while simultaneously all constraints are satisfied (Eschenauer *et al.*, 1997). The classical optimization algorithms are written in such a way that the objective function is minimized. However, if an objective function f is to be maximized, one simply substitutes f by $-f$ in the formulation.

The problem above is formulated mathematically as

$$\min_{\mathbf{x}\in R^n}\{f(\mathbf{x})|h(\mathbf{x}) = 0, \mathbf{g}(\mathbf{x}) \leq 0\} \tag{7.62}$$

where R^n is n-dimensional set of real numbers; $\mathbf{x}$ is vector of the n design variables; $f(\mathbf{x})$ is objective function; $g(\mathbf{x})$ is vector of p inequality constraints; and $h(\mathbf{x})$ is vector of the q equality constraints.

The corresponding feasible domain is defined as

$$X := \{\mathbf{x} \in R^n | h(\mathbf{x}) = \mathbf{0}, \mathbf{g}(\mathbf{x}) \leq \mathbf{0}\} \tag{7.63}$$

Classical optimization methods are the most traditionally and widely used in engineering and the most effective among them are gradient based (Vanderplaats, 1998). Three fundamental steps are usually necessary to implement these methods:

- *Definition of the search direction*: This procedure is the optimization algorithm itself. Gradients of the objective function (in the so-called sequential methods) and both objective and constraint functions (in the direct methods) are manipulated in order to establish search directions along the design space.

- *Definition of the step in the search direction*: Once a search direction **S** is defined in the previous step, the general optimization problem is restricted to a one-dimensional search.

$$\mathbf{x}^{i+1} = \mathbf{x}^i + \alpha \mathbf{S} \tag{7.64}$$

The quantity α in equation (7.64) is the size of the optimizer's move along the search direction in order to update the design configuration from $\mathbf{x}^i$ to $\mathbf{x}^{i+1}$.

- *Convergence check*: Convergence is achieved for the design variable set $\mathbf{x}^*$ upon the satisfaction of the Kuhn–Tucker conditions, expressed by equations (7.65) to (7.67):

$$g_j(\mathbf{x}^*) \leq 0 \tag{7.65}$$

$$\lambda_j g_j(\mathbf{x}^*) = 0 \tag{7.66}$$

$$[\nabla f(\mathbf{x}^*) + \sum_j \lambda_j g(\mathbf{x}^*)] = 0 \tag{7.67}$$

where λ_j are the Lagrange multipliers.

For unconstrained optimization, the most popular approach uses *variable metric methods*, which are also known as *quasi-Newton methods*, because they have convergence characteristics similar to second-order methods. In these methods, the search direction at iteration q is given by equation (7.68):

$$\mathbf{S}^q = -\mathbf{H}\nabla f(\mathbf{x}^q) \tag{7.68}$$

where **H** approaches the inverse of the Hessian matrix during the optimization process. At the initial design point, the **H** matrix is taken as the identity matrix. This means that the first search direction corresponds simply to the steepest descent. The update formula for matrix **H** is given by equation (7.69):

$$\mathbf{H}^{q+1} = \mathbf{H}^q + \mathbf{D}^q \tag{7.69}$$

where **D** is a symmetric update matrix given by equation (7.70):

$$\mathbf{D}^q = \frac{\sigma + \theta\sigma}{\sigma^2}\mathbf{pp}^T + \frac{\theta - 1}{\tau}\mathbf{H}^q\mathbf{y}(\mathbf{H}^q\mathbf{y})^T - \frac{\theta}{\sigma}[\mathbf{H}^q\mathbf{y}\,\mathbf{p}^T + \mathbf{p}(\mathbf{H}^q\mathbf{y})^T] \tag{7.70}$$

The vectors **p** and **y** (change vectors) are defined as

$$\mathbf{p} = \mathbf{x}^q - \mathbf{x}^{q-1} \tag{7.71a}$$

$$\mathbf{y} = \nabla f(\mathbf{x}^q) - \nabla(\mathbf{x}^{q-1}) \tag{7.71b}$$

and the scalars σ and τ are defined as

$$\sigma = \mathbf{p} \cdot \mathbf{y} \tag{7.71c}$$

$$\tau = \mathbf{y}^T\mathbf{H}^q\mathbf{y} \tag{7.71d}$$

A family of variable metric methods uses the same scheme as presented above. The most popular of the variable metric methods are the Davidon–Fletcher–Powell (DFP) and the Broydon–Fletcher–Goldfarb–Shanno (BFGS) methods, the choice depending upon the value of θ used in equation (7.70), i.e., $\theta = 0$ corresponds to DFP method, while $\theta = 1$ corresponds to BFGS method.

To solve constrained optimization problems, the idea is to use unconstrained minimization algorithms as presented above. The general approach is to minimize the objective function as an unconstrained function but to provide some moderate penalty to avoid constraint violations. Penalty increases as the optimization progresses along several unconstrained minimization problems in order to obtain the optimum constrained design. This procedure is known as *sequential unconstrained minimization technique* (SUMT). For this aim a pseudo-objective function is created, according to equation (7.72):

$$\Phi(\mathbf{x}, r_p) = f(\mathbf{x}) + r_p P(\mathbf{x}) \tag{7.72}$$

where $f(\mathbf{x})$ is the original objective function, $P(\mathbf{x})$ is the penalty function, and r_p is a scalar which determines the magnitude of the penalty. To solve the general constrained optimization problem, the *augmented Langrange multiplier method*, which takes into account equality and inequality constraints to create the pseudo-objective function (augmented Lagrangian), as in equation (7.73):

$$A(\mathbf{x}, \lambda, r_p) = f(\mathbf{x}) + \sum_{j=1}^{m} [\lambda_j \Psi_j + r_p \Psi_j^2] + \sum_{k=1}^{1} \{\lambda_{k+m} h_k(\mathbf{x}) + r_p [h_k(\mathbf{x})]^2\} \tag{7.73}$$

where:

$$\Psi_j = \max[g_j(\mathbf{x}), \frac{-\lambda_j}{2r_p}] \tag{7.74}$$

Consequently, the constraints are taken into account in the augmented Lagrangian and it is possible to perform unconstrained minimization by using the methods mentioned earlier in this chapter. The one-dimensional search is done for each search direction using polynomial interpolation techniques combined with the golden-section method, for example.

Besides SUMT, *direct methods* can also be used to perform classical optimization. Possibly the most popular among these methods are the following: the method of feasible directions, the generalized reduced gradient method, and sequential quadratic programming (Vanderplaats, 1998).

7.6 *HEURISTIC OPTIMIZATION METHODS*

Also known as 'random' and 'intelligent' optimization strategies, this group of optimization methods varies the design parameters according to probabilistic rules. It is common to resort to random decisions in optimization whenever deterministic rules fail to achieve the expected success. On the other hand, however, heuristic techniques tend to be more costly, sometimes to the point that certain applications are not feasible unless alternative formulations, designed to spare computational resources, are introduced. Such formulations comprise the response surface meta-modeling method. In this chapter, the most widely known of these methods will be briefly presented.

7.6.1 Genetic Algorithms

Genetic algorithms are random search techniques based on Darwin's 'survival of the fittest' theories, as presented by Goldberg (1989). Genetic algorithms were originated with a binary representation of the parameters and have been used to solve a variety of discrete optimization problems. A basic feature of the method is that an initial population evolves over generations to produce new and hopefully better designs. The elements (or designs) of the initial population are randomly or heuristically generated. A basic genetic algorithm uses four main operators, namely *evaluation, selection, crossover* and *mutation* (Michalewicz, 1996), which are briefly described in the following:

- *Evaluation* – the genetic algorithms require information about the fitness of each population member (fitness corresponds to the objective function in the classical optimization techniques). The fitness measures the adaptation grade of the individual. An individual is understood as a set of design variables. No gradient or auxiliary information is required, only the fitness function is needed.
- *Selection* – the operation of choosing members of the current generation to produce the prodigy of the next one. Better designs, viewed from the fitness function, are more likely to be chosen as parents.
- *Crossover* – the process in which the design information is transferred from the parents to the prodigy. The results are new individuals created from existing ones, enabling new parts of the solution space to be explored. This way, two new individuals are produced from two existing ones.
- *Mutation* – a low probability random operation used to perturb the design represented by the prodigy. It alters one individual to produce a single new solution that is copied to the next generation of the population to maintain population diversity and to avoid premature convergence.

When the parameters (design variables) are continuous, it is more logical to represent them by floating-point numbers. Moreover, since binary genetic algorithm has its precision limited by the binary representation of parameters, using real numbers instead improves the precision of the algorithm. As a single floating-point number represents a continuous parameter, the continuous parameter genetic algorithm requires less storage than the binary genetic algorithm (which needs N_{bits} to represent the same parameter). Consequently, the cost function is also more accurate.

7.6.2 Simulated Annealing

Annealing is a term from metallurgy used to describe a process in which a metal is heated to a high temperature, inducing strong perturbations in the positions of its atoms (Saramago *et al.*, 1999). Providing that the temperature drop is slow enough, the metal will eventually stabilize into an orderly structure. Otherwise, an unstable atom structure arises. Simulated annealing can be performed in direct and inverse problems by randomly perturbing the decision variables and keeping track of the best resulting objective value. After many tries, the most successful design is set to be the center about which a new set of perturbations will take place.

In an analogy to the metallurgical annealing process, let each atomic state (design variable configurations) result in an energy level (objective function value) E. In each step of the algorithm the atomic positions are given small random displacements due to the effect of a prescribed temperature T (standard deviation of the random number generator). As an effect, the energy level undergoes a change ΔE (variation of the objective function value). If $\Delta E \leq 0$, the objective stays the same or is minimized, thus the displacement is accepted and the resulting configuration is adopted as the starting point of the next step. If $\Delta E \leq 0$, on the other hand, the probability that the new configuration is accepted is given by equation (7.75):

$$P(\Delta E) = \bar{e}^{(E/k_b T)} \tag{7.75}$$

where k_b is the Boltzman constant, set equal to 1. This procedure is known as the Metropolis criterion (Kirkpatrick *et al.*, 1983). Since the probability distribution in equation (7.75) is chosen, the system evolves into a Boltzman distribution. The random number r is obtained according to a uniform probability density function in the interval $(0, 1)$. Consequently, if $r < P(\Delta E)$ the new configuration is retained. Otherwise, the original configuration is used to start the next step. The temperature T is simply a control parameter with the same units as the objective function. The initial value of T is related to the standard deviation of the random number generator, whilst its final value indicates the order of magnitude of the desired accuracy in the location of the optimum point. Thus, the annealing schedule starts at a high temperature that is discretely lowered (using a factor $0 < rt < 1$) until the system is 'frozen', hopefully at the optimum, even if the design space is multimodal.

7.6.3 Particle Swarm Optimization

It is well known that the solution of inverse problems by using optimization methods is a difficult task due to the existence of local minima in the design space. This aspect has motivated the authors of this chapter to explore an inverse-problem approach for the determination of external loading in structures, based on the so-called *particle swarm optimization* or PSO (Kennedy and Eberhart, 1995; Eberhart and Kennedy, 1995). The objective function to be minimized is written according to quantities associated with the dynamical behavior of the structure. A similar technique, known as *ant colony optimization* (Dorigo *et al.*, 1991) is also mentioned as another biology-based optimization technique.

PSO is an optimization algorithm based in the swarm theory. The electrical engineer Russell Eberhart and the social psychologist James Kennedy introduced the basic PSO algorithm, in 1995. They were inspired in the 'flocking behavior' models developed by the biologist Frank Heppner.

Heppner's bird model counterbalances the desire to stay in the flock with the desire to roost. Initially, the birds are searching for a roosting area. If a bird searches alone, the success probability is small, but in a flock, the birds can cooperate with each other and this behavior will facilitate the global search.

Using simple rules, a bird when flying is influenced by its own movement, its knowledge of the flight area and the flock performance. In this way, if a bird finds the roosting area, this would result in some birds moving towards the roost. In the socio-cognitive viewpoint this means that mind and intelligence are socials.

Following this principle, each individual learns (and contributes) primarily from the success of his neighbors. This fact requires a balance between exploration (the capacity of individual search) and exploitation (the capacity of learning from the neighbors).

Searching for a roost is an analogous procedure to optimization, and the fly search area has the same meaning as the design or search space in optimization. In the PSO algorithm, the social behavior of birds is modeled by using position and velocity vectors together with parameters like self-trust, swarm trust and inertia of the particle. The outline of a basic PSO is as follows:

1. Define the PSO parameters (inertia, self trust, swarm trust, etc.).
2. Create an initial swarm, randomly distributed throughout the design space (other distributions can be performed).
3. Update the velocity vector of each particle.
4. Update the position vector of each particle.
5. Go to step 3 and repeat until the stop criteria is achieved.

Unlike the genetic algorithms, the PSO algorithm is inherently a continuous optimization tool; however, by using straightforward approaches (by rounding) it is possible to deal with discrete/integer design variables. This technique was successfully applied to the identification of external forces based on the dynamical responses of structural systems (Rojas *et al.*, 2004).

7.6.4 *Meta-Modeling and Response Surface Approximations*

Meta-models are statistical surrogates used to represent cause-and-effect relationships between design parameters and responses of interest within a given design space. They are constructed according to the four fundamental steps whose brief descriptions follow:

(a) *Experimental design* – a design space, including a range of design possibilities, is sampled in order to reveal its contents and tendencies. Each sample is a combination of design variable values.
(b) *Choice of a model* – the nature of the 'meta-model' itself is determined, taking into account that the relations contained in the data gathered in the previous step have to be symbolically represented, with the highest possible accuracy.
(c) *Model fitting* – the model whose shape is defined in (b) is fitted to the data collected in (a). Differences in fitting schemes may affect the efficacy of 'meta-modeling' techniques in the solution of a given problem. In the case of the response surface

method (RSM), the least squares formulation is adopted, as shown in equations (7.76) and (7.77):

$$\mathbf{Y} = \mathbf{EB} + \boldsymbol{\delta} \tag{7.76}$$

where $\mathbf{Y}$ is the vector of responses (dependent variables) obtained for each line of the matrix $\mathbf{E}$ which corresponds to the experimental design stage of meta-modeling. The vector $\boldsymbol{\delta}$ contains free, random error terms. The vector of model parameters $\mathbf{B}$ can be estimated as follows:

$$\mathbf{B} = (\mathbf{E}^T\mathbf{E})^{-1}\mathbf{E}^T\mathbf{Y} \tag{7.77}$$

where the term $(\mathbf{E}^T\mathbf{E})^{-1}$ comes directly from the experimental matrix and is called the variance-covariance matrix, a very important element in evaluating the quality of the meta-model.

(d) *Verification of model accuracy* – the three preceding steps are sufficient to build a first tentative model, whose overall quality and usefulness have to be evaluated by adequate sets of metrics. Each combination of design space sampling, model choice and fitting procedure leads to the use of specific verification procedures.

7.7 MULTICRITERIA OPTIMIZATION

In many practical situations several objective functions have to be minimized simultaneously. As an example of multicriteria structural optimization, we can mention the situation in which the natural frequencies have to be 'safely' away from resonance and, simultaneously, the global strain energy is expected to be minimal. Therefore, multicriteria optimization reflects the idea in which the optimal design represents the minimization of two or more criteria. The solutions of such an optimization problem are called Pareto optimum solutions. The Pareto optimality concept is stated as: 'A vector of x* is Pareto optimal if there exists no feasible vector x which would decrease some objective function without causing a simultaneous increase in at least one objective function'.

This means that problems with multiple objective functions are characterized by the existence of conflict, i.e. none of the possible solutions allows for optimal fulfillment of all objectives for the same design configuration.

To describe mathematically the multicriteria optimization problem, $f(x)$ defined above represents now a vector of r objective functions such that

$$\mathbf{f}(\mathbf{x}) = (f_1(\mathbf{x}) \dots f_r(\mathbf{x}))^T \tag{7.78}$$

Various techniques exist to deal with multicriteria optimization (Osyczka, 1990).

Vanderplaats (1998) presents a compromise optimization formulation as in equation (7.79), which is very effective in the context of multicriteria optimization:

$$f(\mathbf{x}) = \left\{ \sum_{k=1}^{K} \left[\frac{W_k\{f_k(\mathbf{x}) - f_k^*(\mathbf{x})\}}{f_{wk}(\mathbf{x}) - f_k^*(\mathbf{x})} \right]^2 \right\}^{1/2} \tag{7.79}$$

where:

- $f(\mathbf{x})$ is a compromise objective function;
- f_k is the kth response of interest, from a total of K;
- f_k^* is the target value for the kth response;
- f_{wk} is the worst value accepted for the kth response;
- W_k is the weighting factor applied for the kth response of interest.

This formulation is very interesting from the practical point of view because it considers engineering specifications through f_k^* and f_{wk}, which helps in keeping physical insight over the optimization problem. It should be noted that the optimization problem defined by equation (7.79) is unconstrained because the K responses encompass both objective and constraint functions. This is useful when heuristic optimization techniques are used because most of the time it is not trivial to implement the handling of explicit constraints when such methods are used.

7.8 GENERAL OPTIMIZATION SCHEME FOR INVERSE PROBLEMS IN ENGINEERING

Optimization techniques can be used to solve inverse problems in engineering. The two basic ingredients for this aim are the design model and optimization algorithms. Figure 7.2 presents the architecture of general optimization procedure for inverse problem minimization (Luber, 1997).

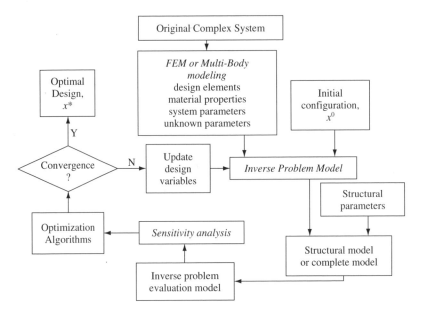

Figure 7.2 Inverse problem architecture for identification

Basically, a design model has to be generated by using FEM or multibody techniques, for example. The unknown parameters are assigned to design variables x, from the optimization point of view. These could be crack position, severity and length, unbalance masses and positions, stiffness and damping coefficients, misalignment characteristics, etc.

The initial configuration x_0 corresponds to the initial values attributed to the design variables. It is important to stress that this is a difficult task in real world inverse problems. The optimal design x^* corresponds to the optimal values obtained for the design variables, hopefully they correspond to the unknown parameters of the system.

The objective function $f(x)$ can be expressed by the difference between measured and simulated quantities associated with the dynamical behavior of the system, such as natural frequencies, eigen-modes, dynamical responses (FRFs and unbalance responses), critical speeds, etc. A combination of these quantities can also be considered for performing multicriteria optimization (Müller-Slany, 1993). Constraint functions $g(x)$ and $h(x)$ can be represented by stresses, strains, flutter speed, buckling load, stress concentration factors and dynamical responses in general. The optimization algorithms can be the classical ones (gradient based) or pseudorandom techniques, such as genetic algorithms and simulated annealing, for example. Similarly, in many practical situations it may be necessary to use hybrid optimization schemes, by combining different techniques in such a way that local minima are avoided.

7.9 APPLICATIONS

In this section various aspects previously addressed are illustrated through applications performed on both numerically simulated and experimentally tested structures.

7.9.1 Application 1: Finite Element Error Localization Based on the Sensitivity of Natural Frequencies and Antiresonance Frequencies

The aim of this application is to evaluate the effectiveness of the error localization method based on the simultaneous use of resonance and antiresonance frequencies, in combination with the model adjustment method based on the sensitivity of eigenvalues and eigenvectors and the best subspace technique for localization of dominant modeling errors. This application was originally developed by Silva (1996). The test structure, depicted in Figure 7.3, is a beam made of steel, which has been tested in laboratory under free–free boundary conditions.

Experimental tests for acquisition of frequency response functions were carried out on the beam in two different situations: the undamaged (error-free) beam and a damaged configuration in which the damage level was simulated by making a saw-cut (see Figure 7.4), as follows: $b = 8.0\,\text{mm}$, $t = 4.0\,\text{mm}$.

For the purpose of error localization, a finite element model of the beam was constructed having 20 two-dimensional Euler–Bernoulli beam elements, 21 nodes, 2 degrees-of-freedom per node (longitudinal degrees-of-freedom eliminated).

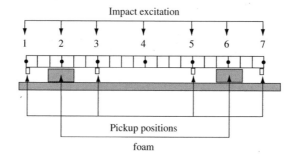

Figure 7.3 Characteristics of the test item and experimental setup

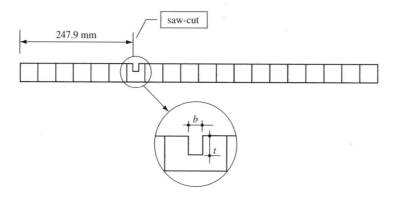

Figure 7.4 Details of saw-cut simulating a structural damage

The following characteristics were adopted in the error localization procedure:

- estimation equations based on the sensitivities of resonance and antiresonance frequencies, combined with the best subspace localization technique;
- poles and zeros of the FRF $H_{76}(\omega)$, in the band [0–200 Hz] were used;
- an error indicator was ascribed to the bending stiffness of each of the 20 finite elements of the model;
- actual damage is located in element number 7.

Table 7.1 Values of FRFs poles and zeros (Hz)

	Poles			Zeros of $H_{76}(\omega)$		
	Undamaged	Damaged	$\Delta f(\%)$	Undamaged	Damaged	$\Delta f(\%)$
2	107.3	104.9	2.2	307.5	291.5	5.2
3	291.8	281.2	3.6	661.4	655.3	0.9
4	569.4	566.2	0.6	1134.2	1106.6	2.4
5	938.1	926.3	1.3	1762.0	1730.9	1.8
6	1393.9	1353.5	2.9		>2000	
7	1939.8	1930.7	0.5			

First identification test using the poles

Localized elements	Representation errors (%)
k_7	7.47
k_{16}	7.30

Second identification test using of the zeros of the FRF $H_{76}(\omega)$

Localized elements	Representation errors (%)
k_7	21.10
k_{14}	6.20

Third identification test using the poles and zeros of the FRF $H_{76}(\omega)$

Localized elements	Representation errors (%)
k_7	16.80
k_{14}	8.72

A difficulty frequently encountered in parameter identification calculations is related to the low sensitivity of the dynamic responses, with respect to some parameters. This is the case, for instance, when a given parameter is related to the stiffness of regions of the structure concentrating a very small fraction of the modal strain energy. In such cases, these parameters could hardly be correctly identified, taking into account that their effect on the dynamic responses is likely to be masked by the inescapable experimental errors. It is common practice simply to discard the 'insensitive' parameters from the identification problem. It should be remembered that this procedure is acceptable only in cases where the adjusted model is to be used to repeat only the experimental data used in the identification process. Otherwise, if the adjusted model is to be used for previsions of the dynamic characteristics of structural configurations other than those used for updating, it should be taken in account that the effect of the parameters can change strongly from one configuration to another. In this regard, the observation of different structural configurations, as suggested herein, can enable the identification of correction parameters that could hardly be identified if a single configuration was exploited. This point is illustrated by Rade and Lallement (1998).

7.9.2   Application 2: Fault Identification in Flexible Rotors

Nowadays the modern design of rotating machinery shows increased use of composite materials and new alloys in various machine components. These materials permit rotating machinery to attain high operational speeds. However, such materials are susceptible to the appearance of faults. These faults lead to a loss of mechanical properties of the materials reducing fatigue life, which can cause the machine to fail or to malfunction of its components. A sudden failure in rotating machinery can cause great economic loss, inconvenience to users and even loss of human lives. To avoid such problems it is necessary to develop a methodology for detecting the faults in their early stages. Several researchers propose mathematical models to represent rotors with cracks (Gasch, 1976; Henry and Okah-Avae, 1976; Mayes and Davies, 1976). Nelson and Nataraj (1986) use the FEM (finite element method) to represent more realistic complex industrial rotors. Cheng and Ku (1991) simulate the dynamic behavior of a damaged rotor, regarding the fault as a source of energy reduction. This reduction entails modification in the element stiffness matrix.

The crack diminishes the area moment of inertia of the rotor cross section in which the fault is localized. Therefore, shaft stiffness decreases and the rotor vibratory pattern alters. The area moments of inertia about the X axis (I_X) and Z axis (I_Z) change from a maximum to minimum value during a rotation cycle, depending on the angular position of the fault. This behavior introduces nonlinearity in the cracked rotor. A simplified fault model considers an average value of the area moment of inertia of the damaged cross section along a rotation cycle (I_d)·I_d is related to the no-fault area moment of inertia, I, by using a parameter ξ, equation (7.80), where ξ represents damage severity. Deeper cracks generate smaller values to ξ.

$$I_d = \xi I \qquad (7.80)$$

This model considers that the shaft possesses a localized region of damage and, for that region, the deterioration is assumed to be uniform per unit length and distributed in such a way that no shift in the line of action of the resultant force occurs (Cheng and Ku, 1991). Then, the corresponding damaged finite element has less capacity to store strain energy with respect to an undamaged element. The damaged element is considered to have a length L_d, which is supposed to be as small as possible in order to represent a realistic fault. As the fault is localized at an arbitrary location along the shaft, its location is associated with the position of the damaged element (p) along the shaft. Thus, three parameters are used to characterize the fault. This way, a vector of design variables, V_p, can be formed for optimization purposes, as given by equation (7.81):

$$V_p = [\xi, p, L_d] \qquad (7.81)$$

Usual methods for identifying faults in structures, such as ultrasound, infrared radiation, magnetic particles, are not effective in dealing with rotors due of the high noise levels found in industrial plants. Moreover, the necessity to stop the plant in order to conduct the tests is a time-consuming operation and causes economic losses. As vibration patterns reflect changes in mechanical properties of a structure, they can be used to identify and localize damage. The identification process is considered as an optimization problem using the minimization of a functional formed by the difference between the dynamic characteristics of the real system and the dynamic characteristics of the FEM model. At the end of the process it is reasonable to expect that the functional will be minimized (global minimum) and the design variable values are such that they correspond to the fault parameters of the real mechanical system.

The dynamic behavior of cracked rotors is studied by analyzing the influence of the three fault parameters (severity, location, length) on the rotor response. Four dynamic features were taken into account, aiming at obtaining meaningful information about fault influence: whirl speeds, vibration modes, unbalance response and critical speeds. The dynamic response of a vertical rotor model, comprising a 10-mm diameter shaft, three discs and two support bearings localized at the ends is shown in Figure 7.5 (Simões and Steffen, 2003).

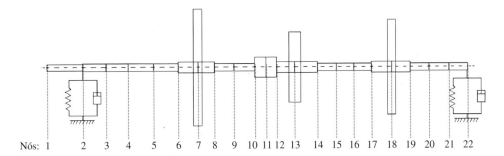

Figure 7.5   Rotor model

The influence of fault severity on whirl speeds is shown in Figure 7.6. It was observed that the more severe the faults, the greater the resulting reductions in whirl speeds, as expected. Figure 7.7 shows the influence of damage location in whirl speeds. It can be observed that the fault localized at the 10th finite element leads to greater changes than do faults localized in another position. It is worth mentioning that this position corresponds to the maximum displacement point for the third mode. The fault does not significantly affect the rotor vibration modes (Figure 7.8), however the modes determine which whirl speed will be the most affected by damage. Table 7.2 shows the results for the third critical speed of the rotor. The position $p = 10$ is the one that most alters the critical speeds in the present example. More detailed analysis about the influence of fault parameters in the rotor dynamic behavior can be found in Simões (2002).

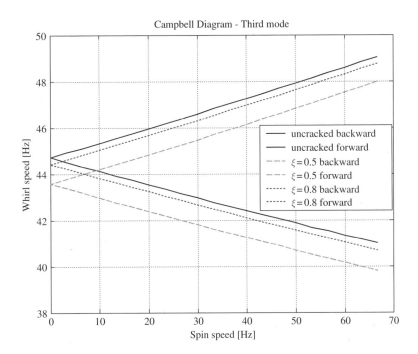

Figure 7.6   Whirl speeds of the cracked rotor, $p = 10$; $L_d = 0.01$ m

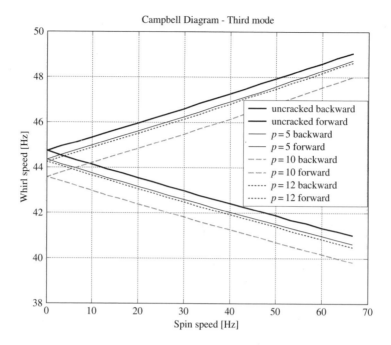

Figure 7.7 Whirl speeds of the cracked rotor, $\xi = 0.5$; $L_d = 0.01$ m

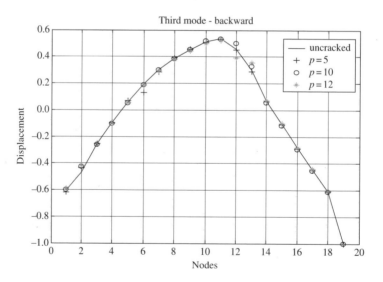

Figure 7.8 Third mode – backward, $\xi = 0.5$; $L_d = 0.01$ m

For simulation purposes, the fault parameters were previously established in such a way that the corresponding results play the role of the real system. The code GAOT version 5 (The Genetic Algorithm Optimization Toolbox for Matlab 5) was used for optimization (this software was developed in the College of Engineering, North Carolina State University, USA). The fault parameters (vector V_p, equation 7.81) are

Table 7.2   Third critical speed (Hz)

Damage severity (ξ)	Damage location		
	$p = 5$	$p = 10$	$p = 12$
0.5	49.02	48.18	48.84
0.8	49.31	49.09	49.29
1		49.41	

optimized by coupling the rotordynamics program with the code GAOT. For this purpose, it is necessary to define lower and upper bounds to the design variables, write the objective function to be minimized, define the initial population and its size, and fix the number of generations. As the optimization code was designed to maximize a fitness function, the negative of that function is set. The fitness function (objective function) is given by equation (7.82):

$$
\text{Fobj} = \sum_{j=1}^{6} w_j^q \cdot \left[\sum_{i=1}^{n} \left(\omega_i^S - \omega_i^m \right)^2 \right]_j + \sum_{i=1}^{3} w_j^r \left(\left(Vc^S - Vc^m \right)^2 \right)_j
$$
$$
+ \sum_{j=1}^{6} w_j^S \cdot \left[\sum_{i=1}^{1} \left(S_i^S - S_i^m \right)^2 \right]_j + w^t \cdot \sum_{i=1}^{k} \left(d^S - d^m \right)^2
$$

(7.82)

where ω^S are vectors corresponding to the first six whirl speeds of the 'real' system, S^S corresponds to the first six modes, d^S is the vector that contains the unbalance response of the system and Vc^S are the first three critical speeds. The vectors ω^m, S^m, d^m and the value Vc^m represent the same dynamic quantities as above, for the mathematical model. w_j^q, w_j^r, w_j^S and w^t are weight coefficients used in the multiobjective function. Finally, in equation (7.82) n corresponds to the number of whirl speeds used, 1 is the number of nodes in the FEM mesh and k is the number of unbalance responses calculated.

The simulation results obtained for the rotor are shown in Table 7.3, using genetic algorithms to solve the inverse problem. The situation in which two simultaneous faults

Table 7.3   Simulation results (two simultaneous faults along the rotor)

	Simulation	Faulty parameters						Objective function
		Fault 1			Fault 2			
		ξ	p	L_d(m)	ξ	p	L_d(m)	
1	Real parameters	0.5	6	0.003	0.8	12	0.003	$1.7 * 10^{-4}$
	Identified parameters	0.53	6	0.0035	0.82	12	0.0034	
2	Real parameters	0.5	5	0.003	0.8	10	0.003	0.004
	Identified parameters	0.7	5	0.006	0.78	10	0.0035	
3	Real parameters	0.5	5	0.005	0.5	13	0.005	0.0058
	Identified parameters	0.64	5	0.0035	0.5	13	0.0044	

are considered together along the rotor is analyzed. The fault location along the shaft was the parameter that was estimated with better accuracy, followed by the fault severity, and finally the fault length, as shown in Table 7.3. This was expected because fault location is the parameter that most influences the dynamical behavior of the system, as shown by the previous results. Evidently, the identification of two simultaneous faults is a difficult task, since the number of parameters required by the optimization process is duplicated as compared with the case in which only one fault is to be determined.

7.9.3 Application 3: Inverse Problem Techniques for the Identification of Rotor-Bearing Systems

The values of bearing parameters of rotating machinery are usually unknown. The problem of determining these unknown parameters is very challenging in rotordynamics as they are needed for various purposes, such as: dynamic analysis and design; control; diagnosis and health monitoring; etc. According to Lalanne and Ferraris (1998), the system equation of motion is given by:

$$\mathbf{M}\ddot{\mathbf{q}} + \mathbf{C}\dot{\mathbf{q}} + \mathbf{K}\mathbf{q} = \mathbf{F}_1 + \mathbf{F}_2 \sin \Omega t + \mathbf{F}_3 \cos \Omega t + \mathbf{F}_4 \sin a \, \Omega t + \mathbf{F}_5 \cos a \, \Omega t \qquad (7.83)$$

where $q = N$ order generalized coordinate displacement vector; $\mathbf{K} =$ stiffness matrix which takes into account the symmetric matrices of the beam and nonsymmetrical matrices of the bearings; $\mathbf{C} =$ matrix consisting of skew-symmetrical matrices due to gyroscopic effects and nonsymmetrical matrices due to bearing viscous damping; $\mathbf{F}_1 =$ constant body force such as gravity; $\mathbf{F}_2$, $\mathbf{F}_3 =$ forces due to unbalance; $\mathbf{F}_4$, $\mathbf{F}_5 =$ forces due to nonsynchronous effects; and $a =$ coefficient.

When optimization techniques are to be used in inverse problems, the design space should be as small as possible to avoid time-consuming search procedures. Identification techniques based on the unbalance response of flexible rotors are used to obtain a first estimate of the unknown parameters of the bearings (Assis and Steffen, 2003; http://www.tandf.co.uk).

A hybrid optimization scheme was designed for performing a parameter estimation procedure, involving classical methods together with pseudorandom ones. For this purpose the following steps should be followed:

(a) The unbalance response of the rotor bearing system (or, alternatively, the FRF) is determined.
(b) A first estimation of the unknown parameters is obtained (Assis and Steffen, 2003). This step allows the initialization of an optimization task in which the design variables correspond to the parameters to be identified.
(c) The following general non-linear optimization problem is defined as:

$$\text{Minimize } F(\mathbf{X}) = \left(D_{\text{measured}} - D_{\text{f.e.m}} \right)^2 \qquad (7.84)$$

$$\text{Subject to } \mathbf{X}_i^l \leq \mathbf{X}_i \leq \mathbf{X}_i^u \qquad (7.85)$$

Neither inequality, nor equality constraints are taken into account in this case.

(d) A sequence of optimization runs are now executed depending on the numerical behavior of the objective function with respect to the existence of local minima:

(d.1) To execute a classical optimization run, the initial design is defined according to the results obtained in step (b) (note that in the applications presented in this chapter this approach proved to be inadequate in many cases because of the low sensitivity of the objective function with respect to the design variables, and also because a significant number of local minima was detected);

(d.2) A GA optimization approach is used when (d.1) fails. For this purpose, an initial population is defined in the neighborhood of the solution found in step (b). The number of generations and the number of individuals are fixed and the evolution process is launched.

(d.3) The best result obtained in (d.2) is used as a starting point for the simulated annealing algorithm. At this point of the procedure, the objective function is expected to be closer to the minimum, and the search space can be reduced. In this case the search space reduction is associated with the initial value attributed to the parameter T (temperature). As the annealing process performs, T is reduced according to the procedure presented before.

(d.4) Eventually step (d.1) can be performed again by using the result obtained in (d.3) to define the initial design for a final classical optimization run.

Using a test rig composed of a vertical rotor supported at the ends by flexible bearings the techniques presented above were validated. Rigid discs could be mounted along the rotor as wished. The test rig was connected to a dedicated workstation and to a modular data acquisition system. Figure 7.9 shows the test rig (a) and its geometry (b). To obtain the finite-element model it was necessary to add supplementary masses at the bearing positions, as shown in Figure 7.9(b). The stiffness of the rotor supports was adjusted as necessary. The adjusting device can be seen in Figure 7.10 for the lower support of the rotor. In this case unbalance was introduced at the central disk, as follows: mass $= 4.5$ g; eccentricity $= 5.46$ cm; angle $= 190°$.

In Table 7.4 the identification history corresponding to the experimental validation procedure is presented. Figure 7.11 shows the unbalance responses for the measured and identified cases.

7.9.4 Application 4: Meta-Modeling Techniques Applied to the Optimization of Test Procedures in Impedance-Based Health Monitoring

The application of statistics techniques, such as the factorial design, has been used in various engineering problems, as reported by Box and Draper (1986), Montgomery (1991) and Barros *et al.* (1995). Through these techniques it is possible to understand the behavior of the system while using a smaller number of experiments. Then, surface response meta-models can be calculated and the optimal design can be obtained. However, the meta-model usually represents a small region of the design space. In the

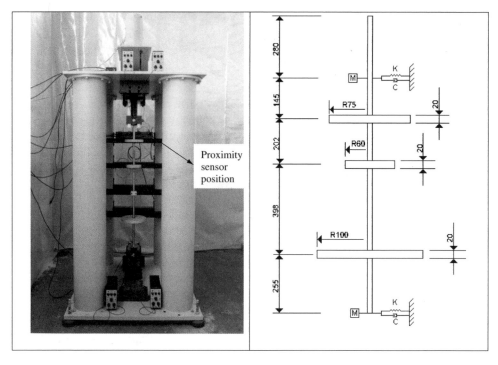

Figure 7.9 (a) Test rig; (b) Test rig geometry (Reproduced by permission of Taylor & Francis from: Inverse Problem Techniques for the identification of rotor-bearing systems, by E.G. Assis and V. Steffen Jr, *Inverse Problems in Engineering*, Vol. 11, No. 1, 2003. http://www.tandf.co.uk.)

area of vibration analysis, it has been shown that the dynamical behavior of mechanical systems can be expressed by representative meta-models (Butkewitsch and Steffen, 2002; Moura and Steffen, 2003).

In the other hand, nondestructive evaluation (NDE) tools, such as the impedance-based health monitoring technique, have been discussed for about 20 years, but only more

Figure 7.10 Lower end support of the rotor (Reproduced by permission of Taylor & Francis from: Inverse Problem Techniques for the identification of rotor-bearing systems, by E.G. Assis and V. Steffen Jr, *Inverse Problems in Engineering*, Vol. 11, No. 1, 2003. http://www.tandf.co.uk.)

Table 7.4 Identification history

Variable	Real value	Step (b)	Step (d.2)	Step (d.3)
K_{xx} (N/m)	8600.0	3557.0	9806.0	9800.0
K_{zz} (N/m)	11 600.0	3557.0	11 521.0	11 700.0
C_{xx} (Ns/m)	83.3	364.0	81.85	81.85
C_{zz} (Ns/m)	83.3	364.0	88.45	88.45
F(X)	—	0.4875	0.3717	0.3717

Source: Reproduced by permission of Taylor & Francis from: Inverse Problem Techniques for the identification of rotor-bearing systems, by E.G. Assis and V. Steffen Jr, *Inverse Problems in Engineering*, Vol. 11, No. 1, 2003. http://www.tandf.co.uk.

recently have the most interesting applications been published. The idea is to compare the monitored vibration level of the system, usually the frequency response function, with a standard vibration level that corresponds to the undamaged structure, aiming at identifying changes in the actual design. For this purpose, piezoelectric patches (PZTs) are bonded to the structure to capture the system response. The PZT is responsible for the conversion of the mechanical impedance of the structure, which is intimately connected with the system stiffness or system integrity. Measuring the electrical impedance of the PZT can determine the impedance variation of the structure due to any structural failure.

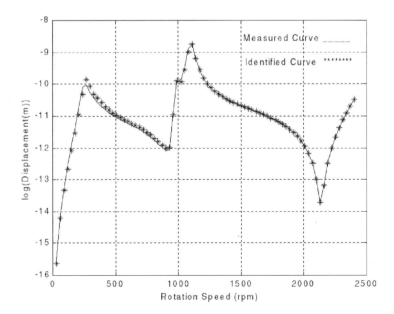

Figure 7.11 Unbalance response at node 10 (Reproduced by permission of Taylor & Francis from: Inverse Problem Techniques for the identification of rotor-bearing systems, by E.G. Assis and V. Steffen Jr, *Inverse Problems in Engineering*, Vol. 11, No. 1, 2003. http://www.tandf.co.uk.)

An important issue with the above mentioned technique is the correct determination of the most sensitive frequency bands for health monitoring purposes. In general, the range for a given structure is determined by a trial-and-error method (Raju, 1997; Giurgiutiu and Zagrai, 2002).

In this application a procedure based on surface responses is developed to determine the most sensitive frequency range for damage identification through the impedance method (Moura and Steffen, 2004)

Figure 7.12 shows a cantilever flexible beam that is used in the present application.

Electrical impedance was measured by using an HP4194A impedance analyzer, as shown in Figure 7.13. A C++ code was written to manage the data acquisition process through a GP-IB card installed in a microcomputer. An ACX PZT was bonded to the beam as depicted in Figure 7.14. An induced fault was introduced in the system by adding a small mass (Figure 7.15, scale in cm) to the structure at its free end. Figure 7.16 shows an overview of the testing apparatus.

The most often used signal in impedance analysis is the R type, which corresponds to the real part of the impedance. In this application the real part, the imaginary part and the modulus of the impedance were evaluated for health monitoring purposes.

(a)   (b)

Figure 7.12   Cantilever beam

Figure 7.13   HP4194A impedance analyzer

Figure 7.14   PZT (ACX) bonded to the beam

Figure 7.15   Adding mass

Figure 7.16   Testing apparatus

This way, in the present contribution three factors were evaluated for impedance-based health monitoring, namely, the lowest frequency to be used, the frequency range, and the type of signal used. The reference values for the experimental tests were the following: OSC Level = 1 volt; average number = 8; sample points = 401. The best frequency range was tentatively found to be between 15 and 33 kHz.

A 4×3^2 factorial design was implemented to determine the optimal conditions for health monitoring. Four frequency levels were defined to obtain the lowest frequency to be used: 15, 19, 23 and 27 kHz. Similarly, for the frequency width, three levels were considered: 2, 4, and 6 kHz. Finally, for the type of signal, three levels were taken into account: $|Z|$, the impedance modulus; R, the real part of the impedance, and X, the imaginary part. All experiments were made randomly, without repetition. Each experiment was made twice: with and without the adding mass. Figure 7.17 shows the differences between the two signals (with and without the mass). It can be seen that experiments Nb. 13, 14, and 15 gave a better performance than the others with respect to the sensitivity of the failure monitoring process. Experiment No. 40 was taken as a reference for comparison purposes in this case. It corresponds to normal test conditions.

Seven meta-models were evaluated in order to choose which of them better represents the system under study within the test conditions domain. For this reason, each model was submitted to a hypothesis test (F test) to compare two consecutive meta-models with regard to their ability in representing the system. For the seventh meta-model the polynomial coefficients are given in Table 7.5, where $f1$, $f2$, and $f3$ are associated respectively with the lowest frequency in the band, the band width, and the type of signal. It can be seen that the resulting curve is distributed around zero, as illustrated in Figure 7.18. This means that this experiment can be considered for statistical evaluation purposes.

Figure 7.19 shows the normal probability distribution for the factors of the final meta-model used in this application (the independent term is not included). Figure 7.20 shows the surface responses for the final meta-model, in which the three factors are depicted. The continuous lines represent the surface responses and the two other curves represent the 95 % confidence intervals. The factors were normalized in the interval -1 to 1.

Table 7.6 shows the ANOVA for the final meta-model. As no repetition was made, pure error is zero, i.e. the impedance signal did not vary (the captured signal is automatically

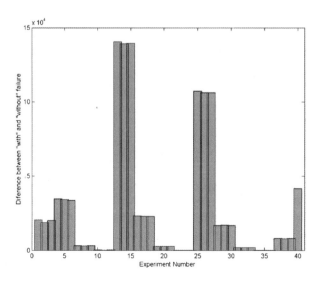

Figure 7.17   Differences between the two signals (with and without the adding mass)

Table 7.5 Polynomial regression coefficients

	Value ($\times 10^4$)
Constant	1.833 011 729 011 32
$f2$	−4.318 864 632 191 67
$f1$	0.853 893 444 270 83
$f3$	−0.014 260 476 192 31
$f2 \times f1$	−1.835 268 995 812 50
$f2 \times f3$	0.021 193 151 175 00
$f1 \times f3$	0.004 081 825 125 00
$f2 \times f1 \times f3$	0.000 268 912 725 00
$f2^2$	3.702 891 193 174 54
$f1^2$	−1.604 194 115 643 00

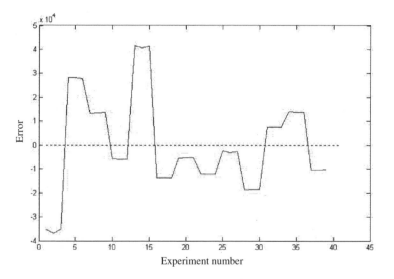

Figure 7.18 Residue distribution for the seventh meta-model

averaged by the impedance analyzer). The resulting explained variation is about 78 %, i.e. this meta-model satisfactorily represents the phenomenon.

As in this case the goal is to obtain a better frequency range for the impedance monitoring, 78 % is satisfactory for representing the system under analysis. It can be observed that MQR/MQr (11.839 502) in this case is five-times the value found in the F hypothesis test (2.222 874).

By observing the surface responses it is possible to say that the best lowest frequency value is the lowest possible one (15 kHz), while the best bandwidth value corresponds to the intermediate one (4 kHz). Finally, the type of signal has a small influence in the tests, however, the $|Z|$ signal (impedance modulus) can be considered as the best. Consequently, the best test configuration corresponds to test No. 13 in Figure 7.17, followed by tests Nb. 14 and 15. This confirms the results obtained by the surface responses for the type of signal used.

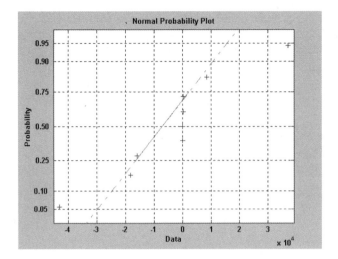

Figure 7.19   Normal probability distribution for the seventh meta-model

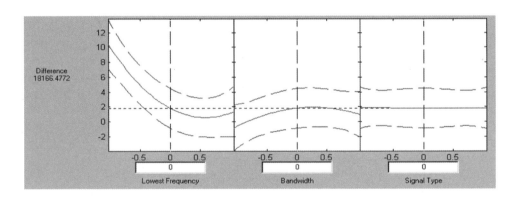

Figure 7.20   Surface responses for the seventh meta-model

Table 7.6   ANOVA for the seventh meta-model

Source of variation	Sum of squares	Degrees of freedom	Mean square
Regression	5.563e + 010	9	6.181e + 009
Residual	1.514e + 010	29	5.22e + 008
Adjustement error	1.514e + 010	29	5.22e + 008
Pure error	0	0	0
Total	7.077e + 010	38	

% explained variation: 0.78607.
% max of explained variation: 1.

From the surface responses it would be possible to optimize the test procedure. Future work on this topic will focus on determining the best test conditions for several fault intensities and positions. This way, by using compromise optimization techniques it is possible to create a robust test procedure for damage monitoring.

ACKNOWLEDGMENTS

Valder Steffen and Domingos Rade are grateful to their former graduate students, Elaine Assis, Sérgio Butkewitsch, Leandro A. Silva and Rodrigo F. Alves Marques, and also to their graduate students Danuza C. Santana, Felipe Chegury Viana, Jhojan Enrique Rojas Flores, José dos Reis V. de Moura Jr., and Ricardo C. Simões, for their contribution to the development and organization of the topics presented, with talent and effort.

REFERENCES

Arora, J.S. (1976) 'Survey of structural reanalysis techniques', *Journal of Structural Division ASCE*, **102**, 783–802.

Assis, E.G. and Steffen Jr, V. (2003) 'Inverse problem techniques for the identification of rotor-bearing systems', *Inverse Problems in Engng.*, **11**, 39–53.

Barros Neto, B., Scarminio, I.S. and Bruns, R.E. (1995) *Planejamento e Otimização de Experimentos*. Unicamp Press, Campinas, Brazil.

Box, G.E.P. and Draper, N. (1986) *Empirical Model Building and Response Surfaces*, John Wiley & Sons, Inc., New York.

Butkewitsch, S. and Steffen Jr, V. (2002) 'Shape optimization, model updating and empirical modeling applied to the design synthesis of a heavy truck side guard', *International Journal of Solids and Structures*, **39**, 4747–4771.

Chen, K., Brown, D.L. and Nicolas, V.T. (1993) 'Perturbed Boundary Condition Model Updating', *Proceedings of the 11 International Modal Analysis Conference*, Kisseemmee, Florida, Society for Experimental Mechanics, pp. 661–667.

Cheng, L.M. and Ku, D.M. (1991) 'Whirl speeds and unbalance response of a shaft-disk system with flaws', *The International Journal of Analytical and Experimental Modal Analysis*, **6**, 279–289.

Dorigo, M., Maniezzo, V. and Colorni, A. (1991) 'Positive feedback as a search strategy', *Technical Report 91–016*, Dipartimento di Elettronica e Informazione, Politecnico de Milano, Italy.

Eberhart, R.C. and Kennedy, J. (1995) 'A New Optimizer Using Particles Swarm Theory', *Proceedings Sixth International Symposium on Micro Machine and Human Science*, Piscataway, New Jersey, IEEE Service Center, pp. 39–43

Eschenauer, H., Olhoff, N. and Schnell, W. (1997) *Applied Structural Mechanics*, Springer-Verlag, Berlin.

Flanelly, W.G. (1971) 'Natural Antiresonances in Structural Dynamics', *Internal Report*, Kaman Aerospace Corporation, Bloomfield, CT.

Friswell, M.I. and Mottershead, J.E. (1995) *Finite Element Model Updating in Structural Dynamics*, Klüwer Academic Publishers, Dordrecht.

Gasch, R. (1976) 'Dynamic Behaviour of a Simple Rotor with a Cross-Sectional Crack', *Institution of Mechanical Engineers Conference Publication*, Vibration in Rotating Machinery, Paper No. C178/76, pp. 123–128.

Giurgiutiu, V. and Zagrai, A.N. (2002) 'Embedded Self-Sensing Piezoelectric Active Sensors for On-Line Structural Identification', *Journal of Vibration and Acoustics*, **124**, 1–10.

Goldberg, D.E. (1989) *Genetic Algorithms in Search Optimization and Machine Learning*, Addison-Wesley, New York.

Haug, E.J., Choi, K.K., Komkov, V. (1986) *Design Sensitivity Analysis of Structural Systems*. Academic Press, Inc., New York.

Henry, T.A. and Okah-Avae, B.E. (1976) 'Vibration in Cracked Shafts', *Institution of Mechanical Engineers Conference Publication*, Vibration in Rotating Machinery, Paper No. C178/76, pp. 15–19.

Kennedy, J. and Eberhart, R.C. (1995) 'Particle swarm optimisation,' *Proceedings of the 1995 IEEE International Conference on Neural Networks*, Perth, Australia, pp. 1942–1948.

Kirkpatrick, S., Gelatt, C.D. and Vecchi, M.P. (1983), 'Optimization by simulated annealing', *Science*, **220**, 671–680.

Lallanne, M. and Ferraris, G. (1998) *Rotordynamics Prediction in Engineering*, John Wiley & Sons, Ltd, Chichester.

Lallement, G. and Cogan, S. (1992) 'Reconciliation Between Measured and Calculated Dynamic Behavior: Enlarging the Knowledge Space of the Structure', *Proceedings, 10th International Modal Analysis Conference*, Society for Experimental Mechanics.

Lallement, G. and Piranda, J., (1990) 'Localisation Methods for Parameter Updating of Finite Element Models in Elastodynamics', *Proceedings of 8th International Modal Analysis Conference*, Orlando, Florida, Society for Experimental Mechanics, pp. 579–585.

Lammens, S., Heylen, W., Sas, P. and Brown, D. (1993) 'Model Updating and Perturbed Boundary Condition Testing', *Proceeding of 11th International Modal Analysis Conference*, Kisseemmee, Florida, Society for Experimental Mechanics, pp. 449–455.

Luber, W. (1997) 'Structural Damage Localization using Optimization Method', *Proceedings of the 15th International Modal Analysis Conference*, Feb. 4–6, Orlando, Florida, pp. 1088–1095.

Maia, N.M.M. and Silva, J.M.M. (1997) *Theoretical and Experimental Modal Analysis*, Research Studies Press Ltd.

Mayes, I.W. and Davies, W.G.R. (1976) 'The Vibration Behaviour of a Rotating Shaft System Containing a Transverse Crack', *Institution of Mechanical Engineers Conference Publication*, Vibration in Rotating Machinery, Paper No. C178/76, pp. 53–64.

Michalewicz, Z. (1996) *Genetic Algorithms + Data Structures = Evolution Programs*, Springer-Verlag, New York.

Montgomery, D.C. (1991) *Design and Analysis of Experiments*, Fourth Edition, John Wiley & Sons, Inc., New York.

Moura Jr, J.R.V. and Steffen Jr, V. (2003) 'Compromise Optimization and Meta-modeling for a Flexible 7 dof Mechanical System Using the Response Surface Method', *Proceedings of the 17th International Congress of Mechanical Engineering – COBEM 2003*, São Paulo, SP, Brazil, November.

Moura Jr, J.R.V. and Steffen Jr, V. (2004) 'Impedance-Based Health Monitoring: Frequency Band Evaluation', *Proceedings IMAC XXII: Conference and Exposition on Structural Dynamics*, Dearborn, Michigan, January.

Müller-Slany, H. (1993) 'A Hierarchical Scalarization Strategy in Multicriteria Optimization Problems', *Proceedings of the 14th Meeting of the German Working Goup* 'Mehrkriterielle Entscheidung', pp. 69–79.

Nalitolela, N.G, Penny, J.E.T. and Friswell, M.I. (1990) 'Updating Structural Parameters of a Finite Element Model by Adding Mass or Stiffness to the System', *Proceedings of the 8th International Modal Analysis Conference*, Orlando, Florida, Society for Experimental Mechanics, pp. 836–842.

Osyczka, A. (1990) *Multi-criterion Optimization in Engineering*, Ellis Horwood Ltd, Chichester, UK.

Nelson, H.D. and Nataraj, C. (1986) 'The dynamics of a rotor system with a cracked shaft', *ASME Journal of Vibration, Acoustics, Stress, and Reliability in Design*, **108**, 189–196.

Rade, D.A. (1994) *Parametric Correction of Finite Element Models: Enlargement of the Knowledge Space*, Doctorate Thesis, University of Franche-Comté, Besançon, France (in French).

Rade, D.A. and Lallement, G. (1996) 'Vibration analysis of structures subjected to boundary condition modifications using experimental data', *Journal of the Brazilian Society of Mechanical Sciences*, **XVIII**, 374–382.

Rade, D.A. and Lallement, G. (1998) 'A strategy for the enrichment of experimenal data as applied to an inverse eigensensitivity-based FE updating method', *Mechanical Systems and Signal Processing*, **12**, 293–307.

Rade, D.A. and Silva, L.A. (1999) 'On the usefulness of antiresonances in structural dynamics', *Journal of the Brazilian Society of Mechanical Sciences*, **XXI**, 82–90.

Raju, V. (1997) *Implementing Impedance-Based Health Monitoring*, MSc dissertation, Virginia Tech, USA.

Rojas, J.E., Viana, F.A.C., Rade, D.A. and Steffen Jr, V. (2004) 'Identification of external loads in mechanical systems through heuristic-based optimization methods and dynamic responses', *Latin American Journal of Solids and Structures*, **1**, 297–318.

Saramago, S.F.P., Assis, E.G. and Steffen Jr, V. (1999) 'Simulated Annealing: Some Applications on Mechanical Systems Optimization', *Proceedings of the XX Iberian Latin-American Congress on Computational Methods in Engineering*, CILAMCE 99, São Paulo, Brazil.

Saldarriaga, M.V. and Steffen Jr, V. (2003) 'Balancing of Flexible Rotors Without Trial Weights by Using Optimization Techniques', *Proceedings of the 17th International Congress of Mechanical Engineering – COBEM 2003*, São Paulo, Brazil, November.

Silva, L.A. (1996) *Identification of Structural Faults Based on the Inverse Sensitivity of Poles and Zeros of Frequency Response Functions*, MSc Dissertation, Federal University of Uberlândia, School of Mechanical Engineering (in Portuguese).

Simões, R.C. (2002) *Fault Identification in Flexible Shaft of Rotors by Using Optimization Techniques* MSc Thesis, Federal University of Uberlandia, Uberlandia , MG, Brazil, October (in Portuguese).

Simões, R.C. and Steffen Jr, V. (2003) 'Optimization Techniques for Fault Identification in Rotor Dynamics', *Proceedings of the 17th International Congress of Mechanical Engineering – COBEM 2003*, São Paulo, Brazil, November.

Vanderplaats, G.N. (1998) *Numerical Optimization Techniques for Engineering Design*, Second Edition, VRAND, Inc. Colorado Springs, CO, USA.

8

Structural Health Monitoring Algorithms for Smart Structures

Vicente Lopes Jr and Samuel da Silva

Universidade Estadual Paulista, Ilha Solteira, São Paulo, Brazil

8.1 INITIAL CONSIDERATIONS ABOUT SHM

There is great interest in the engineering community in the development of real-time condition monitoring of the health of structures to reduce cost and improve safety, based on an effective predictive maintenance program. The fundamental idea is that modal parameters are functions of the physical properties of the system, and any changes in these properties could cause variations in modal parameters. There is no doubt in the engineering community about the importance of damage diagnosis. However, some skeptical authors, still, question the applicability of structural health monitoring (SHM) schemes in practical situations, since the simulated damage is much larger than in real life, or it needs a complex analytical model to describe cracks, for instance, that restrict the application only to simple structures.

These issues will be addressed in the near future, because there are great economic incentives motivating this topic's development. Damage identification is an inverse problem, so, in principle, it is an unsolvable problem. It is a typical optimization problem, and any kind of optimization approach can be used to quantify the damage. Inverse problems are well known in the mathematical community; the challenge now is to define some methods better adapted to predicting structural variations, or damage. There will be no unique method to deal with any kind of machine, or structure. Different applications require different approaches, such as using low or high frequencies, model or nonmodel based techniques, linear or nonlinear models (Doebling, *et al.*, 1998),

Damage Prognosis – For Aerospace, Civil and Mechanical Systems Edited by D.J. Inman, C.R. Farrar, V. Lopes Junior and V. Steffen Junior © 2005 John Wiley & Sons, Ltd

analytical or heuristic methods, global or local approaches (Farris and Doyle, 1991) and so on. It is important, at this stage, to define the advantages and limitations of each technique.

Engineering field problems are defined by governing partial differential or integral equations, shape and size of the space domain, boundary and initial conditions, material properties, and by internal and external forces (Kubo, 1993). If all variables are known, it represents a direct problem, and generally a well-posed system. If any variable is unknown, it becomes an inverse problem, and generally an ill-posed problem. While inverse problems are unsolvable, it can be dealt with if additional information is provided. However, even in this case, there will be no unique solution. There are many techniques that can be applied for solution of inverse problems. Among them, genetic algorithms (GA) and artificial neural network (ANN) are relatively new approaches for optimization and estimation problems. They are considered to be relatively new because there is still much research to be done on the basics of these procedures in order for them to be consolidated and accepted as a defined technology.

Health monitoring is a specific type of nondestructive evaluation (NDE) technique. These methods must allow the integrity of systems to be actively monitored during their operation and throughout their life. For real-time damage detection, the normal variation in the operation conditions must not affect the technique, otherwise it could be difficult to distinguish between these variations and those caused by some kind of fault.

Due to immensely improved available instrumentation, a great change has occurred in the past few years for active self-testing instrumentation. It is now possible to incorporate self-diagnostic technology into the mechanical systems to monitor operating conditions and watch for structural damage. The continually increasing use of new materials to make systems lighter and stronger, and an intelligent maintenance schedule, that only initiates repairs if indeed necessary, could be economically very attractive.

Health monitoring can be formally divided into four sublevels (Doebling *et al.*, 1998): Level 1 – Detect the existence of damage; Level 2 – Detect and locate the damage; Level 3 – Detect, locate and quantify the damage; Level 4 – Detect, locate, and quantify the damage, then estimate the remaining service life. Level 4 requires a more sophisticated model, and represents an emergent area called damage prognosis. In this new area, the inclusion of smart material technology looks to be of definitive importance.

Inman (2001) proposes adding the following sublevels: Level 5 – Combine Level 4 with smart structures to form self-diagnostic structures; Level 6 – Combine Level 4 with smart structures and control to form self-healing structures; Level 7 – Combine Level 1 with active control and smart structures to form simultaneous control and health monitoring. Extensive analysis and investigation have been focused on integrating smart material technology into health monitoring systems (Banks *et al.*, 1996). This is due to the fact that they possess very important characteristics for fault detection and diagnostics, such as that they can serve as sensors as well as actuators for systems that do not contain natural exciting forces, and also, that they come in a variety of sizes, forms, and materials allowing them to be placed in almost anywhere.

A fully adaptive structure could, in principle, be configured to resist a normal loading condition, and also actuate an appropriate system to resist unexpected loads. Another function of interest for an adaptive structure is the provision of an alarm to inform of the necessity for repair. Developing an on-line structural health monitoring system for complex structures is an important issue in the structural community. The goal of this technique is to allow systems and structures to monitor their own integrity while in operation and throughout their lives, thus minimizing maintenance processes and inspection cycles. The proposed methodology for SHM is based on an electric impedance technique utilizing piezoelectric material. Lead-zirconate-titanate, usually abbreviated as PZT, is one of the most commonly used piezoelectric ceramics.

Because of its characteristics, PZT has been widely employed in many different applications both as sensors and as actuators. By using the direct and converse electromechanical properties of the PZT material, the simultaneous actuation and sensing of structural response is possible. The variation in the electrical impedance of a bonded PZT, while driven by a fixed alternating electric field over a frequency range, is analogous to frequency response. The PZT size acting as sensor/actuator is typically small and permits installation without interference in the host structure. Since high-frequency excitation is dominated by local modes, incipient faults can provide measurable variations in impedance curves. Because high frequency also limits the actuating/sensing area, it is possible to conclude that variations in impedance curves are caused by damage in the sensing area.

The main objective of this chapter is to offer a survey of research on solution methods for inverse problems as applied to quantify structural damages. However, the location and number of sensors and actuators for signal monitoring purposes, for active vibration control, or for modal tests, is an extremely important step that can affect the results. Taking into account this consideration, the next section deals with the optimal placement of a sensor/actuator (S/A) based on system norms.

8.2 OPTIMAL PLACEMENT OF SENSORS AND ACTUATORS FOR SMART STRUCTURES

System norm is a measure of 'size' and in this sense it can be applied, for instance, to model reduction, or to find optimal sensor/actuator placement. The problem of the optimal placement of S/A on structures arises from the following considerations (Papatheodorou *et al.*, 1999): use a small number of S/A in order to reduce the cost of instrumentation and data processing; obtain good estimates of modal parameters from noisy data; improve structural control by using valid models; determine efficiently the structural properties and their changes for health monitoring of structure; and ensure visibility of modeling errors.

Many authors present this problem using different optimization methods and objective functions. Some authors apply two optimization-coupled loops: an external loop, corresponding the optimal placement of actuators, and an internal loop, corresponding the obtaining of controller gains. This technique is known as discrete–continuous optimization methods (Lopes, Jr., *et al.*, 2004).

In general, one uses as an objective function the minimization of the mechanical energy and maximization of the energy dissipated by active controller. Among the methods using artificial intelligence to obtain the size and the optimal placement of actuator for multivariable control are 'simulated annealing' and 'genetic algorithms' (GA). On the other hand, some authors use classical optimization techniques and obtain good results. Gabbert *et al.* (1997) present a methodology based on classical techniques for actuator placement. Hiramoto *et al.* (2000) minimize the closed loop transfer function and use the H_∞ controller in a simple beam. The optimal location of two piezoceramics was found using the quasi-Newton method.

It can be shown that the minimum control effort is reached when actuators are placed such that the property of controllability grammians is maximized. These properties can be eigenvalues, traces, norms, etc., and present, in some way, a form to quantify the optimal location. For instance, Costa e Silva and Arruda (1997) considered as a performance index the smallest eigenvalue of controllability grammians for each candidate location. The optimal position is obtained when this index is maximized considering all physical possibilities.

The controllability and observabillity grammians, $\mathbf{W}_c$ and $\mathbf{W}_o$, respectively, are defined as follows:

$$\mathbf{W}_c = \int_0^t \exp(\mathbf{A}t)\mathbf{B}\mathbf{B}^T \exp(\mathbf{A}^Tt)\mathrm{d}t, \quad \mathbf{W}_o = \int_0^t \exp(\mathbf{A}^Tt)\mathbf{C}^T\mathbf{C}\exp(\mathbf{A}t)\mathrm{d}t \qquad (8.1)$$

where $\mathbf{A}$ is the dynamic matrix, $\mathbf{B}$ is the input matrix (related to the actuators) and $\mathbf{C}$ is the output matrix (related to the sensors). They can be determined alternatively from the following differential equations:

$$\dot{\mathbf{W}}_c = \mathbf{A}\mathbf{W}_c + \mathbf{W}_c\mathbf{A}^T + \mathbf{B}\mathbf{B}^T, \quad \dot{\mathbf{W}}_o = \mathbf{A}^T\mathbf{W}_o + \mathbf{W}_o\mathbf{A} + \mathbf{C}^T\mathbf{C} \qquad (8.2)$$

For a stable system, the stationary solutions of the above equations are obtained by assuming $\dot{\mathbf{W}}_c = \dot{\mathbf{W}}_o = 0$. In this case, the grammians are determined from the following Lyapunov equations:

$$\mathbf{A}\mathbf{W}_c + \mathbf{W}_c\mathbf{A}^T + \mathbf{B}\mathbf{B}^T = \mathbf{0}, \quad \mathbf{A}^T\mathbf{W}_o + \mathbf{W}_o\mathbf{A} + \mathbf{C}^T\mathbf{C} = \mathbf{0} \qquad (8.3)$$

Sadri *et al.* (1999) use this approach to maximize the controllability grammian matrix, as an objective function, through genetic algorithms, to obtain the optimal placement and size of actuators in a thin plate.

8.2.1 Optimal Placement Using System Norms

Most of the works use optimization methods in simple structures, such as beams or plates. On the other hand, one can use systems norms as an objective function, mainly when there are a large number of candidate locations. In this sense, only a small number of possible combinations is searched and the time consumed is reduced (Gawronski, 1998). The optimal placement problem, in this case, consists of determining the location of a small set of actuators and sensors such that the system norm is the closest possible to the norms considering a large set of actuators and sensors. We can use any norm, as for instance, H_2, Hankel or H_∞. The generic placement index σ_{ik} evaluates the kth actuator (or sensor) at the ith mode. It is defined in relation to all modes and all admissible actuators (or sensor) as:

$$\sigma_{ik} = \frac{\|G_{ik}\|_g}{\|G\|_g}, i = 1, \ldots, n \quad k = 1, \ldots, S \tag{8.4}$$

where n is the number of the considered modes, S is the number of the candidate position for each actuator (or sensor) and $\|G\|_g$ is a generic norm. It is convenient to represent the placement index of equation (8.4) as a placement matrix:

$$T = \begin{bmatrix} \sigma_{11} & \sigma_{12} & \cdots & \sigma_{1k} & \cdots & \sigma_{1S} \\ \sigma_{21} & \sigma_{22} & \cdots & \sigma_{2k} & \cdots & \sigma_{2S} \\ \cdots & \cdots & \cdots & \cdots & \cdots & \cdots \\ \sigma_{i1} & \sigma_{i2} & \cdots & \sigma_{ik} & \cdots & \sigma_{iS} \\ \cdots & \cdots & \cdots & \cdots & \cdots & \cdots \\ \sigma_{n1} & \sigma_{n2} & \cdots & \sigma_{nk} & \cdots & \sigma_{nS} \end{bmatrix} \Leftarrow i\text{th mode} \tag{8.5}$$

$$\Uparrow$$
$$k\text{th actuator}$$

where the kth column consists of indexes of the kth actuator for all modes, and the ith row is a set of the indexes of the ith mode for all actuators. The actuators with small indexes can be removed as being the least significant. The largest value indexes indicate the optimal position. Similarly, it is possible to determine the optimal placement of sensors. The definition and computation for H_2, H_∞ and Hankel norm is shown in the following.

8.2.2 The H_2 Norm

The H_2 norm of a linear time invariant (LTI) system is defined as (Burl, 1999):

$$\|G\|_2^2 = \frac{1}{\omega} \int_{-\infty}^{+\infty} tr(G(\omega)^* G(\omega)) d\omega \tag{8.6}$$

where $G(\omega)$ is the transfer function of the system and $tr(\cdot)$ is the trace. A convenient way to determine its numerical value is to use the following equations (Gawronski, 1998):

$$\|G\|_2 = \sqrt{tr(\mathbf{C}^T \mathbf{C} \mathbf{W}_c)} \quad \text{or} \quad \|G\|_2 = \sqrt{tr(\mathbf{B} \mathbf{B}^T \mathbf{W}_o)}, \tag{8.7}$$

where $\mathbf{W}_c$ and $\mathbf{W}_o$ are the controllability and observabillity grammians, respectively. Alternatively, one can compute this norm using a convex optimization approach, by solving the problem (Assunção, 2000):

$$\|G\|_2^2 = \min \ tr(\mathbf{C} \mathbf{X} \mathbf{C}^T)$$

$$\text{subject to } \mathbf{A} \mathbf{X} + \mathbf{X} \mathbf{A}^T + \mathbf{B} \mathbf{B}^T \le 0 \tag{8.8}$$

$$\mathbf{X} > \mathbf{0}$$

where $\mathbf{X}$ is a symmetric and positive defined matrix, or in dual form:

$$\|G\|_2^2 = \min \ tr(\mathbf{B}^T\mathbf{Q}\mathbf{B})$$

$$\text{subject to } \mathbf{A}^T\mathbf{Q} + \mathbf{Q}\mathbf{A} + \mathbf{C}^T\mathbf{C} \leq \mathbf{0} \tag{8.9}$$

$$\mathbf{Q} > \mathbf{0}$$

where $\mathbf{Q}$ is a symmetric and positive defined matrix. These problems can be solved using interior-point methods (Gahinet *et al.*, 1995). The H_2 norm of a single-input-single-output (SISO) system corresponds to the area under the frequency response functions (FRF) curve.

8.2.3 The H∞ and Hankel Norm

The H_∞ norm of a stable function $G(\omega)$ is its largest input/output RMS gain:

$$\|G\|_\infty = \max_\omega \sigma_{\max}(G(\omega)) \tag{8.10}$$

where $\sigma_{\max}(G(\omega))$ is the largest singular value of $G(\omega)$. The H_∞ norm of a single-input/single-output (SISO) system is the peak of the transfer function magnitude (in terms of its singular values).

The norm of the ith natural mode can be esteemed in different forms, as for instance, using efficient convex optimization methods. The value of the H_∞ norm in this chapter will be found by solving the optimization problem below (Gonçalves *et al.*, 2002):

$$\|G\|_\infty = \min \mu$$

$$\text{subject to } \begin{bmatrix} \mathbf{A}^T\mathbf{P} + \mathbf{P}\mathbf{A} + \mathbf{C}\mathbf{C}^T & \mathbf{P}\mathbf{B} \\ \mathbf{B}^T\mathbf{P} & -\mu \end{bmatrix} < \mathbf{0} \tag{8.11}$$

$$\mathbf{P} > \mathbf{0}$$

$$\mu > 0$$

where $\mathbf{P}$ is a symmetric and positive defined matrix and μ is a scalar variable. This problem can be solved using interior-point methods (Gahinet *et al.*, 1995).

The Hankel norm of a system is a measure of the effect of its past input on its future outputs, or the amount of energy stored in, and subsequently retrieved from, the system, and given by (Gawronski, 1998):

$$\|G\|_h = \sqrt{\lambda_{\max}(\mathbf{W}_c\mathbf{W}_o)} \tag{8.12}$$

where $\lambda_{\max}(\cdot)$ is relative to the largest eigenvalue. The Hankel norm is approximately one-half of the H_∞ norm.

8.2.4 Application

In this section an example of control design, including the computation norms applied for optimal placement of actuators, is shown. A control system for vibration attenuation of the two first modes in an aluminum plate structure was designed (Figure 8.1). The system was discretized by FEM with 100 elements and 121 nodes (3 d.o.f. per node), making the total number of structural d.o.f. used 363. For the boundary conditions (clamped–free–free–free) the system has 660 states. The properties and dimensions are: Young's modulus 70 GPa; density 2710 kg.m^3; length 0.3 m; width 0.2 m, and thickness 2 mm.

In order to test the proposed optimization methodology, the first step was to find the optimal placement of two pairs of PZT, bonded on both sides of the plate surface. The numbers 1 to 121 show the node positions, and it was considered that the PZT actuators could be positioned among these nodes. In order to reduce the candidate positions, or due to some practical restriction, some physical constraints can be imposed. So, 99 candidates were considered. The placement index of each candidate position for the first and second modes is shown in Figure 8.2.

The largest value index corresponds to the optimal location to control the first two modes. Figure 8.3 shows the experimental setup and the structure with PZT actuators bonded in the optimal position for control of the first two modes. The position of the disturbance input is, also, shown in Figure 8.1.

For analysis purposes, a finite dimensioned system was obtained experimentally in a frequency bandwidth of interest. The test was performed by exciting with two different inputs (one by impact hammer and another one by a pair of PZT actuators). The output signal was obtained at a single point, so there are two frequency response functions (FRF). The output signal was obtained by accelerometer, model 352A10 PCB Piezotronics®. The impact hammer used was model 086C04 from PCB Piezotronics®. In this experiment the software SignalCalc ACE® was used to generate a swept sine signal (bandwidth 0–1000 Hz) for the PZT actuators and to realize the data acquisition. A ten-mode model was identified using an eigensystem realization algorithm (ERA), (Juang and Minh, 2001). The real system has nine modes, but ERA adds a highly damped computation mode to the model.

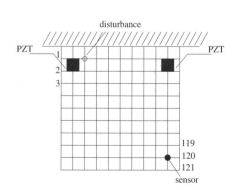
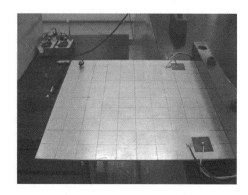

Figure 8.1 FEM model of the plate structure

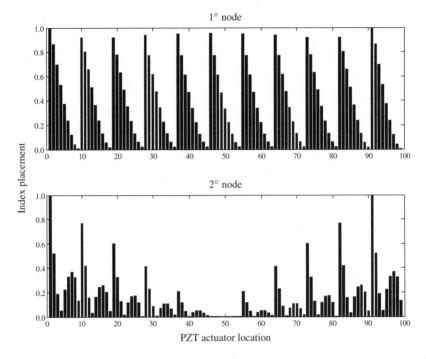

Figure 8.2   Placement indexes for the first and second modes, respectively, versus PZT actuator location

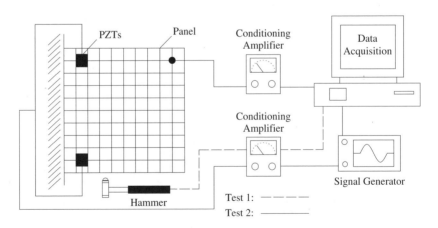

Figure 8.3   Schematic diagram of the measurement setup

The FRFs of the real system and of the identified model are shown in Figures 8.4 and 8.5, for excitation with a impact hammer and with PZT actuators, respectively. The transfer functions H_1 and H_2 are relative to the state-space realization with hammer and PZT actuators, respectively. For the control design a fourth model is considered by

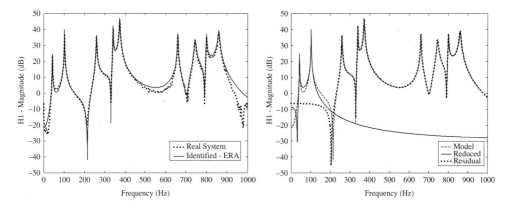

Figure 8.4   FRF of the plate structure excited by impact hammer

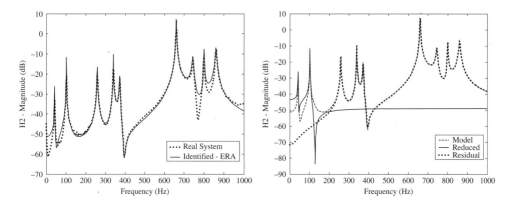

Figure 8.5   FRF of the plate structure excited by PZT actuators

truncating the model. The FRF magnitude plots of H_1 and H_2 for the reduced and residual model are also shown in Figures 8.4 and 8.5.

The regulator is obtained from the solution of the linear matrix inequalities (LMI) problem (Boyd, *et al.*, 1994). Figure 8.6 compares the FRF magnitude plots of the uncontrolled and controlled system.

As result of the control, the resonance peaks of the controlled modes are reduced. Furthermore, the amplitude of other modes, which are not explicitly included in the controller, are reduced occasionally. Nevertheless, some peaks increase the magnitude, because the controller leads to control spillover. Figure 8.7 shows the modal state in time domain for the controlled system considering an impulse disturbance. Clearly, one observes a low influence of the high frequency dynamics in the structural control performance. Comparing the modal magnitude of residual modes, in Figure 8.7, one observes that spillover effect exists there, but it is small when compared with modal magnitude in controlled modes.

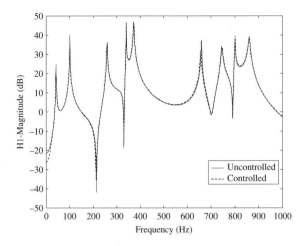

Figure 8.6   FRF for uncontrolled and controlled systems in nominal condition

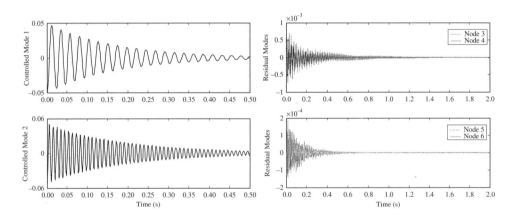

Figure 8.7   Closed-loop for controlled and residual modes in time domain

8.3 PROPOSED METHODOLOGY

Nature is a source of inspiration for engineering. Based on these ideas from nature, the development of new procedures for fault identification in order to prevent catastrophic failures and to increase in-service lifetime are proposed in this chapter. One important aspect in health monitoring that must be analyzed is the life expectation, or prognosis of the life remaining of the mechanical system in normal work conditions. In general, this demands knowledge of the structural model in great details, which is not always possible; in addition, dynamic systems frequently present nonlinear characteristics.

In this proposal the detection and location of faults are accomplished in two stages. Initially, the method of electric impedance is used to determine the location of the faults. Later, the quantification of the faults in two ways, by using genetic algorithms or by using artificial neural networks takes place. Genetic algorithms (GA) are optimization processes founded on principles of natural evolution and selection. A GA takes an initial population of individuals and applies artificial genetic operators to each generation. Neural network is a computational technique inspired by the neuronal structure of intelligent organisms. If the patterns of input data represent the fault appropriately, the neural networks can approach and classify problems associated with nonlinearities. These two methods are compared in this chapter for application to fault identification.

The electromechanical impedance technique is based on high frequency ranges and local vibration modes; therefore, the area of influence of each piezoelectric patch (PSA) is small. This technique can define with good accuracy the region of the fault. It is important to note that this method is not capable of supplying the level of fault severity. More details about this technique are given in Chapter 14. The second part of the proposed methodology supplies quantitative information of the damage severity, Figure 8.8.

A direct problem, which consists of the determination of the modal properties as a function of the physical structural variations, has a unique solution. However, the fault characterization is an inverse problem and it does not present a unique solution. Any optimization method that requires adjusting the model will have a greater chance of

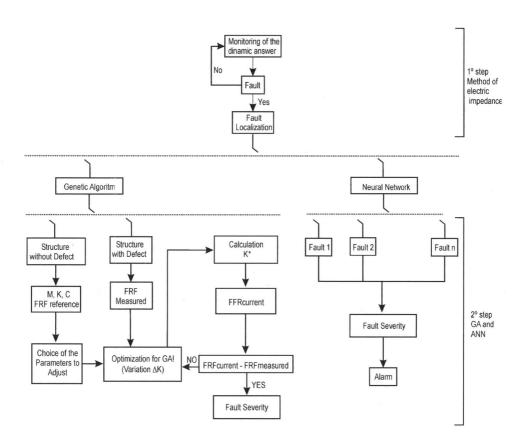

Figure 8.8   Flowchart of the proposed methodology

failing for systems with medium levels of complexity or greater. There exist various methods of model reduction or choices of variables to overcome this difficulty. Among them is sensitivity analysis; however, the fault can occur in positions where the variation of these parameters presents low sensitivity.

This chapter deals with this problem. The main advantage of the proposed methodology is that the method of electrical impedance defines, with accuracy, the location of the fault. Then it is possible drastically to reduce the number of variables that will be used in the optimization process.

8.3.1 Damage Quantification by GA

The choice of the parameters that will be used to quantify the fault is accomplished after the region of the fault is located. After the definition of these parameters, the adjustment of the measured FRF (situation with defect) is done through the optimization technique using GA. When the difference between those curves is smaller than a specified value, the process is finished. The difference between the system matrices without defect, M,K and C, and the matrices M^*, K^* and C^* supply the quantification of the defect.

8.3.2 Damage Quantification Using ANN

The proposed methodology scheme using ANN consists of two steps. In the first step the method of electric impedance is used to identify the fault location. The neural networks are trained for each specific fault using impedance signals from PZT sensors. Therefore, in the second step the number of neural networks necessary is the same as the number of PZT's.

Impedance signals must be processed and normalized in order to represent all conditions of faults that one wants to monitor. One of the largest advantages of this method is that the variation of the signal is local and doesn't affect the other sensors (PZT). Therefore, simultaneous faults, which are difficult to identify fusing conventional methods, can be treated as if they happen independently.

8.4 ARTIFICIAL NEURAL NETWORK AS A SHM ALGORITHM

One important aspect of structural health monitoring is that it will be able to provide information on the life expectancy of the structures, as well as detect and locate damage. In general, this requires an increasing knowledge of the model of the structures that is not always available. Also, several dynamic systems present nonlinear characteristics, which cannot be satisfactorily treated by linear theory. Neural networks may be applied successfully to classify problems associated with nonlinearity, provided that they are well represented by input patterns. Furthermore, the learning capabilities of neural networks are well suited for processing a large number of distributed sensors, which is ideal for smart-structure applications.

The integration of smart structure and neural networks could avoid the complexity introduced by conventional computational methods, since neural networks have been shown to be effective in pertaining to problems in the area of damage detection and system identifications. So, by combining these two technologies, the proposed methodology has the potential of self-diagnosing components and automating and minimizing the maintenance process. Contrary to most health monitoring techniques based on neural networks that utilize frequency response functions (FRF) or time responses to monitor the structures, this technique relies on electrical impedance at various points on the structure, using surface bonded piezoelectric sensor/actuators. To ensure high sensitivity to incipient damage, the impedance is measured at high frequencies and two sets of artificial neural networks were developed in order to detect, locate and characterize structural damages by examining variations on signals.

Artificial neural networks (ANN) are mathematical models inspired by human brain process information. They emulate some of the observed properties of biological nervous systems and draw on the analogy of adaptive biological learning. It is an attempt to simulate within specialized hardware, or sophisticated software, the multiple layers of simple processing elements, called neurons. Each neuron is linked to certain neighbors with varying coefficients of connectivity that represent the strengths of these connections. Learning is accomplished by adjusting these strengths to cause the overall network to output appropriate results. An ANN is configured for a specific application, such as pattern recognition or data classification, through a learning process.

8.4.1   Concepts of Artificial Neural Network

One of the pioneering works in neural network was that of McCulloch and Pitts (1943). McCulloch was a psychiatrist who had spent several years contemplating the representation of an event in the nervous system. Pitts was a mathematician, who joined McCulloch in 1942.

The book *The Organization of Behavior* first presents an explicit statement of a physiological learning rule for *synaptic modification* (Hebb, 1949). In 1958, Rosenblatt developed the original concept of a *perceptron*, a trainable machine for pattern classification. The perceptron computes a weighted sum of inputs, subtracts a threshold, and passes one of two possible values out as the result. Unfortunately, the perceptron is limited, and it was proven to be such during the 'disillusioned years' in Marvin Minsky and Seymour Papert's 1969 book *Perceptrons*. In 1960, Bernard Widrow and Marcian Hoff developed models that they called ADALINE and MADALINE. These models were named for their use of Multiple ADAptive LINear Elements. MADALINE was the first neural network to be applied to a real world problem. It is an adaptive filter that eliminates echoes on phone lines. This neural network is still in commercial use. Werbos (1974) developed and used the back propagation learning method. However, several years passed before this approach was popularized.

Back propagation nets are probably the most well known and widely applied of the neural networks today. In essence, the back propagation network is a perceptron with multiple layers, a different threshold function in the artificial neuron, and a more robust and capable learning rule. Grossberg's (Grossberg, 1976) influence founded a school of thought that explores resonating algorithms. He had a great influence in the development of the

ART (adaptive resonance theory) networks based on biologically plausible models. Significant progress has been made in the field of neural networks, which, in turn, attracts a great deal of attention and funds for further research.

8.4.1.1 Architecture of Neural Network

The manner in which the neurons of a neural network are structured is intimately linked with the learning process. The most common type of artificial neural networks consists of three groups, or layers, of units: a layer of 'input', which is connected to layers of 'hidden' units, which is connected to a layer of 'output'. There are basically two types of connection among the neurons.

Feedforward networks – Feedforward ANNs allow signals to travel one way only, from input to output. There are no feedback loops, i.e. the output of any layer does not affect that same layer. Feedforward nets tend to be straightforward networks that associate inputs with outputs. They are extensively used in pattern recognition.

Feedback networks – Feedback networks can have signals traveling in both directions by introducing loops in the network. Feedback networks are dynamic; their state is changing continuously until they reach an equilibrium point. They remain at the equilibrium point until the input changes and a new equilibrium needs to be found. Feedback architectures are also referred to as interactive or recurrent, although the latter term is often used to denote feedback connections in single-layer organizations.

8.4.1.2 Learning Process

The brain basically learns from experience. Neural networks are sometimes called machine-learning algorithms, because changing of its connection weights (training) causes a network to learn the solution of a problem. The strength of connection between the neurons is stored as a weight value for the specific connection. The system learns new knowledge by adjusting these connection weights, which can be of two types:

Supervised learning – incorporates an external teacher, so that each output unit is compared with its desired response. During the learning process global information may be required. Paradigms of supervised learning include error-correction learning, reinforcement learning and stochastic learning.

Unsupervised learning – uses no external teacher, and is based only upon local information. It is also referred to as self-organization in the sense that it self-organizes data presented to the network and detects their emergent collective properties. Paradigms of unsupervised learning are Hebbian learning and competitive learning.

There are many different types of ANNs. Some of the more popular include the multilayer perceptron, which is generally trained with the error backpropagation algorithm, radial basis function, Hopfield, Kohonen, and, ART.

Back-Propagation Algorithm

The back propagation is a learning algorithm applied in a multilayer perceptron. It is a supervised training based on the *error-correction learning rule*. There are basically two different phases in the training, a forward and a backward pass. In the forward pass, an activity pattern (input vector) is applied to the input neurons, and its effect propagates through the network, for all layers and neurons. Finally, a set of outputs is produced as the actual response of the network. During the forward pass the weights of the network are all fixed. On the other hand, during the backward pass the synaptic weights are all adjusted in accordance with the error-correction rule. The signal error is calculated by subtracting the actual from the desired (target) response. The training process finishes when the error is lower than a predefined value or a set number of iterations is reached. In the normal operating condition, or operating in real time, there is only the forward phase.

Radial Basis Function (RBF)

Radial basis functions are powerful techniques for interpolation in multidimensional space. A RBF is a function that has been built into a distance criterion with respect to a center. Such functions can be used very efficiently for interpolation and for smoothing of data. Radial basis functions have been applied in the area of neural networks where they are used as a replacement for the sigmoidal transfer function. Such networks have three layers, the input layer, the hidden layer with the RBF nonlinearity and a linear output layer. The most popular choice for the nonlinearity is Gaussian. RBF networks have the advantage of not being locked to local minima.

Hopfield Network

A Hopfield network belongs to another important class of neural networks. It has a *recurrent* structure and its development is inspired by different ideas from *statistical physics*. The Hopfield network operates as a nonlinear associative memory, or as a computer solving optimization problems of a combinatorial kind. In a *combinatorial optimization problem* we have a discrete system with a large, but finite, number of possible solutions. The objective is to find the solution that minimizes a cost function providing a measure of system performance. The Hopfield network requires time to settle to an equilibrium condition. It may, therefore, be excessively slow, unless special hardware is used for its implementation (Haykin,1994).

Kohonen Neural Network

The Kohonen neural network is different from the feed-forward back propagation neural network; it is an unsupervised and self-organizing mapping technique that allows projection multidimensional points to a two dimensional network. The Kohonen neural network contains only input and output layers of neurons; there is no hidden layer. Output from the Kohonen neural network does not consist of the output of several neurons. When a pattern is presented, one of the output neurons is selected as a 'winner'. Using a Kohonen network, the data can be classified into groups.

Adaptive Resonance Theory (ART)

Adaptive resonance theory was introduced by Stephen Grossberg in 1976. The term resonance refers to the *resonant* state of the network in which a category prototype vector matches the current input vector within allowable limits. The network learns only in its resonant state. ART is capable of developing stable clustering of arbitrary sequences of input patterns by self-organization. There are several different types of ART network.

Neural networks perform successfully where other methods do not, recognizing and matching complicated, vague, or incomplete patterns. Neural networks have been applied to solving a wide variety of problems. Although one may apply neural network systems for interpretation, prediction, diagnosis, planning, monitoring, debugging, repair, instruction, and control, the most successful applications of neural networks are in categorization and pattern recognition. Therefore, this is a powerful technique for application in varied fields.

8.4.2 Application of ANN for Damage Identification

Figure 8.9 shows the analyzed cubic frame structure built of structural tubing shape material. It is a aluminum structure with 4.462 kg of mass, center of mass of 0.286 m, 0.2515 m, and 0.1645 in the directions x, y and z respectively, and cross section of $25 \times 25 \times 2$ mm. The dimension of the structure is: $597 \times 570 \times 358$ mm. On each edge one PZT element is bonded. After analysis of the undamaged structure by the electrical impedance technique and experimental modal analysis, two elements of the structure were damaged by small cuts (0.8×9 mm) near nodes 4 and 6 in the beam elements 32 and 15, respectively.

The monitoring of each PZT impedance signal is done separately, i.e. each PZT is continuously monitored in a specific frequency range, and the threshold level, above which is an indication of fault in that position, must be defined. Operational conditions and temperature variations can change the modal characteristics of the structure in high frequency range. Hence, these values must be defined by trial and error or by prior

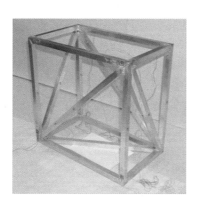

Figure 8.9 Experimental setup and damage details

knowledge of the structural dynamics of the system. The topologies and size of the network depend on the specific application, and the definition of an optimal choice is case dependent.

The training of the network uses a set of inputs for which a specified output is known and in this case, the process finishes when the error E is smaller than a desired value or it reaches a maximum specified number of iterations. In the later case, the training set is considered unsatisfactory. The error E, considered here, is the sum of the squared error between the output value, provided by the trained network, and the value obtained for a specific input. The output could be a combination of zeros and ones, depending on each fault location.

The electrical impedance curves measured at high frequency range show extreme sensitivity to variation in the structure. Correlating the raw electrical impedance signals to each specific damage is very difficult. Therefore, after the acquisition of the electrical impedance, the signals needed to be preprocessed and normalized in order to represent all conditions monitored and used as input patterns for the ANN.

The preprocessed values used to train the ANN were: (i) the area between the undamaged and the damaged impedance curves, (ii) the root mean square (rms) of each curve, (iii) the rms of the difference between the undamaged and damaged curves, (iv) the correlation coefficient between the undamaged and damaged curves. These values were obtained for both real and imaginary parts of the impedance. Therefore, the input vector has eight elements and the input layer of the ANN also consists of eight elements. Theoretically, one can use any data as inputs for ANN, however, the accuracy and generalization capability of the neural network is strongly dependent on the choice of the input patterns.

Table 8.1 shows the natural frequencies, obtained experimentally, for the intact structure and the structure with simultaneous damages in their extreme conditions. Figure 8.10 shows the measurements of 13 damage situations. For the first seven measurements there is continually increased damage only in the element 32 (PZT on node 6). From measurements 8 to 13, there is a continually increased damage only in the element 15 (PZT on node 4). Therefore, the damage situation 13 means simultaneous damage with maximum severity in both elements.

The figures show a clear indication of the damage location. It is possible to verify that there is only damage near PZT 6 for measurement taken from 1 to 7 (Figure 8.10a). After measuring 8, one can see the appearance of simultaneous damage (Figure 8.10b). Although these illustrations can show the location, it is not possible to quantify the damage. Hence, the second set of neural networks, as showed in Section 3, must be applied. Figure 8.11a shows the output of the net trained with signals from PZT 6. Figure 8.11b shows the results from PZT 4. There is an indication of no damage until

Table 8.1 Natural frequencies for undamaged and damaged structure

Mode (Hz)	1	2	3	4	5	6	7	8	9
Undamaged	132.24	214.22	237.11	242.01	250.51	253.64	308.85	327.88	344.28
Damaged	131.87	212.90	230.08	239.45	249.38	251.56	306.46	327.11	342.73
Error (%)	0.28	0.62	2.97	1.06	0.45	0.82	0.77	0.24	0.45

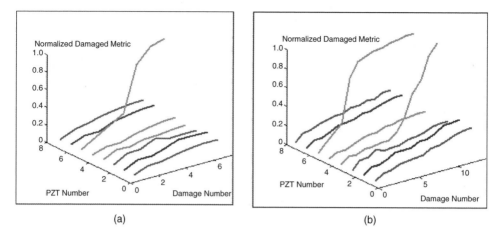

Figure 8.10   Normalized damaged for all eight PZTs

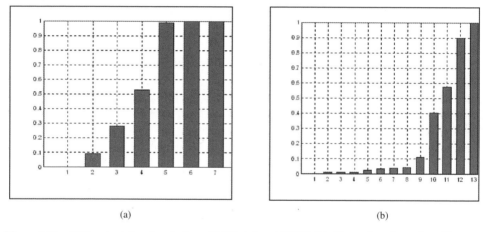

Figure 8.11   NN output for signals from PZT6 (a) and PZT4 (b). The values from 1 to 13 mean damage situations

measurement 7. In these figures, the NN output was set to unity for maximum damage and zero for the undamaged situation.

A sweep frequency range from 18 to 28 kHz was selected after several tests to find a frequency range with good dynamic interactions. After acquiring a damage indication from the impedance-based method, a more detailed inspection was carried out with the trained neural networks. For each test, the signals were taken from all eight PZT sensors.

8.5   GENETIC ALGORITHMS AS A SHM ALGORITHM

Genetic algorithms were, initially, proposed by Holland (1975). Today they are used for a wide variety of problems such as structural analysis, machine learning, cellular manufacturing, combinatorial optimization and game playing. Genetic algorithm

(GA) is a technique based on Darwin's evolution. GA simulates the adaptation process, taking an initial population of individuals and applying artificial genetic operators in each generation. In the optimization process, each individual in the population is coded in a string or chromosome, which represents a possible solution for a certain problem, while the individual's adaptation is evaluated through a fitness function.

Basically, for highly capable individuals (better solutions), larger opportunities are given for reproduction, changing parts of their genetic information in a procedure of crossover. The operator of mutation is used to change some genes in the chromosomes and to cause diversity in the population. The descent of new population can either substitute the whole current population or just substitute for the individuals with smaller adjustment. This evolution cycle, selection, generation, and mutation, is repeated until a satisfactory solution is found.

In order to improve the current population, genetic algorithms commonly use three different genetic operations: *selection, crossover* and *mutation*. Both selection and crossover can be viewed as operations that force the population to converge. They work in a way by promoting genetic qualities that are already present in the population. Conversely, mutation promotes diversity within the population. In general, selection is a fitness-preserving operation, crossover attempts to use current positive attributes to enhance fitness, and mutation introduces new qualities in an attempt to increase fitness.

Mutation is a reproduction operator, which forms a new chromosome by making (usually small) alterations to the values of genes in a copy of a single parent chromosome. There are several kinds of selection including elitism, rank-based selection, tournament selection, roulette-wheel selection, and others. One of the most used is the roulette selection, where the chance of an individual is proportional to its fitness. Individuals are not removed from the source population, so, those with high fitness will be chosen more often than those with a low fitness. Crossover is a reproduction operator that forms a new chromosome by combining parts of each of two parent chromosomes. The simplest form is the single-point crossover, in which an arbitrary point in the chromosome is picked. All the information from parent A is copied from the start up to the crossover point, then all the information from parent B is copied from the crossover point to the end of the chromosome. The new chromosome thus gets the head of one parent's chromosome combined with the tail of the other. Variations exist that use more than one crossover point, or combine information from parents in other ways. The crossover operation is the most important genetic operation.

The general genetic algorithm uses several simple operations in order to simulate evolution. The goal of the genetic algorithm is to come up with a 'good', but not necessarily optimal, solution to the problem.

8.5.1 Application of GA for Damage Identification

The proposed methodology is divided into two parts, (see Figure 8.8). The first one determines the fault location through the electric impedance technique. This technique can define with good accuracy the region of the fault, but it doesn't supply quantitative information of the fault.

The choice of the parameters that will be used to quantify the fault is accomplished after the location of the region of the fault. After the definition of these parameters,

adjustment of the measured FRF (situation with defect) is done through the optimization technique using GA. When the difference between these curves is smaller than a
specified value, the process is finished. The difference between the system matrices
without defect, M, K and C, and the matrices M^*, K^* and C^* supplies the quantification
of the fault. For the analyzed case $M^* = M$ and $C^* = C$ was considered.

The application is carried out on an aluminum beam of 30 mm width, 5 mm thick and
500 mm long. The beam is modeled by finite elements in a code, called 'SmartSys', which
includes the electromechanical coupling of the piezoelectric sensors/actuators and the
host structure. The beam was divided into 20 elements of type 'BEAM', with two
degrees of freedom per node, vertical displacement and rotation in the axis z. It is a
clamped beam and the frequency response function (FRF) of the system was considered
for different combinations of defects and loads.

The presented results consider the fault in elements 5 and 17 simultaneously. The
damage was simulated by decreasing the inertia moment of area of 20 % on element 5
and 15 % on element 17. Figure 8.12 shows the beam with elements 5 and 17 highlighted, as well as the PZT positions. Starting from the finite element models, the FRF
was constructed for the conditions of the mentioned faults up to 2000 Hz, where the
positions of the first seven natural frequencies can be verified.

The objective function of GA used in this work is described below. The GA consists
of the objective function minimization, so, $Ec = 1/Fxe$.

$$fc = f_{current}(1:7) \qquad fm = f_{measured}(1:7)$$

$$Fxe = \sum |fc - fm|$$

where fc is a vector with the first seven natural frequencies of the updating process
(or optimization), and fm is related to the damaged system. The variables used in GA
were: size of the population $= 80$; maximum number of generations $= 80$; crossover
probability $= 0.90$ (90%); and mutation probability $= 0.05$ (5%).

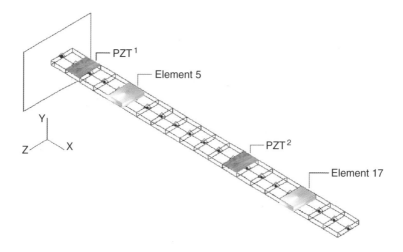

Figure 8.12   Discretized beam with elements 5 and 17 highlighted

Table 8.2   Values of I^4, I^5, I^{16} and I^{17}

Run	1	2	3	4	5	6	7	8	9	10	Average	Error [%}
Generation	80	80	80	80	80	80	80	80	80	80		
I^4 (m^4)	1.000	0.998	0.999	1.000	0.992	1.000	1.000	0.999	0.995	0.999	0.9982	0.266
I^5 (m^4)	0.799	0.806	0.804	0.800	0.802	0.801	0.800	0.813	0.814	0.804	0.8043	0.531
I^{16} (m^4)	1.000	0.991	0.996	0.999	0.987	0.999	1.000	0.987	0.988	0.997	0.9944	0.554
I^{17} (m^4)	0.850	0.845	0.851	0.850	0.846	0.851	0.850	0.851	0.853	0.850	0.8497	0.241

To verify the proposed methodology, several tests with different subsets of parameters were conducted. When updating all elements was attempted, the program did not converge. This demonstrates the importance of the first step of this proposal, and the necessity of choosing a small set of parameters. Table 8.2 shows the results obtained for a small subset of parameters. In this case, the variation on elements 4, 5, 16, and 17 were considered, represented by I^4, I^5, I^{16} and I^{17}, respectively. The ideal values of these parameters to be found by GA are: $I^4 = 1.0$, $I^5 = 0.8$, $I^{16} = 1.0$ and $I^{17} = 0.85$. The error considers the sum of the difference on the seven first natural frequencies. The average of ten runs was considered.

8.6   CONCLUSION

Contrary to most model-based NDE techniques, which rely on the lower-order global modes, an approach utilizing high-frequency structural excitation was developed. This technique would be more useful in identifying and tracking small defects in the sense that damage is a local phenomenon and a high-frequency effect. Neural networks are well suited for structures where a prior analytical model is unknown or the excitation force is not available. Hence, this method should be chosen for situations where FRFs are not available.

Another advantage of this methodology is that multiple damage events, in several different locations, can also be analyzed. If global frequency response functions are used, it is almost impossible to train neural networks for all possible combinations of multiple damage events in different areas. However, in this method, the limited sensing area of each PZT sensor helps to isolate the effect of damage on the signature from other far field changes. Thus, the fault in a remote location has only minor influence on the other PZT sensors, and each PZT sensor reflects only the structural variation occurring in a nearby field.

A combined damage-detection methodology, including the electrical-impedance technique, ANN, and model generation by optimization procedures, was presented. The impedance technique gives clear information about the damage location and, therefore, is used to select a small subset of parameters. The amount of damage is then described by values of parameter variation on the model design level, or by the ANN technique.

This chapter presented the performance of two independent methodologies for SHM, which also can be used in conjunction to certify the results where some doubts could persist after the application of one technique. Both methodologies combine the advantages of smart material technology with neural network, or genetic algorithm features in order to propose a self-diagnostic procedure. High-frequency structural excitations through surface-bonded piezoelectric sensors/actuators were utilized to monitor the structure. The experimental investigation successfully located and identified damage in different structures.

REFERENCES

Assunção, E. (2000) 'Redução H_2 e H_∞ de Modelos Através de Desigualdades Matriciais Lineares: Otimização Local e Global', Tese de Doutorado, UNICAMP, Campinas, SP.

Banks, H.T., Inman, D.J., Leo, D.J. and Wang, Y. (1996) 'An experimentally validated damage detection theory in smart structures', *Journal of Sound and Vibrations*, **191**, 859–880.

Boyd, S., Balakrishnan, V., Feron, E. e El Ghaoui, L. (1994) 'Linear Matrix Inequalities in Systems and Control Theory', SIAM Studies in Applied Mathematics, USA, 193p.

Burl, J.B. (1999) 'Linear Optimal Control: H_2 and H_∞ Methods', Addison-Wesley, New York.

Costa e Silva, V.M. and Arruda, J.R. de F. (1997) 'Otimização do Posicionamento de Atuadores Piezocerâmicos em Estruturas Flexíveis Usando um Algoritmo Genético', *Anais do XIV Congresso Brasileiro de Engenharia Mecânica, COBEM 97*, Bauru, SP, COB. 489, 8p.

Doebling, S.W., Farrar, C.R. and Prime M.B. (1998) 'A summary review of vibration-based damage identification methods', *The Shock and Vibration Digest*, **30**, 91–105.

Farris, T.N. and Doyle, J.F. (1991) 'A global/local approach to cracks in beams: Static analysis', *Journal of Fracture*, 131–140.

Gabbert, U., Schultz, I. and Weber, C.T. (1997) 'Actuator Placement in Smart Structures by Discrete-Continuous Optimization', *ASME Design Eng. Tech. Conferences*, September, Sacramento, CA.

Gahinet, P., Nemirovski, A., Laub, A.J. and Chiliali, M. (1995) *LMI Control Toolbox User's Guide*, The Mathworks Inc., Natick, MA.

Gawronski, W. (1998) '*Dynamics and Control of Structures: A Modal Approach*', Springer-Verlag, New York.

Gonçalves, P.J.P., Lopes Jr, V. and Assunção, E. (2002) 'H_2 and H_∞ Norm Control of Intelligent Structures using LMI Techniques', *Proceedings of ISMA 26 – International Conference on Noise and Vibration Engineering*, Leuven, Belgium, Vol. 1, pp. 217–224.

Gonçalvez, P.J.P., Lopes Jr, V. and Brennan, M.J. (2003) 'Using LMI Techniques To Control Intelligent Structures', In: XXI IMAC – Conference on Structural Dynamics, Kissimmee, Florida, USA.

Grossberg, S. (1976) 'Adaptive pattern classification and universal recoding: I. Parallel development and coding of neural feature detectors', *Biological Cybernetics*, Volume 23.

Haykin, S. (1994) *Neural Networks: A Comprehensive Foundation*, Prentice-Hall, Upper Saddle River, New Jersey.

Hebb, D.O. (1949) *The Organization of Behavior*, John Wiley & Sons, Inc.

Hiramoto, K., Dok, H. and Obinata, G. (2000) 'Optimal sensro/actuator placement for active vibration control using explicit solution of algebraic Riccati equation', *Journal of Sound and Vibration*, **229**, 1057–1075.

Holland, J.H. (1975) Adaption in Natural and Artificial Systems, University of Michigan Press, Ann Arbor.

Hopfield, J.J. (1982) 'Neural Networks and Physical Systems with Emergent Collective Computational Abilities', *Proceedings of the National Academy of Sciences*, Volume 79.

Inman, D.J. (2001) 'Smart Structures: examples and new problems', 16th Congresso Brasiliero de Engenharia Mecânica – COBEM – 2001, Uberlândia, Brazil.

Juang, J. and Minh, Q.P. (2001) *Identification and Control of Mechanical System*, Cambridge University Press, Cambridge, UK.

Kubo, S. (1993) 'Classification of Inverse problem Arising in Field Problems and their Treatments', *IUTAM Symposium on Inverse problem in Engineering Mechanics*, Springer-Verlag, Berlin, pp. 51–60.

Lopes, V. Jr, Müller-Slany, H.H. Brunzel, F. and D.J. Inman (2000) 'Damage Detection in Structures by Electrical Impedance and Optimization Technique', Proceedings: IUTAM Syposium 'Smart Structures and Structronic Systems', Sept. 26–29, Magdeburg.

Lopes Jr, V., Park, G., Cudney, H.H. and Inman, D.J. (2000) 'Structural Integrity Identification Based on Smart Material and Neural Network' *XVIII IMAC – International Modal Analysis Conference, San Antonio, Texas,* , pp. 510–515.

Lopes Jr, V., Steffen Jr, V. and Inman, D.J. (2004) 'Optimal placement of piezoelectric sensor/actuators for smart structures vibration control' *Dynamical Systems and Control,* Ed. Chapman & Hall/CRC, ISBN: 0–415–30997–2, pp. 221–236, by CRC Press LLC.

McCulloch, W.S., Pitts, W.H. (1943) 'A logical calculus of the ideas immanent in neural nets', *Bulletin of Mathematical Biophysics,* Volume 5.

Minsky, M.L. and Papert, S.S. (1969) *Perceptrons: An Introduction to Computational Geometry,* MIT Press, Cambridge, MA.

Papatheodorou, M., Taylor, C.A. and Lieven, N.A.J. (1999) 'Optimal sensor locations for dynamic verification', *Structural Dynamics – Eurodyn 99,* pp. 587–592.

Rosenblatt, F. (1958) 'The perceptron: A probabilistic model for information storage and organization in the brain', *Psychological Review,* Volume 65.

Sadri, A.M., Wright, J.R. and Wynne, R.J. (1999) 'Modelling and optimal placement of piezoelectric actuators in isotropic plates using genetic algorithms', *Smart Materials and Structures,* **8,** 490–498.

Sun, F. (1996) Piezoelectric Active Sensor and Electric Impedance Approach for Structural Dynamic Measurement, Master Thesis, Virginia Polytechnic Institute and State University – CIMSS.

Werbos, P.J. (1974) *Beyond Regression: New Tools for Prediction and Analysis in the Behavioral Sciences,* PhD thesis, Harvard University.

Widrow, B. and Hoff, M. (1960) 'Adaptive Switching Circuits', *1960 IRE WESCON Convention Record,* Part 4.

9

Uncertainty Quantification and the Verification and Validation of Computational Models

François M. Hemez

Los Alamos National Laboratory, Los Alamos, New Mexico, USA

9.1 INTRODUCTION

The material presented in this chapter is largely based on a tutorial from the Los Alamos Dynamics Summer School [1].

In computational physics and engineering, numerical models are developed to predict the behavior of a system whose response cannot be measured experimentally. A key aspect of science-based predictive modeling is to assess the *credibility* of predictions. Credibility, which is usually demonstrated through the activities of model verification and validation (V&V) refers to the extent to which numerical simulations can be analyzed with *confidence* to represent the phenomenon of interest [2].

One can argue, as it has been proposed in recent work [3], that the credibility of a mathematical or numerical model must combine three components: (i) an assessment of fidelity to test data; (ii) an assessment of the robustness of prediction-based decisions to variability, uncertainty, and lack-of-knowledge, and (iii) an assessment of the consistency of predictions provided by a family of models in situations where test measurements are not available. Unfortunately, the three goals are antagonistic, as illustrated in Reference [3] for a wide class of uncertainty models.

Damage Prognosis – For Aerospace, Civil and Mechanical Systems   Edited by D.J. Inman, C.R. Farrar, V. Lopes Junior and V. Steffen Junior   © 2005 John Wiley & Sons, Ltd

The three aforementioned assessments nevertheless require a similar technology in terms of model validation and quantification of uncertainty. Even though V&V in structural dynamics is rapidly evolving, the intent of this contribution is briefly to overview some of the technology developed at Los Alamos National Laboratory (LANL) in support of V&V activities for engineering applications. The discussion focuses on verification, validation, and the quantification of uncertainty for numerical simulations. A common thread to these three activities is to identify where uncertainty arises in numerical simulations and how to quantify it. It is increasingly becoming clear that it may not always be possible to describe uncertainty using the theory of probability. Defining information integration strategies for generalized information theories is an active field of research [4, 5], although not addressed here.

9.2   VERIFICATION ACTIVITIES

Because our intent is to substitute numerical simulations for information that cannot be obtained experimentally, the prediction accuracy of models upon which simulations rely must be established through V&V. *Verification* refers to the assessment that the equations implemented in the computer code are solved correctly, no programming error is present, and the discretization leads to converged solutions both in time and space. *Validation*, on the other hand, refers to the adequacy of a model to describe a physical phenomenon.

Numerical results from, for example, finite element simulations, provide approximations to sets of coupled partial differential equations. Before the validity of the equations themselves can be assessed, verification must take place to guarantee the quality of the solution. Verification is formally defined as '*the process of determining that the implementation of a model accurately represents the conceptual description of the model and its solution*' [6]. The primary sources of errors in computational solutions are inappropriate spatial discretization, inappropriate temporal discretization, insufficient iterative convergence, computer round-off, and computer programming. Verification quantifies errors from these various sources, and demonstrates the stability, consistency, and accuracy of the numerical scheme. The main activities, namely, code verification and calculation verification, are briefly overviewed in this chapter.

9.2.1   Code Verification

Code verification can be segregated into two parts: numerical algorithm verification, which focuses on the design and underlying mathematical correctness of discrete algorithms for solving partial differential equations, and software quality assurance (SQA), which focuses on the implementation of the algorithms.

9.2.1.1   Software Quality Assurance

SQA determines whether or not the code as a software system is reliable (implemented correctly) and produces repeatable results with a specified environment composed of hardware, compilers, and libraries. It focuses on the code as a software product that is sufficiently reliable and robust from the perspective of software engineering.

Successful SQA plans must be defined before and implemented during the development of the product, rather than being viewed as an activity that takes place after the software has been developed. Whether SQA is a legitimate V&V activity is still debated in the structural dynamics community. Our current opinion is that it is not, because SQA activities cannot presume the intended purpose of the software. Code users should nevertheless be cognizant and enforce the implementation of sound SQA practices for the software they use.

9.2.1.2 Numerical Algorithm Verification

Numerical algorithm verification addresses the reliability of the implementation of all of the algorithms that affect the numerical accuracy and efficiency of the code. In other words, this verification process focuses on how correctly the numerical algorithms are programmed in the code. The major goal is to accumulate sufficient evidence to demonstrate that the numerical algorithms in the code are implemented correctly and functioning as intended.

Error estimation and numerical algorithm verification are fundamentally empirical activities. Numerical algorithm verification deals with careful investigations of topics such as spatial and temporal convergence rates, iterative convergence, independence of solutions to coordinate transformations, and symmetry tests related to various types of boundary conditions. This is clearly distinct from error estimation, which deals with approximating the numerical error for particular applications when the correct solution is not known [7].

The principal components of this activity include the definition of appropriate test problems for evaluating solution accuracy, and the determination of satisfactory performance of the algorithms on the test problems. Numerical algorithm verification rests upon comparing computational solutions to the 'correct' answer, which is provided by highly accurate solutions for well-chosen test problems. An example of a verification problem is the nonlinear pendulum. The equation that governs the pendulum angle as a function of time is given by:

$$\frac{d^2\theta(t)}{dt^2} + \lambda^2 \sin(\theta(t)) = 0 \tag{9.1}$$

where $\lambda^2 = (g/L)$, L is the length of the pendulum, and g is the gravitational acceleration constant. A highly accurate solution of equation (9.1) can be compared to the results obtained from a finite element analysis that, for example, represents the pendulum as a rigid body. Comparison between the two solutions verifies the algorithms for this problem only.

It is important to understand that this activity provides evidence of the verification from the *intended purpose* of the code. 'Intended purpose' draws a clear boundary between SQA and numerical algorithm verification. The suite of problems selected to verify the software should exercise all the important dynamics and solution procedures that will be put into play when solving the engineering application of interest.

The main challenge here is to develop test problems for which analytical or highly accurate solutions can be obtained. A technique for developing a special type of analytical solution is the method of manufactured solutions (MMS) [8]. The MMS provides custom-designed verification test problems of wide applicability. Using the MMS in code verification requires that the computed source term and boundary conditions are programmed into the code, and that a numerical solution is computed. Although the intrusive character of the MMS can be viewed as a limitation, this technique nevertheless verifies a large number of numerical aspects in the code, such as the numerical method, differencing technique, spatial-transformation technique for grid generation, grid-spacing technique, and correctness of algorithm coding. Applications in fluid dynamics illustrate that the MSS can diagnose errors very efficiently, but cannot point to their sources, nor does it identify algorithm-efficiency mistakes [7, 8].

9.2.2 Solution Verification

Solution verification basically deals with the quantitative estimation of numerical accuracy of a solution. The primary goal is the attempt to estimate the numerical accuracy of a given solution, typically for a nonlinear system of partial differential equations with singularities or discontinuities. Numerical accuracy is critical in computations used for validation activities, where one should demonstrate that numerical errors are insignificant compared to test-analysis correlation errors.

A first typical issue of a computational grid is to make sure that the resolution provided by the grid must be appropriate to avoid stress concentrations at the 'corners' of a complex geometry. Also, the peak wave speed that can be captured by a given mesh is $C_{max} = (h/\Delta t)$, where h and Δt represent a characteristic element size and time increment. Modeling any phenomenon that could propagate information at a velocity $C > C_{max}$ requires another mesh or time integration, to satisfy $(h/\Delta t) > C$. A third, typical issue involves the degree of distortion of finite elements. Elements with poor aspect ratios tend to provide low-quality approximations.

The two basic approaches for estimating the error in a numerical solution are *a priori* and *a posteriori* approaches. *A priori* approaches use only information about the numerical algorithm that approximates the partial differential operator and the given initial and boundary conditions. *A priori* error estimation is a significant element of numerical analysis for differential equations underlying the finite element and finite volume methods [9, 10].

A posteriori approaches use all of the *a priori* information, plus computational results from numerical solutions using the same numerical algorithm on the same system of partial differential equations and initial and boundary data. Computational results are generally provided as sequences of solutions on consecutively finer grids. The framework upon which solution verification techniques rely is that the true-but-unknown solution of the continuous equations, or y_C, is equal to the solution $y(h)$ of the discretized equations, plus an error assumed to be proportional to the rate of convergence:

$$y_C = y(h) + \alpha h^p + \mathrm{O}(h^{p+1}) \tag{9.2}$$

The discretization parameter h can represent a characteristic mesh size or a time step. Equation (9.2) forms the basis for estimating or verifying the order of convergence p, and verifying that the approximation $y(p)$ converges to the continuous solution y_C. The discussion below focuses on *a posteriori* error estimates because they provide quantitative assessments of numerical error in practical cases of nonlinear equations.

9.2.2.1 The Richardson Extrapolation

A posteriori error estimation has primarily been approached through the use of Richardson extrapolation [7] or estimation techniques with finite element approximations [11]. Richardson's method can be applied to both finite difference and finite element methods. It computes error estimates of dependent variables at all grid points, as well as error estimates for solution functionals. It is emphasized that different dependent variables and functionals converge at different rates as a function of grid size or time step. Error estimation should be carried out mindful of the response outputs of interest.

Three unknowns are shown in equation (9.2): the continuous solution y_C; the order of convergence p; and a constant α. If the convergence order cannot be assumed, a minimum of three equations is needed to estimate the triplet (y_C; p; α). Note that it is good practice to always verify the actual order of convergence that can be severely deteriorated, at least locally, by factors such as programming errors, stress concentrations, and nonlinearity.

The Richardson extrapolation starts by computing three numerical solutions obtained with three (time step or grid size) resolutions h_C, h_M, and h_F:

$$y_C \approx y(h_C) + \alpha h_C^p, \ \ y_C \approx y(h_M) + \alpha h_M^p, \ \ y_C \approx y(h_F) + \alpha h_F^p \tag{9.3}$$

The order of convergence p can then be estimated as:

$$p \approx \log\left(\frac{y(h_M) - y(h_C)}{y(h_F) - y(h_M)}\right) / \log(r) \tag{9.4}$$

where r denotes the ratio between successive refinements, $r = (h_C/h_M) = (h_M/h_F)$, and the subscripts C, M, and F identify the 'coarse', 'medium', and 'fine' resolutions, respectively. It is important that r be kept constant for the estimation (9.4) to stay valid. Finally, the true-but-unknown solution y_C of the continuous partial differential equations can be approximated as:

$$y_C \approx \frac{r^p y(h_F) - y(h_M)}{r^p - 1} \tag{9.5}$$

Posterior error indicators such as the grid convergence index discussed next can be calculated to estimate how far the numerical approximation is from the continuous solution.

9.2.2.2 The Grid Convergence Index

A grid convergence index (GCI) based on Richardson's extrapolation has been developed to assist in the estimation of grid convergence error [7, 12]. The GCI converts error estimates that are obtained from any grid-refinement ratio into an equivalent grid-doubling estimate. Recent studies have established the reliability of the GCI method, even for solutions that are not asymptotically convergent [13, 14].

The definition of the GCI in Reference [7] involves two solutions $y(h_C)$ and $y(h_F)$, obtained with coarse and fine resolutions, respectively:

$$GCI = 100 \left| \frac{y(h_C) - y(h_F)}{y(h_C)} \right| \left(\frac{\beta}{r^p - 1} \right) \tag{9.6}$$

where β is a constant that adds conservatism to the formula, typically, $1 \leq \beta \leq 3$. Small values of the GCI – typically, less than 1 % – indicate that the best approximation obtained from the two resolutions is close to the true-but-unknown continuous solution.

Applying the Richardson extrapolation and GCI concepts starts by performing three finite element calculations, then the order of convergence (9.4) and asymptotic convergence (9.6) are verified. Again, it is emphasized that the convergence study can verify the adequacy of a mesh (when h is defined as the grid size), just like it can be applied to time or frequency-domain convergence (when h is a time step or frequency increment). In the case of nonlinear, transient dynamics simulations, mesh and time convergence should be verified independently of one another, which could lead to significant computational demands.

Another important issue is to assess the validity of the convergence model (9.2). Such error model is generally well suited to analyze numerical methods that solve 'well-behaved' linear problems. Fast dynamics, nonlinear, or shock problems that rely on explicit integration schemes may require an error Ansatz model, as it is termed in physics, where the coupling between time step, grid size, and possibly other parameters, is taken into account [15].

9.2.3 What Can be Expected From Verification?

The rigorous verification of a code requires 'proof' that the computational implementation accurately represents the conceptual model and its solutions. This, in turn, requires proof that the algorithms provide converged solutions of these equations in all circumstances under which the code will be applied. It is unlikely that such proofs will ever exist for general-purpose computational physics and engineering codes. Verification, in an operational sense, becomes the absence of proof that the code is incorrect.

In this definition, verification activities consist of accumulating evidence substantiating that the code does not have any apparent algorithmic or programming errors, and that it functions properly on the chosen hardware and system software. This evidence needs to be documented, accessible, repeatable, and capable of being referenced. The accumulation of evidence also serves to reduce the regimes of operation of the code where one might possibly find such errors.

In the absence of formal proof, what can be expected from the verification activities? Clearly, evidence must be provided that the code does not have any apparent error, and that it provides approximate solutions of acceptable accuracy, consistent with the purpose intended. How much evidence should be provided (in other words, 'How good is good enough?') is not addressed here because such question directly relates to accreditation and the definition of a standard. It is also application specific to a great extent.

9.3 VALIDATION ACTIVITIES

In this section, typical validation activities are presented. These activities are put in the context of a material test instead of being discussed separately.

Materials scientists commonly use an experiment known as the Taylor anvil impact to develop constitutive and equation-of-state models. Because of the regimes for which they are developed, such models generally include plasticity, high strain-rate and temperature dependency. Examples used in engineering mechanics include the Johnson–Cook and Zerilli–Amstrong models [16, 17]. The Taylor anvil experiment consists of impacting a sample of material against a rigid wall and measuring its deformed profile. The measured profiles are compared to numerical predictions and parameters of the constitutive equations can be calibrated to improve the accuracy of the model.

9.3.1 The Validation Domain

Before proceeding with the description of the validation steps, the notion of validation domain must be introduced. Generally speaking, a numerical simulation is always developed to analyze a given operational domain because point predictions, that is, models that cannot be parameterized or modified, are not very useful.

The constitutive model investigated here is developed to run numerical simulations at various combinations of strain rates (S_R) and temperatures (T). For our application, these two inputs define the operational space of interest. The validation domain is simply defined as the region of the operational space where the mathematical or numerical model provides acceptable accuracy for the application of interest. Simply speaking, validation is achieved when the prediction accuracy of the model has been assessed within the operational domain, a consequence of which is the identification of the region – or validation domain – that provides sufficient accuracy.

9.3.2 Validation Experiments

The nature of a validation experiment is fundamentally different from the nature of a conventional experiment. The basic premise is that a validation experiment is a somewhat simpler procedure, which isolates the phenomenon of interest. A suite of validation experiments provides increasing levels of understanding about the mechanics, physics, or coupling between separable phenomena.

In our constitutive modeling example, static material testing comes first, which allows the identification of bulk mechanical properties such as the modulus of elasticity. Because such tests are static in nature, they must be augmented with Hopkinson bar tests. Hopkinson bar tests, however, do not provide sufficient resolution in the regime of interest, that is, at high strain rates and varying temperatures.

To obtain more insight into the behavior of the plasticity throughout the validation domain, Taylor anvil tests are performed next. The validation experiments explore different regions of the validation domain while providing successive material models that are hopefully consistent with each other. The discussion in this work focuses on the definition of an error metric between inferences made from the Hopkinson bar tests and inferences made from the Taylor anvil impact tests (see Section 9.5).

9.3.3 Design vs Calibration Parameters

The Zerilli–Amstrong model estimates the stress resulting from a plastic deformation as:

$$\sigma = C_0 + C_1 e^{-C_3 T + C_4 T \log\left(\frac{d\epsilon_P}{dt}\right)} + C_5 \epsilon_P^N \tag{9.7}$$

where the symbol T represents temperature and ϵ_P denotes plastic strain. The six parameters C_0, C_1, C_3, C_4, C_5 and N are material-dependent constants that can be calibrated to improve prediction accuracy. Because of the range of strain rates for which a validated model is sought (from the quasistatic rate of 10^{-3}/s to $4 \times 10^{+3}$/s), another symbol S_R is introduced that defines the logarithm of the plastic strain rate:

$$S_R = \log_{10}\left(\frac{d\epsilon_P}{dt}\right) \tag{9.8}$$

It is important not to confuse the six calibration variables $(C_0; C_1; C_3; C_4; C_5; N)$ with the two input parameters $(T; S_R)$ that define the operational space or validation domain. The main difference between the two is that calibration variables are introduced by our particular choice of plasticity model. Should another physical model be adopted, the calibration variables would likely change. The dimensionality of the operational space, however, never changes and the plasticity models – whatever they are – must still be validated at various combinations of $(T; S_R)$.

9.3.4 Forward Propagation of Uncertainty

Two key technologies for V&V are the propagation and analysis of uncertainty. This is because model validation is essentially an exercise in assessing and quantifying uncertainty, whether it originates from the model (lack-of-knowledge, modeling assumptions, and parametric variability), computations (mesh and convergence-related numerical errors), physical experiments (variability of the environment and measurement error), or judgments (vagueness and linguistic ambiguity).

In Sections 9.3.4 and 9.3.5, some of the tools employed to propagate uncertainty are briefly illustrated. They include Monte Carlo sampling for propagating uncertainty through forward calculations and Bayesian calibration for backward inference. Other tools, which include the design of experiments and analysis of variance, are illustrated in Section 9.4 [18–20].

The Monte Carlo simulation is one of many sampling techniques that can propagate uncertainty from the calibration variables ($C_0;C_1;C_3;C_4;C_5;N$) through the numerical simulation of Taylor impact. In this illustration, each of the six variables is assumed to vary according to a Gaussian probability density function (PDF). Because the variables are assumed to be independent and uncorrelated, sampling the six individual PDF laws is straightforward. Each combination of variables ($C_0;C_1;C_3;C_4;C_5;N$) defines a specific material model, as shown by equation (9.7). Once the material model selected, it is implemented to simulate numerically the impact experiments using the finite element package HKS/Abaqus^{TM} [21]. An axisymmetric mesh is impacted against a perfectly rigid surface, which produces large deformations (over 260 %) and significant plastic strain at the crushed end of the cylinder. The number of Monte Carlo runs is typically selected based on the time necessary to perform a single analysis and the available computing resource.

Two features of the response are defined to characterize the deformed profiles, the ratios of final-to-initial lengths (L/L_o) and the ratios of initial-to-final radii or footprints (R/R_o). Each point in Figure 9.1(a) represents the pair of output features ((L/L_o); (R/R_o)) of an impact simulation for a particular material model. A total of 1000 finite element calculations are performed. It can be observed that there is a significant correlation between the two output features, as one would expect because the shorter the cylinder, the larger its footprint.

The histograms shown in Figure 9.1(b) are obtained by projecting the distribution of output features on the horizontal and vertical axes. Each axis is then discretized in 20 bins and the histograms simply show how many features are counted within each bin. The histograms approximate the output PDF. It can be observed from their asymmetries and long tails that the probability laws of response features (L/L_o) and (R/R_o) are not normal, illustrating the well-known result that a Gaussian PDF propagated through a nonlinear system such as this finite element simulation does not stay Gaussian.

Monte Carlo simulations are popular because of their simplicity and well-established convergence properties. Our simple application illustrates the propagation of uncertainty from inputs to outputs and the estimation of the response's probability structure, from which statistics can be calculated. To guarantee convergence, however, large numbers of samples may be required in which case other sampling strategies (for example, stratified sampling such as the Latin hypercube sampling [22] or orthogonal arrays [23]) offer alternatives.

9.3.5 Inverse Propagation of Uncertainty

The calibration of model parameters is a technique often employed to improve the fidelity-to-data. Calibration is generally formulated as a deterministic inverse problem. A cost function is defined as the 'distance' in some sense between measurements and predictions. Model parameters are then optimized to minimize the cost function. When the calibration parameters and response features are considered to be realizations of random processes, a mechanism must be found to propagate uncertainty from the measurements back to the inputs. This is here referred to as the inverse propagation of uncertainty.

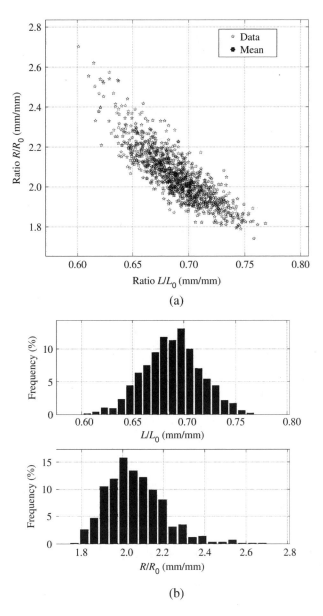

Figure 9.1 Output features (L/L_o) and (R/R_o) obtained from the Monte Carlo simulation. (a) Distribution of (L/L_o) and (R/R_o); (b) Histograms of (L/L_o) and (R/R_o)

   Although many formulations are possible, the concept of Bayesian inference is illustrated here for the Taylor impact application [24]. Like in the deterministic case, a procedure for inverse propagation of uncertainty starts with the definition of a cost function. The main difference is to take into account the fact that the input parameters p and output features y are random variables, which generally implies that the cost function becomes a statistical test. In the case of Bayesian inference, the cost function defines the posterior probability that the model parameters $p = (C_0;C_1;C_3;C_4;C_5;N)$ are correct given evidence provided by the physical measurements y^{Test}. The posterior **PDF** is the conditional probability law of the calibration variables p:

$$e^2 = -2\log(\text{Prob}(p|y^{Test}))\tag{9.9}$$

The Bayes Theorem states that the posterior probability (e^2) is the product of the likelihood function – likelihood to predict the measurements based on a given model – multiplied by the prior probability of p. Under the assumption of Gaussian probability laws, the likelihood function becomes the mean square error between measurements and predictions. One advantage is that the cost function obtained is a closed-form expression:

$$e^2 = \sum_{k=1...N} \left(y_k^{Test} - y_k(p)\right)^T \left(\sum\nolimits_{y_k}^{Test}\right)^{-1} \left(y_k^{Test} - y_k(p)\right) + (p - p_o)^T \left(\sum\nolimits_p\right)^{-1} (p - p_o)$$

$$\tag{9.10}$$

where the inverted matrices are formed from variance and covariance values, and the symbol p_o denotes the nominal material coefficients. The quantify $y = \{(L/L_o);(R/R_o)\}$ collects the output features. The sum aggregates information obtained from different experiments. When Gaussian probability distributions are assumed, the cost function (9.10) becomes the well-known chi-square statistical test that attempts to reject the null hypothesis that measurements y^{Test} and model predictions y are sampled from the same parent population.

The general procedure for calculating the chi-square statistics goes as follows. First, the simulation is analyzed for a given experimental configuration defined by the temperature and strain-rate parameters $(T;S_R)$, and given material coefficients (p). The features (L/L_o) and (R/R_o) are calculated from the final deformed profile. These two predictions are compared to the measurements. Figure 9.2 shows a typical

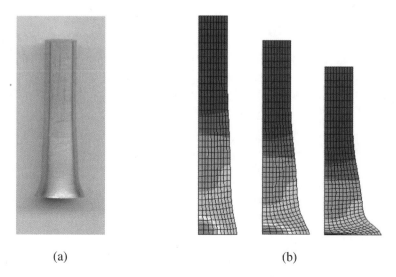

(a)   (b)

Figure 9.2 Test-analysis comparison for a Taylor anvil experiment. (a) Measured profile; (b) simulations at 17 ms, 33 ms, and 50 ms after impact

test-analysis comparison. The procedure is repeated for each combination of temperature and strain-rate $(T;S_R)$ to accumulate the chi-square metric.

Once a numerical procedure has been defined to compute the cost function, the calibration variables $p = (C_0;C_1;C_3;C_4;C_5;N)$ are optimized to search for the lowest possible chi-square value. Because the prior and posterior PDF laws have been assumed to be Gaussian, a deterministic optimization solver can be used to optimize the posterior mean and covariance. In the case where no evidence is available to suggest a particular distribution, the main difficulty becomes the estimation of a posterior PDF whose functional form is unknown. This can be resolved with a Markov chain Monte Carlo optimizer that exhibits the attractive property of being able to sample an unknown probability law [20, 25].

9.4 UNCERTAINTY QUANTIFICATION

In Section 9.3, the propagation of uncertainty from input variables to output features or vice-versa (forward propagation or inverse propagation, respectively) is discussed. Another type of analysis, namely the *screening experiment*, is addressed here. Effect screening refers to the identification of interactions between inputs to explain the variability of outputs.

Effect screening addresses questions such as: 'Which inputs or combination(s) of inputs explain the variability of outputs?' Screening is typically performed to truncate the list of input variables by determining which ones most influence the output features over the entire design domain. Identifying the effects most critical to an input–output relationship is also the basis for replacing physics-based models by fast-running surrogates (see Section 9.4.2).

9.4.1 Effect Screening

Effect screening is typically achieved using the concept of analysis-of-variance [26] or other techniques such as Bayesian screening [20], not discussed here. Our experience with these methods is that a successful screening of effects should always provide consistent results when techniques of different nature are employed [18].

To understand the importance of screening, it is useful to go back to Figure 9.1(a). Each datum results from a numerical simulation performed with a specific material model (9.7). The '*spread*' of predictions $((L/L_o);(R/R_o))$ comes from varying the six material coefficients C_0, C_1, C_3, C_4, C_5, and N. Figure 9.1(a), however, does not explain which input (C_0, C_1, C_3, C_4, C_5, or N) or combination of inputs (such as C_1C_3, C_1N, or N^2) explains the variability of (L/L_o) and (R/R_o). Screening answers such question by performing multiple regression analyses and estimating correlations between the input effects and output features.

The screening of main effects (also known as linear screening) attempts to identify the inputs that control the output variability, without accounting for higher-order effects. This can be extended to higher-order effect screening to assess the effect of combinations of inputs such as C_1C_3, C_1N, or N^2. Analyses of variance rely on the calculation of the R^2 statistics that estimate coefficients of correlation between inputs and outputs. A large R^2 relative to the other values indicates that the corresponding effect (such as C_0, C_1, C_3, etc., for linear screening or C_1C_3, C_1N, N^2, etc., for higher-order screening) produces a significant variability of the output feature. Each output feature must be analyzed individually. Runs of the finite element model are selected with a design of computer experiments that must be carefully chosen to avoid introducing bias in the results of the screening experiments.

The analysis of Taylor impact simulations indicates that the variability of features (L/L_o) and (R/R_o) is controlled, for the most part, by the material coefficients C_1, C_3, and C_4. The other factors do not produce significant output variability, which means that they can be kept constant and equal to their nominal values. The number of calibration variables is therefore reduced from $(C_0;C_1;C_3;C_4;C_5;N)$ to $(C_1;C_3;C_4)$, which can lead to significant computational savings when performing parameter calibration or propagating uncertainty through the code.

9.4.2 Surrogate Modeling

Many analyses (such as the propagation of uncertainty, parameter calibration, and reliability) become computational prohibitive when the number of inputs and outputs increases. Instead of arguing for approximate and less expensive solutions, the approach can be taken to replace physics-based models by surrogates.

Surrogate models (also known as meta-models, emulators, or response surfaces) capture the relationship between inputs and outputs without providing a detailed description of the physics, geometry, loading, etc. The advantage of surrogate models is that they can be analyzed at a fraction of the cost of performing the physics-based simulations. Examples include polynomials, exponential decays, neural networks, principal component decomposition, and statistical inference models. An example of meta-model is presented in Section 9.5 for assessing the prediction accuracy of the Taylor simulations. Techniques for the design of experiments, screening, and meta-modeling are overviewed in References [27, 28].

Meta-models must be trained, which refers to the identification of their unknown functional forms and coefficients. Their quality must be evaluated independently from the training step. Because analyzing finite element models at every combination of input variables is generally a combinatorial impossibility, training can be based on a subset of carefully selected runs that, statistically speaking, provide the same amount of information.

9.4.3 Design of Computer Experiments

Design of experiments (DOE) techniques have been developed for exploring large design spaces when performing complex, physical experiments. They can be brought to bear to select judicious subsets of finite element runs. Such simulations provide the input–output data used, for example, to fit surrogate models.

It is important to realize that meta-modeling and effect screening can both be performed with the same DOE because identifying which effects and interactions

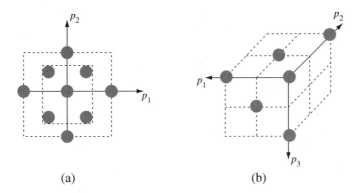

Figure 9.3 Designs for computer experiments. (a) Two-factor central composite design; (b) Three-factor orthogonal array design

capture a particular input–output relationship essentially delivers the functional form of the surrogate model.

Examples of popular designs include the Morris method, fractional factorials, the central composite design (CCD), and orthogonal arrays. Figure 9.3 illustrates a two-factor CCD and a three-factor orthogonal array. In both cases, finite element analyses are performed at the combinations $(p_1;p_2)$ or $(p_1;p_2;p_3)$ shown. It can be observed that the CCD and orthogonal array require nine and six finite element runs, respectively. Using the CCD, a fully quadratic polynomial can be identified to serve as surrogate to the physics-based simulations. The orthogonal array shown in Figure 9.3b can only identify a few interactions (such as p_1p_2 or p_1p_3). Quadratic effects (such as p_1^2 or p_2^2) cannot be screened because there are simply not enough data points to identify these higher-order effects without aliasing.

Characteristics of a good design are that it relies on as few runs as possible; provides screening capabilities; leads to high-quality surrogate models; and minimizes aliasing, which refers to the compounding of effects that cannot be captured by the design [27, 28].

Although generally not recommended, DOE such as orthogonal arrays can be used to propagate uncertainty from inputs to outputs. A statistically more rigorous approach would be to first perform a number of effect screening experiments based on the appropriate DOE. After some of the input variables have been eliminated, sampling or stratified sampling techniques can be used to propagate the uncertainty at any required level of accuracy.

9.5   ASSESSMENT OF PREDICTION ACCURACY

In Section 9.3, the calibration of model parameters under uncertainty has been illustrated. Calibration, however, is only a tool in support of prediction accuracy assessment. A calibrated model is likely to provide small prediction errors in the neighborhood of the points used for calibrating its parameters but the question of adequacy in other regions of the operational space remains. In the remainder, an assessment of prediction accuracy for the plasticity model is illustrated. The illustration is purposely simplified for the sake of clarity.

The concept of operational space (or validation domain) introduced in Section 9.3.1 is essential to the discussion. The operational domain, that is, the set of conditions for which a validated model of plasticity is sought, is defined by the combination of temperatures T and strain rates S_R, as shown in Figure 9.4(a).

The dots in Figure 9.4(a) symbolize the settings $(T;S_R)$ at which physical experiments have been performed. Data collected during these seven tests can be used to calibrate the coefficients $(C_0;C_1;C_3;C_4;C_5;N)$ of the material model, see the discussion in Section 9.3.5, but the central question remains: 'What is the prediction accuracy of the model throughout the validation domain?' Providing an answer necessitates estimating prediction accuracy away from settings $(T;S_R)$ at which experiments have been performed.

The assessment of prediction accuracy starts by calculating a quantitative metric of test-analysis correlation at those settings $(T;S_R)$ in the validation domain where experiments have been performed. Test-analysis metrics define fidelity-to-data, that is, the ability of numerical predictions to match the data collected experimentally. It is helpful to keep a clear distinction between *features* of the response and correlation *metrics*; this avoids confusion between what the model needs to predict and how good a job it does of predicting it.

In the case of our Taylor impacts, the Mahalanobis distance is calculated between measurements and predictions:

$$e^2 = \left(y^{Test} - y(p)\right)^T \left(\sum\nolimits_y^{Test}\right)^{-1} \left(y^{Test} - y(p)\right) \tag{9.11}$$

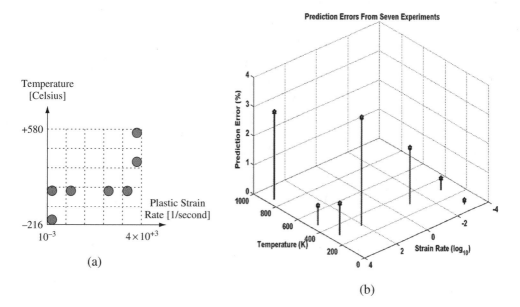

(a)

(b)

Figure 9.4 Metrics for prediction accuracy assessment shown in the validation domain. (a) The validation domain; (b) Mahalanobis error metrics

where y^{Test} represents the mean of measured response features (L/L_o) and (R/R_o) and $y(p)$ is the mean prediction of the same features obtained with the calibrated model of plasticity. The covariance matrix is initialized with the measurement variance.

The Mahalanobis statistic is adopted here because it assesses fidelity-to-data relative to the experimental variability. However, the choice of a metric for test-analysis comparison is generally application-specific. Figure 9.4(b) shows the values of the error metric (9.11) computed at the seven settings in the two-dimensional space $(T;S_R)$ where Taylor anvil impacts have been performed. Note that the question remains of knowing how well the model performs at locations in the operational space where no physical experiment is available.

This question is addressed by developing a meta-model of prediction accuracy. Clearly, assessing fidelity-to-data away from settings that have been tested must involve some sort of extrapolation. Although it may not seem very rigorous, extrapolation is justified whenever it is believed that the prediction error at a setting $(T;S_R)$ that has not been tested experimentally will not differ significantly from the prediction error at a setting $(T;S_R)$ that has been tested as long as the two are '*close*' to each other.

The model of prediction accuracy is referred to as a meta-model because it is not based on physical principles. Its only purpose is to capture an input–output relationship between the input parameters $(T;S_R)$ and the prediction error e^2, and extrapolate beyond the available data with reasonable accuracy. A quadratic polynomial is selected here for simplicity. Other choices may include polynomials with orthogonal basis functions, neural networks, principal component analysis, Kriging, or statistical inference.

It is emphasized that an arbitrary choice of meta-model expresses *lack-of-knowledge*. Even though this question is not discussed here, the robustness of decisions based on an

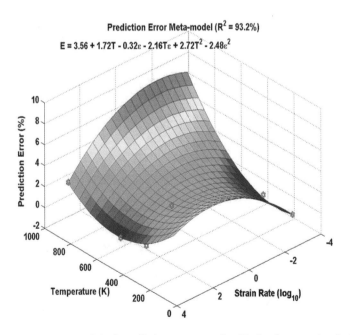

Figure 9.5 Meta-model of prediction accuracy for Taylor impact simulations

assessment of prediction accuracy should be studied to ensure that they are not vulnerable to such lack of knowledge. A theory and practical tools are proposed in References [29] and [30] to analyze robustness under severe uncertainty. Studying the relationship between fidelity-to-data, robustness to uncertainty, and the consistency between predictions made by multiple models is at the core of science-based decision making [3].

Figure 9.5 illustrates the estimation of prediction accuracy over the entire validation domain $(T; S_R)$. The significance of Figure 9.5 is that it estimates the adequacy of the Zerilli–Amstrong model of plasticity throughout the operational domain, without having to perform any additional physical experiment or simulation. Should the accuracy be insufficient, it can then be decided to replace this model by something else. Here, 'validation' does not mean that the model should be 'perfect' everywhere, which is a significant shift of paradigm compared to the model calibration approach. The model is validated because its prediction accuracy has been assessed throughout the design space, and away from tested conditions.

9.6 CONCLUSION

This chapter offers a brief overview of the tools brought to bear at Los Alamos National Laboratory in support of engineering Verification and Validation (V&V) programs. The material is based to a large extent on a tutorial taught yearly at the Los Alamos Dynamics Summer School [1]. In this publication, the tools developed are applied to the validation of a nonlinear plasticity model. They include code verification, solution verification, statistical sampling, design of experiments, effect screening, feature extraction (although not discussed here), metrics for test-analysis correlation, surrogate modeling, and calibration.

The ultimate objective of verification and validation is to quantify the level of confidence in predictions made by numerical models, possibly away from settings that have been tested experimentally. Obtaining this information over an entire operational space is essential to answer questions such as: 'How appropriate is the model overall?' 'Which one of several competing models is more appropriate for a particular application?' 'Which physical tests would be useful to improve the prediction accuracy of the model?'

To conclude, we briefly discuss three issues that have not been addressed, yet are key issues, we believe, in establishing the credibility of decisions based on modeling and simulation: the breakdown of total uncertainty; the quantification of modeling uncertainty; and the robustness of science-based decisions.

(i) ***Total uncertainty*** refers to the aggregation of all potential sources of uncertainty, originating from measurements, models, or expert knowledge. Clearly, one difficulty is to identify these sources of uncertainty and lack of knowledge. Another is to model uncertainty with an appropriate mathematical theory, which may not always be probability. Information integration and total uncertainty assessment are currently being studied at Los Alamos and elsewhere. These concepts are critical for validation, reliability, and margin assessment.

(ii) *Modeling lack of knowledge* refers to the uncertainty that results from somewhat arbitrary choices made when selecting modeling rules, assumptions, and functional forms. A credible and rigorous assessment of accuracy should account for such lack of knowledge. One serious roadblock is that much of this uncertainty may not always be quantified probabilistically.

(iii) *The robustness of science-based decisions* refers to the ability to demonstrate that predictions and decisions based on the analysis of numerical models are not vulnerable to our lack of knowledge about some of the processes being analyzed. It is especially important to assess the robustness of predictions in cases where not all the sources of uncertainty can be modeled and propagated through the analysis. Hence, it appears that robustness is the missing link between fidelity-to-data and confidence in predictions.

REFERENCES

[1] The Los Alamos Dynamics Summer School, ext.lanl.gov/projects/dss/home.htm.

[2] Doebling, S.W. 'Structural Dynamics Model Validation: Pushing the Envelope', *International Conference on Structural Dynamics Modeling: Test, Analysis, Correlation and Validation*, Madeira, Portugal, June 3–5, Instituto Superior Técnico, Lisbon, 2002.

[3] Hemez, F.M. and Ben-Haim, Y. 'The Good, the Bad, and the Ugly of Predictive Science', *Fourth International Conference on Sensitivity Analysis of Model Output*, Santa Fe, New Mexico, March 8–11, Los Alamos National Laboratory, 2004.

[4] Booker, J., Ross, T., Hemez, F.M., Anderson, M.C. and Joslyn, C. 'Quantifying Total Uncertainty in a Validation Assessment Using Different Mathematical Theories', *Fourteenth Biennial Nuclear Explosives Design Physics Conference*, Los Alamos National Laboratory, Los Alamos, New Mexico, October 20–24, 2003.

[5] Ross, T., Booker, J. and Hemez, F.M. 'Quantifying Total Uncertainty and Performance Margin in Assessing the Reliability of Manufactured Systems', *Fifth Biennial Tri-lab. Engineering Conference*, Santa Fe, New Mexico, October 21–23, Los Alamos National Laboratory 2003.

[6] AIAA. *Guide for the Verification and Validation of Computational Fluid Dynamics Simulations*, American Institute of Aeronautics and Astronautics, AIAA-G-077-1998, Reston, Virginia, 1998.

[7] Roache, P.J. *Verification and Validation in Computational Science and Engineering*, Hermosa Publishers, Albuquerque, New Mexico, 1998.

[8] Roache, P.J. 'Code Verification by the Method of Manufactured Solutions', *Journal of Fluids Engineering*, **114**, 4–10 (2002).

[9] Ferziger, J.H. and Peric, M. *Computational Methods for Fluid Dynamics*, Springer-Verlag, New York, 1996.

[10] Oden, J.T. 'Error Estimation and Control in Computational Fluid Dynamics', in *The Mathematics of Finite Elements and Applications*, Whiteman, J.R. (Ed.) John Wiley & Sons, Inc., New York, 1993, pp. 1–23.

[11] Ainsworth, M. and Oden, J.T. *A Posteriori Error Estimation in Finite Element Analysis*, John Wiley & Sons, Inc., New York, 2000.

[12] Roache, P.J. 'Perspective: A method for uniform reporting of grid refinement studies', *Journal of Fluids Engineering*, **116**, 405–413 (1994).

[13] Cadafalch, J., Perez-Segarra, C.C., Consul, R. and Oliva, A. 'Verification of finite volume computations on steady state fluid flow and heat transfer', *Journal of Fluids Engineering*, **124**, 11–21 (2002).

[14] Chen, C.-F., Lotz, R.D. and Thompson, B.E. 'Assessment of numerical uncertainty around shocks and corners on blunt trailing-edge supercritical airfoils', *Computers and Fluids*, **31**, 25–40 (2002).

[15] Brock, J.S. 'Isolating Temporal Discretization Errors for Separate Verification Activities', *Technical report LA-UR-03-9160*, Los Alamos National Laboratory, Los Alamos, New Mexico, December 2003; *2004 AIAA Aerospace Sciences Conference*, paper number AIAA-2004-0741, January 2004.

[16] Adessio, F.L., Johnson, J.N. and Maudlin, P.J. 'The effect of void growth on Taylor cylinder impact experiments', *Journal of Applied Physics*, **73**, 7288–7297 (1993).

[17] Zerilli, F.J. and Amstrong, R.W. 'Dislocation mechanics-based constitutive relations for material dynamics calculations', *Journal of Applied Physics*, **61**, 1816–1825 (1987).

[18] Cundy, A.L., Schultze, J.F., Hemez, F.M., Doebling, S.W. and Bingham, D. 'Variable Screening in Metamodel Design for a Large Structural Dynamics Simulation', *Twentieth International Modal Analysis Conference*, Los Angeles, California, February 4–7, 2002, Society for Experimental Mechanics, pp. 900–903.

[19] Hemez, F.M., Wilson, A.C. and Doebling, S.W. 'Design of Computer Experiments for Improving an Impact Test Simulation', *Nineteenth International Modal Analysis Conference*, Kissimmee, Florida, February 5–8, 2001, Society for Experimental Mechanics, pp. 977–985.

[20] Kerschen, G., Golinval, J.-C. and Hemez, F.M. 'Bayesian model screening for the identification of non-linear mechanical structures', *Journal of Vibration and Acoustics*, **125**, 389–397 (2003).

[21] *HKS/Abaqus^{TM} Explicit User's Manual*, Version 5.8, Hibbitt, Karlsson & Sorensen, Pawtucket, Rhode Island, 1998.

[22] McKay, M.D., Beckman, R.J. and Conover, W.J. 'A comparison of three methods for selecting values of input variables in the analysis of output from a computer code', *Technometrics*, **21**, 239–245 (1979).

[23] Hedayat, A.S., Sloane, N.J.A. and Stufken, J. *Orthogonal Arrays: Theory and Applications*, Springer-Verlag, New York, 1999.

[24] Hanson, K.M. 'A framework for assessing uncertainties in simulation predictions', *Physica D*, **133**, 179–188 (1999).

[25] Carlin, B.P. and Chib, S. 'Bayesian model choice via Markov chain Monte Carlo', *Journal of the Royal Statistical Society Series B*, **77**, 473–484 (1995).

[26] Saltelli, A., Chan, K. and Scott, M. *Sensitivity Analysis*, John Wiley & Sons, Inc., New York, 2000.

[27] Myers, R.H. and Montgomery, D.C. *Response Surface Methodology: Process and Product Optimization Using Designed Experiments*, Wiley Interscience, New York, 1995.

[28] Wu, C.F.J. and Hamada, M. *Experiments: Planning, Analysis, and Parameter Design Optimization*, John Wiley & Sons, Inc., New York, 2000.

[29] Ben-Haim, Y. 'Robust rationality and decisions under severe uncertainty', *Journal of the Franklin Institute*, **337**, 171–199 (2000).

[30] Ben-Haim, Y. *Information-Gap Decision Theory: Decisions Under Severe Uncertainty*, Academic Press, New York, 2001.

10
Reliability Methods

Amy Robertson and François M. Hemez

Los Alamos National Laboratory, Los Alamos, New Mexico, USA

10.1 INTRODUCTION

As stated in the introduction to this book, *damage prognosis* is the process of estimating a system's remaining useful life. The goal is to forecast system performance by measuring the current state of the system, estimating the future loading environments for that system, and predicting through simulation and past experience the remaining useful life of the system. Damage prognosis faces numerous sources of variability, uncertainty, and lack of knowledge. Examples are experimental variability, parametric uncertainty, unknown functional forms of the mathematical models, and extrapolated future loading and environments. The discussion of damage prognosis would therefore be incomplete without addressing the issue of decision making under uncertainty.

In the presence of damage, decisions must be made about whether a system can continue to function as required for a given amount of time. For instance, if a wing of an airplane has been damaged during flight, one might need to decide whether that plane can continue its flight mission, or needs to abort. These decisions are based on an understanding of how probable it is that a system might fail during the period of time that it takes to complete the mission. Reliability analysis is the procedure used to provide an estimate of the failure probability. Reliability can be defined as the probability that a system, at a given point in time, will be able to perform a required function without failure. Through an assessment of the current reliability, as well as information about how that reliability will be changing over time, a decision of what to do in the presence of damage can be made.

To demonstrate the concepts associated with performing a reliability analysis, this chapter outlines the probabilistic analysis used to predict the remaining useful life of a

Damage Prognosis – For Aerospace, Civil and Mechanical Systems   Edited by D.J. Inman, C.R. Farrar, V. Lopes Junior and V. Steffen Junior   © 2005 John Wiley & Sons, Ltd

composite plate after initial damage caused by a foreign body impact. This problem ismotivated by aerospace applications where fiber reinforced composite wing and fuselage components often experience foreign object impact both in flight and as a result of maintenance activities. Currently, the industry does not have a method to quantify the damage introduced by such events, nor does the industry have the ability to predict the growth of such damage when the aircraft is subjected to fatigue loading produced by flight operations. The type of damage that will be focused on is delamination, because this is the most common form of damage in composites and is also one of the most difficult to assess. Recent surveys show that 60 % of all observed damage in composite parts of civil aircrafts is in the form of delamination [1]. Once damage is initiated in the composite panels, it is propagated using cyclic loading, which represents the normal in-service loading a composite structure might endure. At any point in this loading, it is important to determine what the reliability of the system is and how many more loading cycles it can endure before failure.

This chapter is organized as follows. First, the steps needed to perform a reliability analysis are summarized and applied to the composite delamination problem described above. The next section discusses the computational approaches commonly used in reliability analysis for estimating the probability that a structure will fail. This discussion includes the role of surrogate modeling (also known as meta-modeling) for integrating the experimental diagnostics, modeling and simulation, data interrogation, and reliability assessment into a practical damage prognosis solution. Finally, time-based reliability is discussed to show how the estimate of probability of failure at a given point in time can be extrapolated to making decisions about the remaining useful life of the system.

10.2 RELIABILITY ASSESSMENT

The reliability of a system (R) is defined as the probability that the system in its current condition will not fail. The probability of failure (P_F) is directly related to the reliability and is the value that is sought after in a reliability analysis:

$$P_F = 1 - R \tag{10.1}$$

In basic terms, failure is defined as the demand (stress, strain) needed to exceed the capacity of that system. Of course, determination of these two values at any given system state is a difficult task and must involve estimation. Invariably, there will be uncertainty in these values and therefore a probabilistic analysis must be performed, whereby these values are represented in terms of probability density functions (PDFs). Examining the figure below, one can see that failure occurs in the region where the capacity and demand PDFs overlap.

If the PDFs of the demand and capacity are known, the probability of failure can be calculated as follows [2]:

$$P_F = P(C < D) = P(C - D < 0) \tag{10.2}$$

where C denotes capacity and D denotes demand. If the demand and capacity are statistically independent, this probability is calculated as:

$$P_F = \int\limits_{0}^{\infty} F_D(x)f_C(x)\mathrm{d}x \qquad (10.3)$$

where F_D is the cumulative distribution of the demand and f_C is the probability distribution of the capacity (Figure 10.1).

In general, the reliability problem will not be so simple. Demand and capacity are the fundamental concepts behind the failure of a system, but they are not always the best way of representing the reliability. The more broad view of reliability is that it answers the fundamental question of 'what is the probability that the response of the system will not exceed some critical level?' Throughout this chapter we will be focusing on one example to help demonstrate the concepts of reliability. The problem addressed is the reliability of a composite material that has been impacted by a projectile. A reliability statement for this application could be 'what is the probability that the damage to the plate has not exceeded a critical size, beyond which the composite will no longer perform as desired?' If desired performance is the ability of the composite to carry a load, one can see that this problem is just another way of stating whether the demand on the plate, which could be the ambient loading to the structure, has exceeded the capacity, in this case the strength of the composite.

10.2.1 Failure Modes

The first step in assessing the reliability of a system begins with identifying what the failure modes of the system are. For composites, there are generally three different ways the material can fail: matrix cracking, fiber breakage and delamination (see Figure 10.2). For our example problem, failure will be defined only in terms of delamination even though the experiments and simulations performed clearly involve all three types of failure.

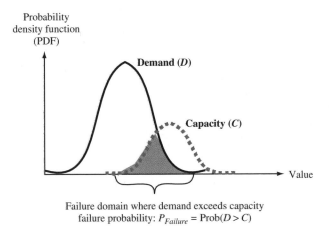

Figure 10.1   Illustration of the concepts of demand, capacity, and failure of a system

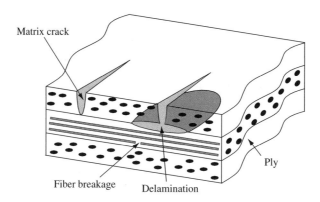

Matrix crack

Fiber breakage   Delamination

Ply

Figure 10.2   Schematic demonstrating the three failure modes of a [90°/0°/90°] composite

Failure is defined as the exceedance of the delamination size beyond some critical value:

$$Failure: \ a_N > a_C \tag{10.4}$$

where a_N is the size (area) of the delamination after N loading cycles and a_C is the critical delamination size. As cyclic loading is applied, the delamination will grow, therefore increasing the probability of failure.

The structural performance of the system, which will affect the delamination size, is subject to uncertainty in many areas including the natural variability of the material properties and loads, the statistical uncertainties due to lack of data, and modeling uncertainties due to idealizations. These parameters are the random variables of the analysis. When thinking of reliability in terms of demand and capacity, these parameters are what constitute the uncertainty in the demand and capacity values.

To reduce the number of random variables in the analysis, parameter screening may be performed to determine which parameters most influence the delamination growth. This is accomplished by a method called 'analysis of variance' [3], which looks at how the variation of each parameter affects the variation of the output through means of the finite-element model. The influence of each parameter is largely dependent on the range over which that parameter may vary. Experiments can be performed to provide information about the variability of some of the parameters, but the remainder must be approximated, which decreases the level of confidence in the analysis. In this example, since we will be applying the loading, we will assume that there is no uncertainty associated with it. In general, however, characterizing the present and future loading of a structure can be a difficult task. A common approach is to use data recorded from operational and environmental sensors and, from that data, to develop models that are able to predict the future loading on the system.

Once the list of random variables is reduced to the most significant, a distribution form (PDF) for the uncertainty of the parameters must be chosen. This information comes from the experiments discussed above, or can be chosen based on what is accepted in the literature for the variable of interest. For instance, Young's modulus is commonly thought to have a lognormal distribution. Using experimental data, the type of distribution is chosen through chi-square or Komogorov–Smirnov tests, which examine which distribution type results in the least amount of fit error [4].

Once the distributions are formed, all the variables are combined into a joint probability density function (JPDF), which defines the parameter space of the system. This can be done directly if all variables are normally distributed and independent, otherwise they must first be transformed so that they are no longer correlated.

10.2.2 Limit State

The failure state of a system for a given failure mode is represented by a function of the response known as the performance function g. This function can be used to divide the parameter space into regions where the structural conditions or operating environments lead to either safe operation, or to system failure. Figure 10.3 simplistically defines a limit state $g = 0$ as the plane that divides these two regions. In terms of capacity and demand, the limit state is expressed as $g = C - D$ such that negative values ($g < 0$) of the function g denote the failure or unsafe operating mode of the system:

$$g = 0 \quad \text{Limit State}$$
$$g < 0 \quad \text{Failure}$$
$$g > 0 \quad \text{Safe}$$

A significant challenge arises in expressing the performance function g in terms of a mathematical equation or set of equations. For our example, the performance function is formed by rewriting the failure mode (equation 10.4) such that when it is less than zero, failure will occur:

$$g = a_C - a_N \tag{10.5}$$

As shown by our example, this formulation is generally straightforward when failure is defined through a physical criterion, such as a critical delamination size, or perhaps a critical speed of an aircraft that could lead to the occurrence of wing flutter. When the

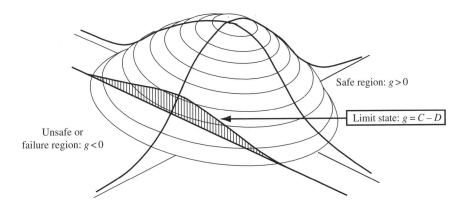

Figure 10.3 Joint probability density function of demand (D) and capacity (C) and limit state $g = C - D$ that separates the safe region from the failure region

definition of the limit state involves nonphysical criteria, subjective performance evaluations or linguistic ambiguities, its translation into a set of equations that can be implemented in a reliability code can be more challenging. Clearly the derivation of a limit state or multiple limit states is application specific.

Once the limit state has been defined, reliability analysis consists of estimating the probability of failure, that is, the probability that demand exceeds the system's capacity. For our example, we want to determine the probability that the delamination size has exceeded the critical value after N load cycles. This probability is calculated based on the distribution of a_N, whose variability is due to the uncertainty of the system parameters.

10.2.3 Probability of Failure

The probability of failure $P_{Failure}$ is defined as the integral of the joint probability density function (JPDF) over the failure or unsafe region. As mentioned previously, the failure region is mathematically defined as $g < 0$, and is the region bounded by the limit state $g = 0$ as shown in Figure 10.3. For our example, where failure is defined in terms of delamination size, reliability analysis consists of estimating the probability of reaching the critical delamination size given uncertainties about the predictive model, current health of the system, and expected loading.

Reliability analysis requires the incorporation of the failure model into a finite-element model or other model to build a relationship between system response and damage level. Interfacing the analysis model and the calculation of a limit state can be a significant computational and software integration challenge. For instance, our delamination example involves a fairly simple limit state, which states that failure occurs when the delamination size reaches a critical value (equation 10.5). However, determining the size of the delamination after a given amount of loading is not an easy task. A model must be created that represents the physics of the delamination damage, as well as how that damage will grow. Also, the composite, damage model, and loading could have uncertainty associated with them. Determining the distribution of the delamination size therefore requires incorporating all these different aspects of the problem together, which can be a difficult or at least very time-consuming task.

Once a model is formed, integration of the JPDF across all random variables for the failure region $g < 0$ of the parameter space gives the probability of failure $P_{Failure}$:

$$P_F = P\{g < 0\} \tag{10.6}$$

$$P_F = \int\limits_{g<0} f_X(x)\mathrm{d}x \tag{10.7}$$

where $f_X(x)$ is the JPDF of the response domain.

The calculation of the probability of failure can be difficult due to the complexity of integrating a function in a multiple-dimension space. Also, in most cases the number of random variables will be large. The integration can also be difficult because it involves computing rare events located in the tails of the statistical distributions. In addition, often closed form representations of the performance function g and the failure domain ($g < 0$) are not available for complex systems, therefore an approximation must be sought. For all these reasons, computational strategies based on a system model are commonly used to estimate the reliability. Three of these computational approaches are briefly discussed next.

10.3 APPROXIMATION OF THE PROBABILITY OF FAILURE

10.3.1 Monte Carlo Simulation

There are many different methods for estimating the integral in equation (10.7), and they can be categorized into three different types. The first that will be discussed is a sampling-based approach that consists of numerically integrating the probability of failure through statistical sampling. A series of simulations of the system model are run to try to find the percentage that falls within or outside the failure region of the response space. The simplest sampling method is a Monte Carlo simulation that chooses which simulations to run by randomly sampling the probability distributions of the input parameters.

A Monte Carlo simulation was performed for our composite delamination example and is represented in Figure 10.4. Samples are taken from two parameter distributions and used to evaluate the performance function, g. In most cases, each evaluation of g requires running a system model, such as a finite-element simulation. The probability of

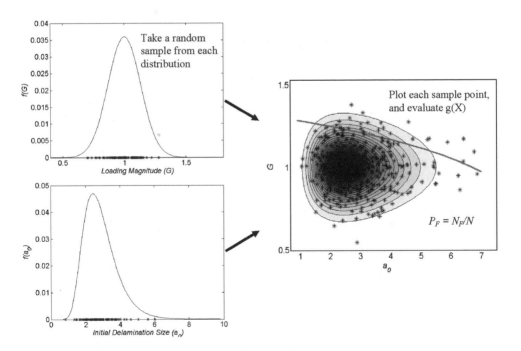

Figure 10.4 Monte Carlo approximation of probability of failure. Two parameter distributions are sampled to determine the value of $g(X)$

failure is determined by finding the ratio of the results that are negative (N_F), meaning failure has occurred, to the total number of evaluations (N):

$$P_F = \frac{N_F}{N} \tag{10.8}$$

A theoretical estimate of the error associated with using this approach is available, and can be taken advantage of to estimate convergence of the probability of failure [5]:

$$\%\text{error} = 200 * \sqrt{\frac{1 - P_F}{N * P_F}} \tag{10.9}$$

This equation is based on assuming a 95 % confidence interval.

The main limitation of Monte Carlo sampling is that it can require more simulation runs than is reasonable to perform, especially when the number of random variables is large (typically, more than ten) or when the probability of failure is low. Stratified sampling techniques such as the latin hypercube sampling (LHS) and fractional factorial designs of experiments can provide a trade-off between convergence of the numerical integration and the number of computational simulations. Another way of limiting the number of simulations is by intelligently choosing the computer runs. Adaptive sampling techniques are available that can sample simulations mainly around the limit state and failure domain, hence, leading to an accelerated rate of convergence compared to purely random sampling [6].

10.3.2   Analytical Approximation of the Limit State

The second strategy for estimating the probability of failure focuses on simplifying the integral in equation (10.7) by approximating the limit state using a Taylor series expansion. There are a number of methods available for forming this approximation, and most can be divided into two categories: those that approximate the limit state about the mean values of the random variables, and those that approximate the limit state about the system's *most probable point* (MPP). The limit state is usually located near the boundary of the parameter space, which makes approximation about the mean values not a very accurate approach. For that reason, the MPP approach will be focused on here.

The system's MPP is the location on the limit state $g = 0$ that is closest to the origin, if the limit state is transformed to standard normal space. In standard normal space, the JPDF is centered about the origin and decays as it progresses out from the origin, so the closest location to the origin on the limit state is the 'most probable' point. From this point, an approximation of the probability of failure can be determined. This calculation is illustrated in Figure 10.5. First, the random variables X of the simulation are transformed into an uncorrelated, standard normal space described by the random variables u. The standard normal multivariate JPDF of normalized variables u is denoted by Φ. Second, the MPP is estimated by solving a minimum distance optimization problem defined as:

$$\begin{aligned}
\text{Minimize}: \ & D = \sqrt{X^T X} \\
\text{Constrained by}: \ & g(X) = 0
\end{aligned} \tag{10.10}$$

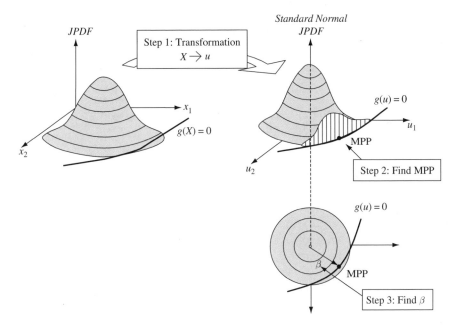

Figure 10.5   Concept of fast probability integration for reliability analysis

Third, the distance β between the MPP and the origin is calculated. Finally the probability of failure is estimated by finding the standard normal cumulative distribution value at β, that is, $P_{Failure} = \Phi(-\beta)$.

Expansion-based approximations such as the first order reliability method (FORM), second order reliability method (SORM), and advanced mean value (AMV) follow the principle of these calculations, and are collectively referred to as fast probability integration (FPI) methods. FPI methods differ by the order of the polynomial expansion used in the neighborhood of the MPP and other details [7]. Generally, the limit state $g = 0$ is approximated using either a first-order or second-order Taylor series expansion about the MPP (FORM and SORM). Expansions about the mean values of the input random variables, rather than the MPP, are also possible with the AMV method. In this context, runs of the simulation code are performed to estimate the derivatives needed for the Taylor series expansion, rather than sampling the probability distributions. This explains the significant computational savings that can be achieved compared to sampling-based approaches. However, convergence error estimates, such as the one provided in equation (10.9) for a Monte Carlo simulation, are no longer available.

The computational savings of expansion-based FPI methods are provided at the expense of formal proofs of convergence. Another potentially devastating limitation is the accuracy with which linear and quadratic polynomials can approximate nonlinear limit state functions. A solution that mandates further research and development could be to provide more flexibility in the way implicit limit state functions (such as those generally provided by finite-element simulations) are approximated. Libraries of response-surface methods, fractional functions, exponential decays, neural networks, and radial-basis functions, for example, could be implemented to augment the limited capability of first-order and second-order polynomial approximations.

Finally the value of hybrid methods for reliability analysis is recognized. Hybrid methods refer to the integration between sampling-based and expansion-based approaches to concentrate the effort of statistical sampling in areas of the reliability domain where it is most needed, that is, along the limit state. More information about the limit state, in turn, makes it possible to fit with better accuracy the polynomial models that approximate the true-but-unknown limit state function.

10.3.3 *Metamodeling for information integration*

Approximations implemented to reduce the cost of a reliability assessment do not, however, reduce the computational burden of a single numerical simulation. When the prognosis of damage involves the analysis of large finite-element models or other physics-based models, the number of runs that can be executed in a reasonable time might be limited. It may then be advantageous to spend the available computational resources on developing fast-running surrogates to the computational mechanics simulation, rather than attempting to estimate directly the reliability of the system. Reliability assessments based on fast-running meta-models then become computationally efficient.

Meta-models commonly take the form of polynomial response surfaces as illustrated in Figure 10.6. Other functional forms, such as fractional models, exponential decays, neural networks, statistical-based models, and radial basis functions, to name only a few, can be considered depending on the type of input–output relationship considered. The design of experiment techniques discussed in previous chapters can greatly improve the efficacy of the training step by choosing intelligently a subset of simulation runs,

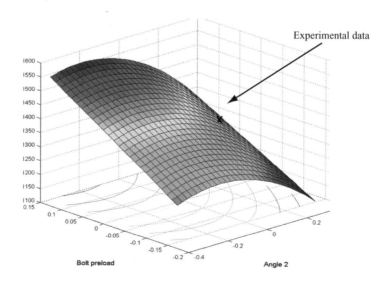

Figure 10.6 Response surface developed as a surrogate to a finite element analysis

depending on the type of meta-model trained and the level of fidelity required. Meta-models make it feasible to perform sampling-based reliability, thereby guaranteeing a converged estimate of the probability of failure. The accuracy then depends on how well the meta-models estimate the response of the computational model at parameter values not used for training. Fortunately, validation criteria and goodness-of-fit indicators are available to control the quality of training and level of fidelity. The adequacy of meta-models can be assessed prior to their deployment on the damage prognosis nodes and their integration with the sensing system.

Because reliability analysis is applicable to many engineering problems, general-purpose software is being developed to perform sampling-based and expansion-based sensitivity, uncertainty quantification, and reliability assessments. The software packages NESSUS [6] and DAKOTA [8] are two examples that can be interfaced with a variety of general-purpose, finite-element, packages. These and other uncertainty quantification, reliability, and numerical optimization packages also include limited capabilities for meta-modeling that can be taken advantage of.

10.4 DECISION MAKING

After the probability of failure is estimated, decision making is used to answer such questions as 'is the probability of failure acceptable?' or 'which scenario, configuration of the system or operating condition leads to an acceptable probability of failure?' Decision-making relies on the estimation of reliability, as well as a quantification of its confidence, to decide which course of action should be taken. Many times the end goal is to estimate the remaining useful life of the system, which is based on an understanding of how the reliability of the system will be changing in time.

For our composite plate example, the initial delamination in the plate will grow as cyclic loading is applied, resulting in a decrease in the effective reliability of the plate. To determine how the reliability is changing with time, the limit state must be reformed after each loading cycle, and the probability of critical delamination size exceedance assessed using a computational method such as FORM. As shown in Figure 10.7, the delamination size at a given point in time will actually be a distribution due to the uncertainty in the system parameters, and FORM determines what percentage of the distribution lies above the critical value, a_C. The resulting time-based reliability curve is shown in Figure 10.8.

From the time-based reliability plot, the remaining useful life of the system can be determined. 'Remaining useful life' is defined as the time remaining until the reliability drops below a defined lower operating threshold. This will be important to the operation of a system in use. If an operator is only given information about the level of damage that has been induced, he/she will not know exactly when the system will fail. Providing this predictive information will allow the operator to know the appropriate time that a system needs to be taken out of service to be inspected.

More information about the damage state can be gained if active monitoring of the structure is being performed. This information can then be used to update the time-based reliability curve. A sensing system could determine the size of the delamination at a given point in time. However, this will not be an exact evaluation of the damage size,

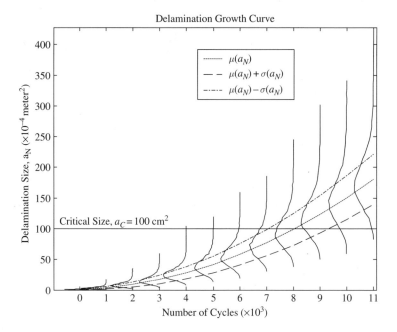

Figure 10.7 Delamination growth curve

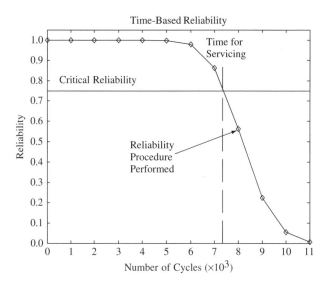

Figure 10.8 Time-based prediction of reliability

as it will have uncertainty associated with it just as the finite element model will have. The difference here is that the uncertainty is not based on a lack of knowledge of the system parameters, but rather the inability to obtain a precise measurement of the delamination size. The difficulty comes in understanding how to combine the uncertainty in the measurement with the uncertainty in the predicted delamination size based on the simulations. One approach would be to use Bayesian statistical methods to combine the different types of uncertainty [9]. In the end, the ability actively to diagnose a system along with a general understanding of how that damage might grow in time can be a valuable asset in determining the probability of completing a mission, or the need to adopt a new flight profile.

While this estimate is of high value in systems where damage accumulates gradually and at predictable rates, it is of less value in more extreme conditions such as an aircraft in combat, where the users of the system (the pilot and mission commander) would prefer to know the probability of completing the current mission, given the current assessment of the damage state. Because predictive models typically have more uncertainty associated with them when the structure responds in a nonlinear manner, as will often be the case when damage accumulates, an alternate goal might be to estimate how long the system can continue to safely perform in its anticipated environments before one no longer has confidence in the predictive capabilities of the models that are being used to perform the prognosis.

10.5 SUMMARY

Reliability methods provide a framework for predicting the remaining useful life of a system with quantified uncertainty, and are an integral part of the damage prognosis process. This chapter has presented the methodologies used, and issues related to performing a structural reliability analysis. These procedures were demonstrated through an example problem that sought to determine the life expectancy of a composite structure that is subject to delamination damage resulting from a foreign body impact.

Several issues were shown to be of importance when performing a reliability analysis. Foremost is the problem of uncertainty, including the issues of identifying all areas where uncertainty may occur, accurately modeling the levels of uncertainty, and understanding what types of model are best for describing the uncertainty. A major concern with this last issue is how to incorporate the nonparametric sources of uncertainty that are likely to influence the reliability assessment. Examples are the lack of knowledge associated with the functional form of the models, interactions between damage modes, or the field of residual stresses. Such sources of uncertainty may not be amenable to a probabilistic treatment in which case other theories to represent the uncertainty must be brought to bear. In the end, a clear understanding of the underlying uncertainty in an analysis must be had in order to have any confidence in the conclusions drawn from it.

Another important issue in performing a reliability analysis is the integration of the parameter space over the region where failure will occur. The term *reliability method* is commonly used to describe the computational approaches used to approximate this integral. Central to this problem is the developing of a mathematical representation of the mode of failure, which incorporates the uncertain parameters that contribute to the response of the system. For many complex systems, a finite-element model is used to estimate the system response, and no explicit form of the performance function is available. This chapter showed three approaches for approximating the probability of failure in this situation, using either a linearized approximation of the limit state, a sampling based approach, or a meta-model. Meta-models are becoming an important resource in many areas of diagnosis and prognosis, and can allow for an easier integration of the individual components in an analysis.

Reliability analysis then plays the important role of being able to answer the pertinent question of 'what is the remaining useful life of the system.' This is done through the assessment of the current reliability of the system as well as prediction of how that reliability will change in time, which is based on the prediction of future loading conditions of the structure. With this knowledge, decisions can be made as to how to proceed in the presence of damage, for instance, 'is there enough time to complete a flight mission?' or 'does one need to change the mission profile?'

REFERENCES

[1] Kruger, R. and Konig, M. (1996) 'Investigation of Delamination Growth Between Plies of Dissimilar Orientations', *ISD-Report 96/5*, November.

[2] Ang, A.H. and Tang, W.H. (1984) *Probability Concepts in Engineering Planning and Design; Volume II: Decision, Risk, and Reliability*, John Wiley & Sons, Inc., New York.

[3] Myers, R.H. and Montgomery, D.C. (1995) *Response Surface Methodology: Process and Product Optimization Using Designed Experiments*, Wiley Interscience, New York.

[4] Haldar, A. and Mahadevan, S. (2000) *Probability, Reliability, and Statistical Methods in Engineering Design*, John Wiley & Sons, Inc., New York.

[5] Long, M.W. and Narciso, J.D. (1997) *Probabilistic Design Methodology for Composite Aircraft Structures*, US Department of Transportation, DOT/FAA/AR-99/2, October.

[6] *NESSUS User's Manual*, Version 2.4, Southwest Research Institute, San Antonio, TX, 1998.

[7] Wu, Y.T. (1990) 'Advanced probabilistic structural analysis method for implicit performance functions', *AIAA Journal*, **28**, 1663–1669.

[8] Eldred, M.S., Giunta, A.A., Van Bloemen Waanders, B.G., Wojtkiewicz, S.F., Hart W.E. and Alleva M.P. (2002) *DAKOTA User's Manual, A Multilevel Parallel Object-Oriented Framework for Design Optimization, Parameter Estimation, Uncertainty Quantification, and Sensitivity Analysis*, Version 3.0, Report SAND-2001–3796, Sandia National Laboratories, Albuquerque, NM, April 2002.

[9] Hanson, K.M. (1999) 'A framework for assessing uncertainties in simulation predictions', *Physica D*, **133**, 179–188.

11

Lamb-Wave Based Structural Health Monitoring

Ajay Raghavan and Carlos E.S. Cesnik[1]

Department of Aerospace Engineering, The University of Michigan, USA

11.1 INTRODUCTION

A structural health monitoring (SHM) system typically consists of an onboard network of sensors for data acquisition and some central processor employing an algorithm to evaluate structural health. The system utilizes stored knowledge of structural materials, operational parameters, and health criteria. The schemes available for SHM can be broadly classified as:

(a) passive schemes, i.e. schemes that do not involve the use of actuators;
(b) active schemes, i.e. schemes that do involve the use of actuators.

The most significant passive schemes are acoustic emission (AE) and strain/loads monitoring, which have been demonstrated with some success [1–8]. However, they suffer the drawback of requiring high sensor density on the structure. They are typically implemented using fiber-optic sensors and, for more benign environments, foil strain gages.

Foremost among active schemes are Lamb-wave testing methods. Unlike passive methods, active schemes are capable of 'checks-on-demand' within seconds. In particular, Lamb-wave testing can offer a reliable method of estimating information about

[1] Corresponding author. E-mail: cesnik@umich.edu; Phone +1 (734) 764-3397

Damage Prognosis – For Aerospace, Civil and Mechanical Systems Edited by D.J. Inman, C.R. Farrar, V. Lopes Junior and V. Steffen Junior © 2005 John Wiley & Sons, Ltd

damage in a structure in terms of location, severity and type of damage. They have been used in the nondestructive evaluation and testing (NDE/NDT) industry for two decades. There, Lamb waves are excited and received in a structure using handheld transducers for scheduled maintenance. They have also demonstrated suitability for SHM applications having an onboard, preferably built-in, sensor and actuator network to assess the state of a structure during operation. The advantage of Lamb-wave testing over conventional structural health monitoring approaches, such as AE and strain-loads monitoring, is that an actuator–sensor pair has a large coverage area, implying that the density of sensors and actuators can be coarse (typically less than 10 sensor–actuator pairs per square meter). In general, Lamb waves are excited in structures using surface-bonded/embedded piezoelectric (hereafter referred to as 'piezo') actuators. For sensing, piezos or other high-frequency strain devices can be used.

Elastic waves in solids have been a subject of much study [9–12] owing to their wide-ranging applications such as in seismology, inspection, material characterization, delay lines, etc. Lamb waves are a class of elastic perturbation that can propagate in a solid plate (or shell) with free surfaces, for which displacements occur both in the direction of wave propagation and normal to the plane of the plate (or shell). There are other classes of elastic waves that have been examined in detail. One such class is that of Rayleigh waves [13], which propagate close to the free surface of elastic solids. Other examples of such classes are Love [14], Stoneley [15] and Scholte [16] waves that travel at material interfaces. Lamb waves were first predicted mathematically and described by Horace Lamb [17]; however he was unable to produce them experimentally. Worlton [18] was probably the first person to recognize the potential of Lamb waves for NDE. Lamb-wave generation using conventional handheld transducers is now well understood and commonly used in NDE, where the transducers operate by 'tapping' the surface or generating normal stresses on the surface. Analytical formulations giving fairly good models for the resultant Lamb wave generation are covered in Reference [19], which is a standard text in this field. Lamb-wave excitation using surface-bonded/embedded piezos for SHM is a relatively nascent field. In the use of piezos for SHM, Lamb waves are excited when the piezo actuator 'pinches' the surface (causes in-plane strains). Piezos are inexpensive devices that can be surface mounted for SHM on existing plate or shell-type structures, inserted between the layers of lap joints, or inside composite materials. Such configurations are found abundantly in aerospace structures and other mechanical systems, which become potential application areas for this technology.

The conventional Lamb-wave diagnostic methods are the pulse–echo method and the pitch–catch method. In either approach, the actuator generating Lamb waves is excited by a pulse signal (typically a sinusoidal toneburst of some limited number of cycles). In the former, after exciting the structure with a pulse, the actuator is also used as a sensor to 'listen' for echoes of the pulse coming from discontinuities such as damage areas and structural boundaries. Since the boundaries are known and the wave speed for a given center actuation frequency of the toneburst is known, the signals from the boundaries can be filtered out, and signals from the defects remain. From these signals, the defects can be located. In the pitch–catch approach, a pulse signal is sent across the specimen under interrogation and a sensor at the other end of the specimen receives the signal. From various characteristics of the received signal, such as delay in time of transit, amplitude, frequency content, and by using certain signal processing techniques such as wavelet transforms, pattern recognition and neural networks, information about the damage can be obtained.

Lamb-wave testing can become very complex due to the dispersive nature of these waves. Hence a fundamental understanding of wave propagation fundamentals and the theory of Lamb waves is essential for the successful application of this method. The first section that follows presents these concepts. It is followed by a detailed discussion of how these basic wave propagation concepts can be used to analyze Lamb-wave excitation and sensing using surface-bonded piezos.

11.2 FUNDAMENTALS OF ELASTIC WAVE PROPAGATION

11.2.1 Isotropic Llinear Elastic 3-D Solid

The general 3-D equilibrium equations of motion for elastic solids are:

$$\nabla \cdot \boldsymbol{\sigma} + \boldsymbol{f} = \rho \ddot{\boldsymbol{u}} \tag{11.1}$$

or equivalently,

$$\sigma_{ij,j} + f_i = \rho \ddot{u}_i \tag{11.2}$$

where indices i, j, k can assume values 1,2,3, $\boldsymbol{\sigma}$ is the stress tensor, ρ is the density of the solid, $\boldsymbol{f}$ is the body force per unit volume and $\ddot{\boldsymbol{u}}$ is the particle acceleration at the point.

For simplicity, consider the case of a linear elastic isotropic medium. For that, the constitutive relations are given by:

$$\sigma_{ij} = \lambda \varepsilon_{kk} \delta_{ij} + 2\mu \varepsilon_{ij} \tag{11.3}$$

where λ and μ are Lamé's constants for the isotropic medium, δ_{ij} is the Kronecker delta, and ε_{ij} are the components of the strain tensor.

To complete the set of elasticity equations, the kinematical relations for the case of geometrically linear deformations are given by:

$$\varepsilon_{ij} = \frac{1}{2}(u_{i,j} + u_{j,i}) \tag{11.4}$$

Combining these three sets of equations, one can reach the Navier's equation for a 3-D isotropic body:

$$(\lambda + \mu)\nabla\nabla \cdot \boldsymbol{u} + \mu\nabla^2\boldsymbol{u} + f = \rho\ddot{\boldsymbol{u}} \tag{11.5}$$

11.2.2 1-D Wave Propagation

To understand some basic concepts on wave propagation, consider the 1-D wave propagation problem along the x_1 direction. Therefore,

$$\frac{\partial}{\partial x_2} = \frac{\partial}{\partial x_3} = 0; \quad u_2 = u_3 = 0 \tag{11.6}$$

This reduces the above equations to:

$$(\lambda + 2\mu)u_{1,11} + f_1 = \rho\ddot{u}_1 \tag{11.7}$$

or

$$\frac{\partial^2 u_1}{\partial x_1^2} - \frac{1}{c^2}\frac{\partial^2 u_1}{\partial t^2} = -f_1 \tag{11.8}$$

where

$$c^2 = \frac{\lambda + 2\mu}{\rho} \tag{11.9}$$

which represents the square of the propagation speed.

Further, consider the free vibration problem with zero body force, $f_1 = 0$. The problem of waves in a taut string serves as a good model problem for 1-D wave propagation, as illustrated in Figure 11.1. Also for the rest of the 1-D analysis, x_1 is replaced by x for convenience.

Consider the following assumptions:

(a) The restoring force in the string is due to the tension T.
(b) The deflections w are small and so also are the angles of the string with respect to the horizontal at various points.
(c) T is much bigger than the dynamic strain fluctuations.

From the equilibrium of an infinitesimal piece of the string, one obtains:

$$T\frac{\partial^2 w}{\partial x^2} + q = \rho_L\frac{\partial^2 w}{\partial t^2} \tag{11.10}$$

For the free vibrations case, $q = 0$ and equation (11.10) simplifies to:

$$\frac{\partial^2 w}{\partial x^2} - \frac{1}{c^2}\frac{\partial^2 w}{\partial t^2} = 0 \tag{11.11}$$

where:

$$c^2 = \frac{T}{\rho_L} \tag{11.12}$$

which is mathematically equivalent to equation (11.8) derived above.

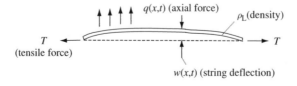

Figure 11.1   Representation of the 1-D wave problem in a string (deflections exaggerated)

11.2.3 D'Alembert's Solution for an Infinite String

One can show that:

$$w(x, t) = f(x - ct) + g(x + ct) \tag{11.13}$$

is a solution of equation (11.11), where f and g are arbitrary functions. This can be simply done by direct substitution:

$$\frac{\partial^2 f(x - ct)}{\partial x^2} = f'' \tag{11.14}$$

and

$$\frac{\partial^2 f(x - ct)}{\partial t^2} = c^2 f'' \tag{11.15}$$

$$\therefore \text{LHS} = \frac{1}{c^2}(c^2 f'' + c^2 g'') = f'' + g'' = \text{RHS} \tag{11.16}$$

This solution for the 1-D wave equation, known as D'Alembert's solution, can thus be used to solve initial value problems on infinite strings. For example, consider the following initial conditions:

$$w(x, 0) = U(x) \tag{11.17}$$
$$\dot{w}(x, 0) = V(x) \tag{11.18}$$

Inserting these conditions into D'Alembert's solution, equation (11.13) yields:

$$w(x, t)|_{t=0} = f(x) + g(x) = U(x) \tag{11.19}$$

$$\dot{w}(x, t)|_{t=0} = c(-f'(x) + g'(x)) = V(x) \tag{11.20}$$

These two equations yield:

$$w(x, t) = \frac{1}{2}[U(x - ct) + U(x + ct)] + \frac{1}{2c} \int\limits_{x-ct}^{x+ct} V(s)\mathrm{d}s \tag{11.21}$$

To aid visualization, consider the simple case $V(x) = 0$; $U(x) = H(x + a) - H(x - a)$, where $H(\ )$ is the Heaviside function. This case is illustrated in Figure 11.2.

11.2.4 Reflection from a Fixed Boundary

Consider a string fixed at one end and a wave pulse $f(x - ct)$ approaching the fixed end from $-\infty$. Then one can solve for the reflected wave from the fixed end at $x = 0$, which imposes the condition $w(0, t) = 0$, by creating an imaginary wave in the $+x$ axis. For this, as illustrated in Figure 11.3, imagine the string to be extended symmetrically to $+\infty$

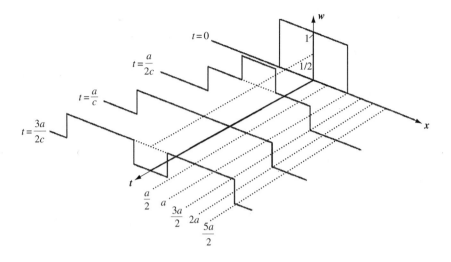

Figure 11.2   Simple example of transient wave propagation in a 1-D string

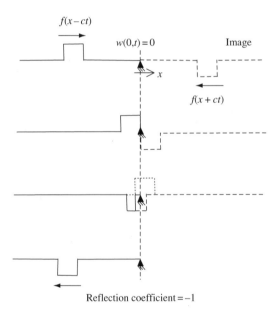

Figure 11.3   Reflection of a wave pulse in a string from a fixed boundary

and further consider a negative (of equal magnitude to the original pulse, but opposite
in sign) wave pulse approaching the fixed end simultaneous with the original pulse. As
the two waves approach the fixed end, their superposition will ensure the satisfaction of
the fixed boundary condition at that end. Since the superposition of the two waves is a
feasible solution, by the uniqueness condition, it is indeed the solution. This is
illustrated in Figure 11.3. As shown in Figure 11.3, for a fixed end, the incident wave

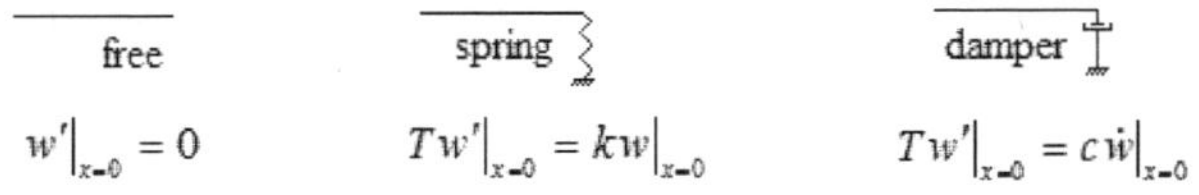

free
spring
damper
$w'|_{x=0} = 0$
$Tw'|_{x=0} = kw|_{x=0}$
$Tw'|_{x=0} = c\dot{w}|_{x=0}$

Figure 11.4 Other boundary conditions for the string problem

packet simply bounces off with its sign reversed. Other boundary conditions are described in Figure 11.4.

11.2.5 Reflection and Build-Up of Standing Waves in Finite Domains

Consider a finite string of length L, with one end fixed and the other being harmonically excited, as shown in Figure 11.5. Thus, its boundary conditions are:

$$\begin{aligned} w(0,t) &= w_0 e^{-i\omega t} \\ w(L,t) &= 0 \end{aligned} \tag{11.22}$$

Before the rightward wave arrives at the fixed end, it propagates as if it were in an infinite string. Therefore, for the transient wave, D'Alembert's solution for an infinite string applies for this case for $t < \frac{L}{c}$.

The reflection coefficients γ_s at $x = 0$ and γ_e at $x = L$ are defined as follows:

$$\gamma_s = \left. \frac{w_{reflected}}{w_{incident}} \right|_{x=0} \tag{11.23}$$

and

$$\gamma_e = \left. \frac{w_{reflected}}{w_{incident}} \right|_{x=L} \tag{11.24}$$

Since the boundary at $x = L$ is fixed, as seen in Section 11.2.4.

$$\gamma_e = -1 \tag{11.25}$$

Since only rightward propagating waves are sought,

$$w(x,t) = f(x - ct) \tag{11.26}$$

Now using the boundary condition at $x = 0$:

$$w(0,t) = f(-ct) = w_0 e^{-i\omega t} \tag{11.27}$$

String
$w(0,t) = w_0 e^{-iwt}$
x
L

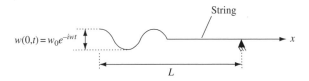

Figure 11.5 Build up of waves in a finite string

One can identify the unknown function f as:

$$f() = w_0 e^{\frac{i\omega}{c}()} \tag{11.28}$$

and, therefore, until $ct < L$,

$$w(x, t) = w_0 e^{i(\gamma_0 x - \omega t)}, \quad \gamma_0 = \frac{\omega}{c} \quad \text{for } x < ct$$
$$= 0 \qquad\qquad\qquad \text{for } x > ct \tag{11.29}$$

The symbols ω, γ_0, and c introduced here are commonly encountered when analyzing harmonic waves in a medium. ω is called the angular frequency, and is related to the frequency of harmonic oscillations f by the relation:

$$\omega = 2\pi f \tag{11.30}$$

γ_0 is the wavenumber (this is also sometimes denoted by ξ) and is related to the spatial wavelength λ_w of the wave by the following relation:

$$\gamma_0 = \frac{2\pi}{\lambda_w} \tag{11.31}$$

As can be seen from (11.31), the wavenumber is proportional to the number of wavelengths per unit distance. c (also denoted by c_p) is called the phase velocity and is the speed at which the harmonic wave propagates through the medium. It is related to ω and γ_0 by the equation:

$$\frac{\omega}{c} = \gamma_0 \tag{11.32}$$

When the disturbance reaches the fixed end at $x = L$ (at $t = \frac{L}{c}$) for the first time, the solution must be augmented by a leftward wave according to what was described in Section 11.2.4. Therefore, the total solution for $t > \frac{L}{c}$ is:

$$w^1_{total} = \left(w_0 e^{i\gamma_0 x} + w_0 \gamma_e e^{i\gamma_0 L} e^{-i\gamma_0(x-L)} \right) e^{-i\omega t} \tag{11.33}$$

$$\underbrace{\qquad\qquad}_{\text{Rightward wave}} \qquad\qquad \underbrace{\qquad\qquad}_{\text{Leftward wave}}$$

The first factor $e^{i\gamma_0 L}$ of the new second term appearing in the expression for w is needed to cancel out the first term (i.e. maintain $w = 0$ at $x = L$ as imposed by the boundary condition). The second factor of the second term $e^{-i\gamma_0(x-L)}$ is the resultant leftward wave that has just started.

Once the wave travels all the way back to the left (length L), it will arrive at $x = 0$ at $t = \frac{2L}{c}$. Again, in order to satisfy the boundary condition at this point, yet another wave is introduced, so that the new resultant wave is:

$$w^2_{total} = w^1_{total} + \gamma_s \gamma_e w_0 e^{2i\gamma_0 L} e^{i(\gamma_0 x - \omega t)} \tag{11.34}$$

This goes on and each time the wave hits one of the string boundaries, a new term is added. The terms of this expression are in geometric progression, and can be written as:

$$w(x, t) = w_0(e^{i\gamma_0 x} + \gamma_e e^{i\gamma_0(2L-x)} + \gamma_s \gamma_e e^{i\gamma_0(2L+x)}$$
$$+ \gamma_s \gamma_e^2 e^{i\gamma_0(4L-x)} + \gamma_s^2 \gamma_e^2 e^{i\gamma_0(4L+x)} + \ldots)e^{-i\omega t} \quad (11.35)$$

For the steady-state solution, the sum of an infinite number of these terms is needed, and this can be solved using the formula for geometric series, yielding:

$$w(x, t) = w_0 \left[\frac{e^{i\gamma_0 x} + \gamma_e e^{i\gamma_0(2L-x)}}{1 - \gamma_e \gamma_s e^{2i\gamma_0 L}} \right] e^{-i\omega t} \quad (11.36)$$

Note that the same *steady-state* solution could have been reached by a much simpler approach. Assume a solution of the form:

$$w(x, t) = (Ae^{i\gamma_0 x} + Be^{-i\gamma_0 x})e^{-i\omega t} \quad (11.37)$$

and impose $w(0, t) = w_0 e^{-i\omega t}, w(L, t) = 0$ to find the constants A and B. However the earlier approach was adopted to provide an insight into how transients build up into the steady-state solution. A key lesson to learn from the above analysis is that *the solution for the infinite string holds for the finite string for analyzing the transient waves*. This will eventually prove useful in the analysis for Lamb-wave testing, where essentially transient waves are used, particularly the transmitted pulse and first few echoes. Hence an infinite plate wave solution can be used as part of the transient wave analysis in finite plates.

11.2.6 The Helmholtz Decomposition for Isotropic Media

Consider a decomposition for the displacement vector $\boldsymbol{u}$ into a scalar potential ϕ and a vector $\mathbf{H}$. The physical meaning of these will become evident in due course. This decomposition eases the solution of the problem of wave propagation in a 3-D elastic linear isotropic medium. Thus, let the displacement vector be expressed as follows:

$$\boldsymbol{u} = \nabla\phi + \nabla \times \mathbf{H} \quad (11.38)$$

Furthermore, for uniqueness of the solution, impose:

$$\nabla \cdot \mathbf{H} = 0 \quad (11.39)$$

Such decomposition is always possible and unique, since there are four equations and four unknowns (ϕ and the three components of $\mathbf{H}$) for a given displacement field. Again recall Navier's elastodynamic equation of motion for zero body force, equation (11.5):

$$(\lambda + \mu)\nabla\nabla \cdot \boldsymbol{u} + \mu\nabla^2\boldsymbol{u} = \rho\ddot{\boldsymbol{u}} \quad (11.40)$$

Using equations (11.38) and (11.39) into (11.40) yields:

$$(\lambda + \mu)\nabla[\nabla \cdot (\nabla\phi + \nabla \times \mathbf{H})] + \mu\nabla^2(\nabla\phi + \nabla \times \mathbf{H}) = \rho(\nabla\ddot{\phi} + \nabla \times \ddot{\mathbf{H}}) \quad (11.41)$$

This can be rewritten as:

$$\nabla\{(\lambda + 2\mu)\nabla^2\phi - \rho\ddot{\phi}\} + \nabla \times \{\mu\nabla^2\mathbf{H} - \rho\ddot{\mathbf{H}}\} = 0 \quad (11.42)$$

Applying the operators $\nabla \cdot (\ )$ and $\nabla \times (\ )$ on equation (11.42), one gets:

$$(\lambda + 2\mu)\nabla^2\phi - \rho\ddot{\phi} = 0 \tag{11.43}$$

$$\mu\nabla^2\mathbf{H} - \rho\ddot{\mathbf{H}} = 0 \tag{11.44}$$

or

$$\nabla^2\phi = \frac{1}{c_1^2}\ddot{\phi} \tag{11.45}$$

and

$$\nabla^2\mathbf{H} = \frac{1}{c_2^2}\ddot{\mathbf{H}} \tag{11.46}$$

Therefore, the original elastodynamic vector equation (11.40) is decomposed into a scalar wave equation and a vector wave equation, (11.45) and (11.46) respectively.

11.2.7 Rayleigh–Lamb Wave Equation for an Infinite Plate

Consider an infinite plate of thickness $2b$, i.e. the domain $\Omega: \{(x_1, x_2, x_3) = (-\infty, \infty) \times (-b, b) \times (-\infty, \infty)\}$ with free surfaces (Figure 11.6). Waves of interest are in the x_1–x_2 plane, and there are no variations along x_3, i.e. $\frac{\partial}{\partial x_3} = 0$.

For this case, Helmholtz's decomposition yields:

$$u_1 = \frac{\partial\phi}{\partial x_1} + \frac{\partial\mathrm{H}_3}{\partial x_2} \tag{11.47}$$

$$u_2 = \frac{\partial\phi}{\partial x_2} - \frac{\partial\mathrm{H}_3}{\partial x_1} \tag{11.48}$$

$$u_3 = -\frac{\partial\mathrm{H}_1}{\partial x_2} + \frac{\partial\mathrm{H}_2}{\partial x_1} \tag{11.49}$$

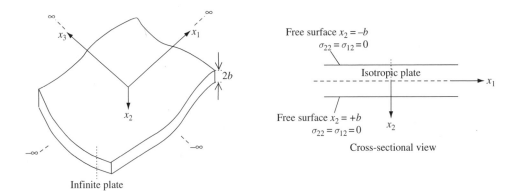

Figure 11.6 Infinite plate with free surfaces

$$\frac{\partial H_1}{\partial x_1} + \frac{\partial H_2}{\partial x_2} = 0 \tag{11.50}$$

Since the boundaries $x_2 = +b$ and $x_2 = -b$ are stress free, one has:

$$\sigma_{22} = 0; \sigma_{12} = \sigma_{21} = 0; \sigma_{23} = \sigma_{32} = 0 \text{ at } x_2 = \pm b \tag{11.51}$$

Using equations (11.3), (11.4) and (11.47) to (11.50), these stress components can be written as:

$$\sigma_{22} = (\lambda + 2\mu)\nabla^2\phi - 2\mu\left(\frac{\partial^2\phi}{\partial x_1^2} + \frac{\partial^2 H_3}{\partial x_1 \partial x_2}\right) \tag{11.52}$$

$$\sigma_{21} = \mu\left(2\frac{\partial^2\phi}{\partial x_1 \partial x_2} + \frac{\partial^2 H_3}{\partial x_2^2} - \frac{\partial^2 H_3}{\partial x_1^2}\right) \tag{11.53}$$

$$\sigma_{23} = \mu\left(-\frac{\partial^2 H_1}{\partial x_2^2} + \frac{\partial^2 H_2}{\partial x_1 \partial x_2}\right) \tag{11.54}$$

which will be eventually equated to zero at the plate boundaries to find the governing equations. It is interesting to note here that u_1, u_2, σ_{22}, and σ_{21} depend only on ϕ and H_3, while u_3 and σ_{23} depend only on H_1 and H_2. Thus, it suffices to consider the following two cases separately:

Case I: In this case $u_3 = 0$ and the governing equations are

$$\nabla^2\phi = \frac{1}{c_1^2}\ddot{\phi} \tag{11.55}$$

and

$$\nabla^2 H_3 = \frac{1}{c_2^2}\ddot{H}_3 \tag{11.56}$$

The surface conditions are:

$$\sigma_{22}|_{x_2=0} = 0 \tag{11.57}$$
$$\sigma_{12}|_{x_2=0} = 0 \tag{11.58}$$

Case II: In this case the governing equation is

$$\nabla^2 u_3 = \frac{1}{c_2^2}\ddot{u}_3 \tag{11.59}$$

The surface condition is:

$$\sigma_{23}|_{x_2=0} = 0 \tag{11.60}$$

As indicated by the previous equations, while equations (11.55) and (11.56) are uncoupled, their solutions are coupled through the surface traction-free conditions. As a result, the combination of waves governed by equations (11.55) and (11.56) in a plate (known as Lamb waves) is always 'dispersive', i.e. the wave velocity depends on the frequency of the traveling wave, whereas with the waves governed by equation (11.59) there exists a fundamental dispersionless mode at all frequencies.

To derive the equations for Lamb waves, consider first the governing equations in terms of the Helmholtz scalar and vector potentials, equations (11.55) and (11.56), respectively. Seeking plane wave solutions in the x_1–x_2 plane for waves propagating along the $+x_1$ direction, assume solutions of the form:

$$\phi = f(x_2)e^{i(\xi x_1 - \omega t)} \tag{11.61}$$

$$H_3 = h_3(x_2)e^{i(\xi x_1 - \omega t)} \tag{11.62}$$

Using equations (11.61) and (11.62) into equations (11.55) and (11.56), one gets:

$$\frac{d^2 f}{dx_2^2} + \left(\frac{\omega^2}{c_1^2} - \xi^2\right)f = 0 \tag{11.63}$$

$$\frac{d^2 h_3}{dx_2^2} + \left(\frac{\omega^2}{c_2^2} - \xi^2\right)h_3 = 0 \tag{11.64}$$

Let,

$$\frac{\omega^2}{c_1^2} - \xi^2 \equiv \alpha^2; \quad \frac{\omega^2}{c_2^2} - \xi^2 \equiv \beta^2 \tag{11.65}$$

The solutions to the differential equations (11.63) and (11.64) are:

$$f(x_2) = A \sin \alpha x_2 + B \cos \alpha x_2 \tag{11.66}$$

$$h_3(x_2) = C \sin \beta x_2 + D \cos \beta x_2 \tag{11.67}$$

Substituting these solutions in equations (11.52), (11.53), (11.61) and (11.62), and enforcing the conditions given by equations (11.57) and (11.58), two possible solutions result:

(i) *Symmetric modes*

$$\begin{bmatrix} -(\xi^2 - \beta^2)\cos \alpha b & 2i\xi\beta\cos\beta b \\ -2i\xi\alpha\sin\alpha b & (\xi^2 - \beta^2)\sin\beta b \end{bmatrix}\begin{bmatrix} B \\ C \end{bmatrix} = \begin{bmatrix} 0 \\ 0 \end{bmatrix} \tag{11.68}$$

Nontrivial solutions exist if and only if

$$\det\begin{bmatrix} -(\xi^2 - \beta^2)\cos \alpha b & 2i\xi\beta\cos\beta b \\ -2i\xi\alpha\sin\alpha b & (\xi^2 - \beta^2)\sin\beta b \end{bmatrix} = 0 \tag{11.69}$$

This gives

$$\frac{\tan \beta b}{\tan \alpha b} = \frac{-4\alpha\beta\xi^2}{(\xi^2 - \beta^2)^2} \tag{11.70}$$

(ii) *Antisymmetric modes*

$$\begin{bmatrix} -(\xi^2 - \beta^2)\sin\alpha b & -2i\xi\beta\sin\beta b \\ 2i\xi\alpha\cos\alpha b & (\xi^2 - \beta^2)\cos\beta b \end{bmatrix} \begin{bmatrix} A \\ D \end{bmatrix} = \begin{bmatrix} 0 \\ 0 \end{bmatrix} \tag{11.71}$$

Nontrivial solutions exist if and only if

$$\det \begin{bmatrix} -(\xi^2 - \beta^2)\sin\alpha b & -2i\xi\beta\sin\beta b \\ 2i\xi\alpha\cos\alpha b & (\xi^2 - \beta^2)\cos\beta b \end{bmatrix} = 0 \tag{11.72}$$

This gives

$$\frac{\tan\beta b}{\tan\alpha b} = \frac{-(\xi^2 - \beta^2)^2}{4\alpha\beta\xi^2} \tag{11.73}$$

Therefore, given a certain isotropic material, equations (11.70) and (11.73) can be solved numerically to find the relation between the driving frequency ω and the wavenumber ξ from which the corresponding phase velocity c_p can be found. This is plotted in Figure 11.7. below for an aluminum alloy (material properties used: Young's modulus $E = 70 \times 10^9$ N/m^2; Poisson's ratio $\nu = 0.33$; density $\rho = 2700$ kg/m^3).

The resulting displacement components are given by:

$$u_1 = \{i\xi(A\sin\alpha x_2 + B\cos\alpha x_2) + \beta(C\cos\beta x_2 - D\sin\beta x_2)\}e^{i(\xi x_1 - \omega t)}$$
$$u_2 = \{\alpha(A\cos\alpha x_2 - B\sin\alpha x_2) - i\xi(C\sin\beta x_2 + D\cos\beta x_2)\}e^{i(\xi x_1 - \omega t)} \tag{11.74}$$

For $A = D = 0$; $B, C \neq 0$, one has symmetric Lamb modes, i.e. symmetric u_1 about $x_2 = 0$ and antisymmetric u_2 about $x_2 = 0$.

For $A, D \neq 0$; $B = C = 0$, one has antisymmetric Lamb modes, i.e. antisymmetric u_1 about $x_2 = 0$ and symmetric u_2 about $x_2 = 0$.

11.2.8 Lamb-Wave Excitation Using Surface-Bonded Piezos

Piezos operate on the piezoelectric principle that couples the electrical and mechanical behavior of the material and can be modeled as:

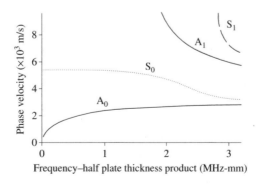

Figure 11.7 Phase velocity dispersion curve for an aluminum alloy

$$S_{ij} = s^E_{ijkl} T_{kl} + d_{kij} E_k$$
$$D_j = d_{jkl} T_{kl} + \varepsilon^T_{jk} E_k$$

(11.75)

where S_{ij} is the mechanical strain, T_{kl} is the mechanical stress, E_k is the electrical field, D_j is the electrical displacement, s^E_{ijkl} is the mechanical compliance of the material measured at constant (zero) electric field, ε^T_{jk} is the dielectric permittivity measured at constant (zero) mechanical stress, and d_{kij} represents the piezoelectric coupling effect. Typically for SHM, piezos are actuated in the d_{31} mode, i.e. voltage is applied in the 3-direction direction (or in the case of anisotropic piezocomposite actuators, in the d_{11} mode). When bonded onto a substrate and actuated, it tends to contract or expand, depending on the polarity of the applied electric field. As it deforms, a bending motion is induced in the substrate. As pointed out in Reference [20], piezos can be effectively modeled as causing line-shear forces along their edges.

In this section, the nature of Lamb waves excited by piezoelectric actuators causing line shear forces along their edges is analyzed. The objective is to derive the longitudinal normal strain and displacement equations first presented by Giurgiutiu [21]. Consider an infinite isotropic plate with similar coordinate system and geometry as in Section 11.2.7, with an infinitely long piezo actuator along the x_3-direction,[2] as illustrated in Figure 11.8. The situation where a piezo actuator is bonded only on one side ($x_2 = +b$) between $x_1 = +a$ and $x_1 = -a$ is examined. The governing equations for this problem are still equations (11.55) and (11.56), that is:

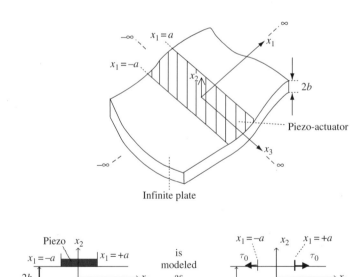

Figure 11.8   Infinite plate with infinitely wide surface-bonded piezo actuator

---

[2] So that the $\frac{\partial}{\partial x_3} = 0$ assumption used in the derivation of the Lamb waves equation is still valid.

$$\nabla^2\phi = \frac{1}{c_1^2}\ddot{\phi}$$

and

$$\nabla^2 H_3 = \frac{1}{c_2^2}\ddot{H}_3$$

However in this case the surface conditions are different. Since the piezo actuator is modeled as causing line shear forces along its edges, one has:

$$\sigma_{22} = 0, \quad x_2 = \pm b$$

$$\sigma_{12} = \tau(x_1)e^{-i\omega t}, \quad x_2 = +b \tag{11.76}$$

$$\sigma_{12} = 0, \quad x_2 = -b$$

where

$$\tau(x_1) = \tau_0[\delta(x_1 - a) - \delta(x_1 + a)] \tag{11.77}$$

$\delta(\ )$ being the Kronecker delta function, for ideal bonding between the piezo and the substrate. Thus the forcing function is introduced through the surface condition here. Also notice that sinusoidal excitation at frequency ω is considered, which is reflected in the $e^{-i\omega t}$ factor in the shear force at $x_2 = +b$. Since this is a forced vibration problem, general solutions are sought in the form:[3]

$$\phi = \int_{-\infty}^{\infty} (A(\xi)\sin\alpha x_2 + B(\xi)\cos\alpha x_2)e^{i(\xi x_1 - \omega t)}\,d\xi \tag{11.78}$$

$$H_3 = \int_{-\infty}^{\infty} (C(\xi)\sin\beta x_2 + D(\xi)\cos\beta x_2)e^{i(\xi x_1 - \omega t)}\,d\xi \tag{11.79}$$

The displacements are then given by

$$u_1 = \int_{-\infty}^{\infty} \{i\xi(A(\xi)\sin\alpha x_2 + B(\xi)\cos\alpha x_2) + \beta(C(\xi)\cos\beta x_2$$

$$- D(\xi)\sin\beta x_2)\}e^{i(\xi x_1 - \omega t)}\,d\xi \tag{11.80}$$

$$u_2 = \int_{-\infty}^{\infty} \{\alpha(A(\xi)\cos\alpha x_2 - B(\xi)\sin\alpha x_2) - i\xi(C(\xi)\sin\beta x_2$$

$$+ D(\xi)\cos\beta x_2)\}e^{i(\xi x_1 - \omega t)}\,d\xi \tag{11.81}$$

---

[3] Eventually it will be seen that only terms with wavenumbers corresponding to the natural modeshapes contribute to the response.

Using the stress-displacement relations established by equations (11.52) and (11.53), along with the surface stress conditions from equation (11.76), one gets:

$$\int_{-\infty}^{\infty} \{(\lambda + 2\mu)\alpha^2 + \lambda\xi^2\}(A(\xi)\sin\alpha b + B(\xi)\cos\alpha b) + 2i\mu\xi\beta(C(\xi)\cos\beta b$$
$$- D(\xi)\sin\beta b)e^{i(\xi x_1 - \omega t)}d\xi = 0 \qquad (11.82)$$

$$\int_{-\infty}^{\infty} \{(\lambda + 2\mu)\alpha^2 + \lambda\xi^2\}(-A(\xi)\sin\alpha b + B(\xi)\cos\alpha b) + 2i\mu\xi\beta(C(\xi)\cos\beta b$$
$$+ D(\xi)\sin\beta b)e^{i(\xi x_1 - \omega t)}d\xi = 0 \qquad (11.83)$$

$$\mu\int_{-\infty}^{\infty} 2i\xi\alpha(A(\xi)\cos\alpha b - B(\xi)\sin\alpha b) + (\xi^2 - \beta^2)(C(\xi)\sin\beta b$$
$$+ D(\xi)\cos\beta b)e^{i(\xi x_1 - \omega t)}d\xi = \tau(x)e^{-i\omega t} \qquad (11.84)$$

$$\mu\int_{-\infty}^{\infty} 2i\xi\alpha(A(\xi)\cos\alpha b + B(\xi)\sin\alpha b) + (\xi^2 - \beta^2)(-C(\xi)\sin\beta b$$
$$+ D(\xi)\cos\beta b)e^{i(\xi x_1 - \omega t)}d\xi = 0 \qquad (11.85)$$

Moreover, equation (11.84) can be rewritten as:

$$\mu\int_{-\infty}^{\infty} (2i\xi\alpha(A(\xi)\cos\alpha b - B(\xi)\sin\alpha b) + (\xi^2 - \beta^2)(C(\xi)\sin\beta b$$
$$+ D(\xi)\cos\beta b))e^{i(\xi x_1 - \omega t)}d\xi = e^{-i\omega t}\int_{-\infty}^{\infty} \bar{\tau}(\xi)e^{i\xi x_1}d\xi \qquad (11.86)$$

where $\bar{\tau}(\xi)$ is the spatial Fourier transform of $\tau(x_1)$.

By harmonic balancing of the terms of the previous equations, the following relations result:

$$\begin{bmatrix} -(\xi^2 - \beta^2)\cos\alpha b & 2i\xi\beta\cos\beta b \\ -2i\xi\alpha\sin\alpha b & (\xi^2 - \beta^2)\sin\beta b \end{bmatrix}\begin{bmatrix} B(\xi) \\ C(\xi) \end{bmatrix} = \begin{bmatrix} 0 \\ \frac{\bar{\tau}(\xi)}{\mu} \end{bmatrix} \qquad (11.87)$$

and

$$\begin{bmatrix} -(\xi^2 - \beta^2)\sin\alpha b & -2i\xi\beta\sin\beta b \\ 2i\xi\alpha\cos\alpha b & (\xi^2 - \beta^2)\cos\beta b \end{bmatrix}\begin{bmatrix} A(\xi) \\ D(\xi) \end{bmatrix} = \begin{bmatrix} 0 \\ \frac{\bar{\tau}(\xi)}{\mu} \end{bmatrix} \qquad (11.88)$$

It is interesting to note here that the excitation term appears in the equations for both modes, implying that excitation using actuators bonded on only one side excites both symmetric and antisymmetric modes. Solving for $A(\xi)$, $B(\xi)$, $C(\xi)$ and $D(\xi)$ and substituting in equation (11.80) and the expression for strain in terms of the Helmholtz components [which can be obtained using equations (11.3), (11.4), (11.47)–(11.50), (11.78) and (11.79)], the following expressions are obtained:

$$\varepsilon_{11}(x_1, t) = \frac{1}{4\pi\mu} \int_{-\infty}^{\infty} \left(\frac{\overline{\tau} N_S}{D_S} + \frac{\overline{\tau} N_A}{D_A} \right) e^{i(\xi x_1 - \omega t)} d\xi \tag{11.89}$$

$$u_1(x_1, t) = \frac{1}{2\pi} \frac{-i}{2\mu} \int_{-\infty}^{\infty} \frac{1}{\xi} \left(\frac{\overline{\tau} N_S}{D_S} + \frac{\overline{\tau} N_A}{D_A} \right) e^{i(\xi x_1 - \omega t)} d\xi \tag{11.90}$$

where

$$\begin{aligned} N_S &= \xi\beta(\xi^2 + \beta^2) \cos \alpha b \cos \beta b \\ D_S &= (\xi^2 - \beta^2)^2 \cos \alpha b \sin \beta b + 4\xi^2 \alpha\beta \sin \alpha b \cos \beta b \\ N_A &= \xi\beta(\xi^2 + \beta^2) \sin \alpha b \sin \beta b \\ D_A &= (\xi^2 - \beta^2)^2 \sin \alpha b \cos \beta b + 4\xi^2 \alpha\beta \cos \alpha b \sin \beta b \end{aligned} \tag{11.91}$$

The integrals in equations (11.89) and (11.90) are singular at the roots of D_A and D_S and are solved by the residue theorem, using a contour consisting of a semicircle in the upper half of the complex ξ plane and the real axis. The equations $D_s = 0$ and $D_A = 0$ are exactly the Rayleigh–Lamb wave equations for symmetric and antisymmetric modes, equations (11.70) and (11.73). Now, for $\tau(x_1) = \tau_0[\delta(x_1 - a) - \delta(x_1 + a)]$, the corresponding Fourier transform is:

$$\overline{\tau}(\xi) = \tau_0(-2i \sin \xi a) \tag{11.92}$$

From equations (11.89), (11.90), (11.91) and (11.92), one can obtain the strain and displacement along the wave propagation direction as:

$$\varepsilon_{11}(x_1, t) = \frac{\tau_0}{\mu} \sum_{\xi^S} \sin \xi^S a \frac{N_S(\xi^S)}{D_S'(\xi^S)} e^{i(\xi^S x_1 - \omega t)} + \frac{\tau_0}{\mu} \sum_{\xi^A} \sin \xi^A a \frac{N_A(\xi^A)}{D_A'(\xi^A)} e^{i(\xi^A x_1 - \omega t)} \tag{11.93}$$

$$u_1(x_1, t) = -\frac{i\tau_0}{\mu} \sum_{\xi^S} \frac{\sin \xi^S a}{\xi^S} \frac{N_S(\xi^S)}{D_S'(\xi^S)} e^{i(\xi^S x_1 - \omega t)} - \frac{i\tau_0}{\mu} \sum_{\xi^A} \frac{\sin \xi^A a}{\xi^A} \frac{N_A(\xi^A)}{D_A'(\xi^A)} e^{i(\xi^A x_1 - \omega t)} \tag{11.94}$$

11.3 APPLICATION OF LAMB-WAVE FORMULATION TO SHM

Having established a basic understanding of elastic-wave propagation and generation using piezoelectric actuators, some results based on the above formulation are examined below [22].

11.3.1 Actuator/Sensor Selection

11.3.1.1 Sensor Modeling and Design

When using active methods for SHM, an important practical requirement is the minimization of power drawn, while ensuring that the signal-to-noise ratio is still high to allow detection of weaker signal reflections from possible defects. Power considerations significantly affect the final design of the SHM system. For this reason, it is crucial that the piezo sensor/actuator geometry, configurations, and materials are optimally chosen. Suppose a piezo sensor is used to detect Lamb waves generated in a plate by a different piezo actuator. Let the length of the piezo sensor be $2c$. For the response of the piezo sensor, consider the charge accumulated by the sensor per unit width. This is given by:

$$Q_S = E_S g_{31} k \varepsilon_0 \int\limits_{x_S-c}^{x_S+c} \varepsilon_{11} dx_1 \qquad (11.95)$$

where E_s is the Young's modulus of the sensor material, k is the dielectric constant of the piezo material, ε_0 is the in-vacuum permittivity, g_{31} is the voltage constant of the piezo material, and x_s is the sensor location. However from the strain-displacement relation, equation (11.4),

$$\int\limits_{x_S-c}^{x_S+c} \varepsilon_{11} dx_1 = u_1(x_S + c) - u_1(x_S - c) \qquad (11.96)$$

Also the capacitance of the sensor per unit width is:

$$C_S = \frac{(k\varepsilon_0 \cdot 2c)}{h_S} \qquad (11.97)$$

where h_s is the sensor thickness, and the sensor response is:

$$V_S = \frac{Q_S}{C_S} \qquad (11.98)$$

Combining equations (11.96), (11.97) and (11.98), the sensor voltage response is proportional to:

$$V_S \propto \frac{u_1(x_S + c) - u_1(x_S - c)}{2c} \qquad (11.99)$$

Substituting equation (11.94) into this equation, we obtain:

$$V_S \propto \left[\sum_{\xi^S} \frac{\sin \xi^S a}{\xi^S} \frac{N_S(\xi^S)}{D'_S(\xi^S)} \frac{\sin \xi^S c}{2c} e^{i(\xi^S x_1 - \omega t)} + \sum_{\xi^A} \frac{\sin \xi^A c}{\xi^A} \frac{N_A(\xi^A)}{D'_A(\xi^A)} \frac{\sin \xi^A c}{2c} e^{i(\xi^A x_1 - \omega t)} \right]$$

$$(11.100)$$

Assuming that all parameters are fixed except the sensor length $2c$, the sensor voltage response is simply:

$$V_S \propto \frac{\sin \xi c}{2c} \tag{11.101}$$

Since the function $\frac{\sin t}{t}$ is maximum at $t = 0$, and its subsequent peaks rapidly drop off, it is concluded that for maximum sensor response V_S, $2c$ should be as small as possible, preferably much smaller than the half-wavelength of the traveling wave, $\frac{\lambda_w}{2}$. Note, however, that only if the sensor length is much smaller than $\frac{\lambda_w}{2}$ does it make sense to maximize the local strain as given by equation (11.93). This is due to the fact that in the limit of $2c$ tending to zero, the sensor response V_s becomes proportional to the local strain ε_{11}. To illustrate this, consider the following example: a 1-cm length actuator is exciting S_0 Lamb waves in a 1-mm thick and uniform plate. The sensor response [equation (11.100)] over a range of operating frequencies (0–1.25 MHz) for different sensor lengths (various values from $2c = 2$ cm to $2c = 0.125$ cm), same sensor material, and given actuator length is plotted in Figure 11.9. It may be seen from Figure 11.9 that the smaller the sensor length, the higher the amplitude of the sensor response. The conventional practice of using the same dimensions for sensor and actuator obviously is a poor choice from a sensitivity point of view as can be seen from the curve for $2c = 2a = 1$ cm. It can also be seen that below a certain value, the sensor response tends to a limit. At the limit of $2c$ tending to zero, the sensor response (V_s) tends to the local strain (ε_{11}). Only then does it make sense to maximize local strain using equation (11.93). This conclusion regarding sensor dimension also holds for other strain sensors such as fiber-optic strain sensors or resistive strain gauges, since those too have a response proportional to the average strain over their lengths. Furthermore, as can be seen from this analysis, the piezo-sensor material chosen should have the highest possible voltage constant g_{31} value (assuming we are using standard piezo wafers; if piezo composites are being used, then they should be chosen so as to maximize the g_{33} value). Other possibilities for increasing sensor response by material optimization are

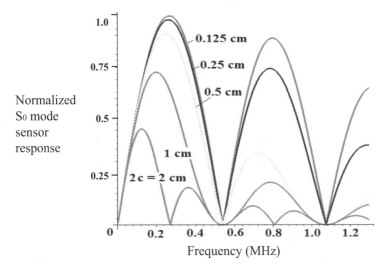

Figure 11.9   Sensor response for various sensor lengths, actuator length fixed ($2a = 1$ cm)

increasing E_s, the sensor material Young's modulus and h_s, the sensor thickness. Increasing these values (i.e., E_s and h_s), however, may adversely increase the local plate stiffness and affect the Lamb-wave propagation.

11.3.1.2 Actuator Design

Consider next the effect of actuator length on the sensor response. If all parameters with the exception of $2a$ are kept constant, from equation (11.100):

$$V_S \propto \sin \xi a \tag{11.102}$$

Thus the response is maximum for the actuator length of:

$$2a = \left(n + \frac{1}{2} \right) \lambda_w, \quad n = 0, 1, 2, 3, \ldots \tag{11.103}$$

Thus as far as maximizing sensor response goes, any of the values of $2a$ given by equation (11.103) are equally optimal. However, consider the expression for power drawn by the actuator (denoted by P):

$$P = 2\pi f C_a V^2 \tag{11.104}$$

where V is the voltage applied to the actuator, and f is the operating frequency in Hz. As before for the sensor case, the capacitance of the actuator per unit width is:

$$C_a = \frac{k\varepsilon_0 \cdot 2a}{h_a} \tag{11.105}$$

Therefore the power consumption is directly proportional to the actuator length, i.e.:

$$P \propto 2a \tag{11.106}$$

Since all lengths suggested by equation (11.103) are equally optimal for maximizing sensor response, the smallest value of $2a$ is desirable for minimal actuator power consumption. This is achieved by $n = 0$ in equation (11.103), which also reduces the area occupied by the actuator on the structure.

As discussed in Reference [20], the magnitude of the shear force applied by the piezo actuator [τ_0 in equation (11.77)] is directly proportional to the coupling constant (or d-value) of the piezo actuator. Hence the actuator material must be chosen such that the appropriate coupling constant d_{31} or d_{11} (depending on whether standard piezo wafers or piezo composites are being used) is maximum. It is also important to realize that in the analyses in Sections 11.2.8 and 11.3.1 above, ideal bonding between the piezo and the structure has been assumed. In practice there is some finite *shear lag* introduced due to the elasticity of the bond layer.

Due to shear lag, the assumption that the piezo causes shear forces along its edges degenerates into the reality that the force is transmitted to the structure over a finite length close to the edges. While the equation derived is less sensitive to the fact that the actuator forces are transmitted over a small finite length rather than a point, in using a very short length sensor, this may be a limiting factor in reducing the sensor length beyond a point. However it is difficult to analyze this effect since bond strength may be difficult to quantify and even for the same epoxy used to bond the piezo onto a structure, it depends significantly on the amount of epoxy applied, and curing conditions. There is an analytical discussion on this in Reference [20].

11.3.2 Comb Transducers

Comb transducers using piezos for SHM is a concept that has been borrowed from NDE. The concept of comb structures for exciting Rayleigh and Lamb waves was already presented in 1958 [19]. In NDE, the comb structure transducer consists of a metal plate with a corrugated, comb-shaped profile consisting of alternating projections and slots of width $\frac{\lambda_w}{2}$ and a quartz plate resting on top. These transducers used in NDE have long been known for their selective wavenumber excitation, which is very useful in reducing the dispersiveness of the generated Lamb waves. Similarly, consider an array of piezos spaced regularly and excited in phase to excite Lamb waves. Figure 11.10 shows the cross-sectional view of such a transducer attached to a plate. The equation for comb-transducer induced-strain is obtained by modeling the comb transducer as line shear forces along the edges of each actuator in the array, similar to the derivation for a single actuator derived in Section 11.2.8. For example, the final equation for induced local strain obtained for the A_0 mode by a transducer array with n (n being even) actuators each of length $2a$ spaced by p from each other is

$$\varepsilon_{11}^n(x_1, t) = \frac{2\tau_0}{\mu} \sum_{\xi^A} \sin(\xi^A a) \{\cos(\xi^A p) + \cos(\xi^A 2p) + \cdots + \cos(\xi^A \frac{n}{2} p)\} \frac{N_A(\xi^A)}{D'_A(\xi^A)} e^{i(\xi^A x_1 - \omega t)}$$

(11.107)

If $2a$ and p are chosen to be $\frac{\lambda_w}{2}$, the comb transducer exhibits a very narrow and selective wavenumber excitation region as compared to a single actuator, as illustrated in Figure 11.11. The plot compares the local strain induced by a single piezo transducer and a 8-piezo set comb transducer, with each element being the same size as in the case of excitation with a single piezo. For both cases, the power consumption is kept constant. The wavenumber selectivity of the comb transducer increases with increasing n, i.e., the peaks appearing in Figure 11.11 get narrower, and correspondingly a narrower wavenumber bandwidth is excited when using a sinusoidal toneburst signal. The resulting generated Lamb waves are less dispersive.

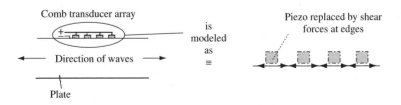

Figure 11.10 Comb transducer array model

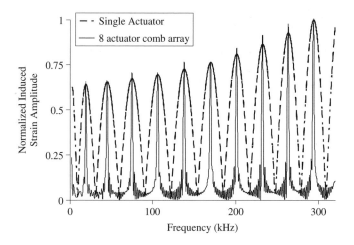

Figure 11.11   Comparison of induced strain in A_0 mode between an 8-array piezo comb transducer and that of a single piezo actuator (power is kept constant)

11.4 EPILOGUE

Besides what was presented in this chapter, the authors are actively engaged in research on other aspects related to Lamb-wave based SHM technologies. Current efforts are directed towards the development of Lamb-wave based SHM in support to integrated systems health management of spacecraft and space habitats. These structures present a very challenging application area for SHM, due to the harsh environment of outer space. Investigations have been conducted to examine the behavior of piezoelectric actuators and sensors over a range of temperatures. As an example, an aluminum plate specimen was instrumented with two PZT-5A piezo patches for a simple pitch–catch configuration Lamb-wave test and subjected to three thermal cycles up to 150 °C. The variation of sensor response magnitude with temperature is shown in Figure 11.12. It describes a significant drop in sensor response magnitude after the first thermal cycle. However, the degradation after the second and third cycles is much smaller, and the sensor response asymptotes to a certain response curve. Other investigations to characterize and model thermal behavior of piezos are also being conducted at the University of Michigan. Other recent efforts [22] have focused on further development of models that describe the excitation and sensing of transient diagnostic waves in plate structures using surface-bonded piezo transducers, part of which has been covered in Section 11.3 of this chapter. An analysis for excitation of circular-crested Lamb-waves using piezo discs based on the 3-D elasticity equations is also presented in Reference [22]. Future efforts are planned towards modeling rectangular piezo actuators and developing novel variable length actuator designs which would enable optimal operation at different points on the Lamb-wave dispersion curves with the same actuator element. Finally, work towards a surface scanning scheme that would enable rapid scanning of a whole structure from a compact array of centrally placed actuators is also underway at the University of Michigan.

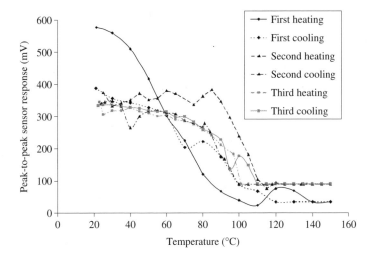

Figure 11.12   Thermal behavior of sensor response in pitch–catch Lamb-wave test

REFERENCES

[1] Schoess, J.N. and Zook, J.D. 'Test results of resonant integrated microbeam sensor (RIMS) for acoustic emission monitoring', *Proceedings of the SPIE Conference on Smart Electronics and MEMS*, **3328**, pp. 326–332 (1998).

[2] Marantidis, C., Van Way, C.B. and Kudva, J.N. 'Acoustic-Emission Sensing in an On-board Smart Structural Health Monitoring System for Military Aircraft', *Proceedings of the SPIE Conference on Smart Structures and Integrated Systems*, **2191**, pp. 258–264 (1994).

[3] Seydel, R.E. and Chang, F.K. 'Implementation of a Real-time Impact Identification Technique for Stiffened Composite Panels', *Proceedings of the 2nd International Workshop on Structural Health Monitoring*, Stanford University, California, Technomic Publishing Company, Lancaster, PA, pp. 225–233 (1999).

[4] Kollar, L. and Steenkiste, R.J. 'Calculation of the stresses and strains in embedded fiber optic sensors', *Journal of Composite Materials*, **32**, 1647–1679 (1998).

[5] Rees, D., Chiu, W.K. and Jones, R. 'A numerical study of crack monitoring in patched structures using a piezo sensor', *Smart Materials and Structures*, **1**, 202–205 (1992).

[6] Chiu, W.K., Galea, S.C., Koss, L.L. and Rajic, N. 'Damage detection in bonded repairs using piezo-ceramics', *Smart Materials and Structures*, **9**, 466–475 (2000).

[7] Hautamaki, C., Zurn, S., Mantell, S.C. and Polla, D.L. 'Experimental evaluation of MEMS strain sensors embedded in composites', *Journal of Microelectromechanical Systems*, **8**, 272–279 (1999).

[8] Ellerbrock, P. 'DC-XA Structural Health Monitoring Fiber-optic Based Strain Measurement System', *Proceedings of the SPIE*, **3044**, 207–218 (1997).

[9] Rose, J.L. *Ultrasonic Waves in Solid Media*, Cambridge University Press, Cambridge, UK, 1999.

[10] Auld, B.A. *Acoustic Fields and Waves in Solids*, Volume I, Second Edition, R.E. Kreiger Publishing Co., Florida, 1990.

[11] Graff, K.F. *Wave Motion in Elastic Solids*, Dover Publications, New York, 1991.

[12] Achenbach, J.D. *Wave Propagation in Elastic Solids*, North-Holland, New York, 1984.

[13] Rayleigh, J.W.S. 'On waves propagated along the plane surface of an elastic solid', *Proceedings of the London Mathematical Society*, **17**, 4–11 (1887).

[14] Love, A.E.H. *Some Problems of Geodynamics*, Cambridge University Press, Cambridge, UK, 1926.

[15] Stoneley, R. 'Elastic waves at the surface of separation of two solids', *Proceedings of the Royal Society of London, Series A*, **106**, 416–428 (1924).

[16] Scholte, J.G. 'On the Stoneley wave equation', *Proc. Kon. Nederl. Akad. Wetensch*, **45**, (20/5), 159–164 (1942).

[17] Lamb, H. 'On waves in an elastic plate', *Proceedings of the Royal Society of London Series A*, **93**, 293–312 (1917).

[18] Worlton, D.C. 'Experimental confirmation of Lamb waves at megacycle frequencies', *Journal of Applied Physics*, **32**, 967–971 (1961).

[19] Viktorov, I.A. *Rayleigh and Lamb waves*, Plenum Press, New York, 1967.

[20] Crawley, E.F. and Lazarus, K.B. 'Induced strain actuation of isotropic and anisotropic plates', *AIAA Journal*, **29**, 944–951 (1991).

[21] Giurgiutiu, V. 'Lamb wave generation with piezoelectric wafer active sensors for structural health monitoring', *Proceedings of the SPIE*, **5056**, 111–122 (2003).

[22] Raghavan, A. and Cesnik, C.E.S. 'Modeling of piezoelectric-based Lamb-wave generation and sensing for structural health monitoring', *SPIE Symposium on Smart Structures and Materials/NDE 2004*, Paper 5391-42, San Diego, California, March 14–18 (2004).

12
Structural Energy Flow Techniques

José Roberto de F. Arruda

Departamento de Mecânica Computacional, Universidade Estadual de Campinas, Brazil

12.1 INTRODUCTION

Most of the time, the effect of damage in the dynamic behavior of structures is of a local nature. Therefore, damage usually has little effect on global properties such as natural frequencies and low-frequency mode shapes. The variability due to environmental changes often has a more important effect than do changes due to damage, thus making most fault detection methods based upon modal properties and dynamic response at low frequencies deceptive. On the other hand, predicting the structural behavior at higher frequencies is usually out of reach of numerical methods such as finite and boundary elements. Nevertheless, it is true that damage does change the dynamic behavior. Therefore, there must exist analytical and experimental dynamic analysis techniques that are able detect, localize and evaluate damage. Due to the localized effect of damage, methods that use extensive spatial data have the potential for detecting and localizing damage. Such spatially dense data can be obtained using optical techniques, such as laser Doppler vibrometry and pulse holography, and acoustic techniques, such as near-field acoustic holography. In this chapter, techniques for processing spatially dense dynamic response data, aimed at detecting and localizing damage, are discussed. Because damage can cause local energy dissipation and change the spatial distribution of dynamic forces, power-flow techniques that are able to localize energy sources and energy sinks in structures are presented. On the other hand, damage can cause nonlinearity, which can be detected using frequency-domain methods and specialized excitation signals, which are briefly reviewed. This chapter ends with a

Damage Prognosis – For Aerospace, Civil and Mechanical Systems Edited by D.J. Inman, C.R. Farrar, V. Lopes Junior and V. Steffen Junior © 2005 John Wiley & Sons, Ltd

discussion about the limitations of deterministic numerical models for structures and the problem of dynamic modeling at higher frequencies.

12.2 POWER AND INTENSITY CONCEPTS

In mechanical systems, instantaneous power is defined as the inner product of force times velocity. For harmonic behavior, the signals may be represented by a complex amplitude. It can be easily shown that the average value of the instantaneous power is equal to one-half the real part of the product of the complex force amplitude and the complex conjugate of the velocity amplitude:

$$I = \overline{f(t)\nu(t)} = \frac{1}{2}\Re\{F^*(\omega)V(\omega)\} \tag{12.1}$$

where $f(t) = \Re\{F(\omega)e^{i\omega t}\}$, $\nu(t) = \Re\{V(\omega)e^{i\omega t}\}$ and $*$ denotes the complex conjugate. The reactive power in this case will be simply half the imaginary part of the same product:

$$P_r = \frac{1}{2}\Im\{F^*(\omega)V(\omega)\} \tag{12.2}$$

Therefore, the instantaneous power at steady state will be the active power plus the oscillating power:

$$f(t)\nu(t) = I + \sqrt{P_r^2 + I^2}\cos(2\omega t + \varphi) \tag{12.3}$$

The plot in Figure 12.1 shows the power components of an underdamped SDOF system initially at rest excited with a sinusoidal force. The plots show the steady-state active and reactive components. The envelope of the instantaneous power is the sum of the active component and the magnitude of the oscillating component, which is the mean square value of the active and reactive components, while the average is the active power component.

Structural dynamicists are usually unfamiliar with the concept of reactive intensity. However, this concept is current in electrical engineering, especially in electric power systems. The electrical energy consumption is measured in terms of active power, power being voltage times current. However, active power does not fully characterize the electrical charge. A purely inductive charge, for instance, demands energy to be started but does not demand power in the steady-state condition. Electrical power companies adjust the consumption bill whenever the ratio between active and total power is above an established limit, usually of the order of 85 %. This is done because the company needs to have a higher installed power in order to be able to supply the transient demand of the reactive charges. It also needs to have enough energy storage capacity in the network to absorb the oscillating part of the power, which is continuously exchanged back and forth between the power generation network and the consumer's charge.

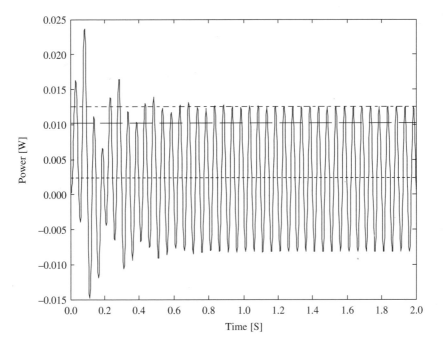

Figure 12.1 Power components for an underdamped SDOF system excited by a sinusoidal force. (—) Instantaneous power; (---) active part; (– –) reactive part; (- - -) magnitude of the oscillating part

In a distributed-parameter system, such as a beam or a plate, a steady-state intensity vector can be defined at a plane passing through a point in the continuum as the power per unit area oriented according to the surface normal:

$$I_i = \frac{1}{2}\Re\left\{\sum_j \sigma_{ij}(\omega) V_j^*(\omega)\right\} \tag{12.4}$$

where $\sigma_{ij}(\omega)$ is the amplitude of the stress tensor and $V_j(\omega)$ is the amplitude of a component of the point velocity. By analogy, the reactive power can be expressed as:

$$R_i = \frac{1}{2}\Im\left\{\sum_j \sigma_{ij}(\omega) V_j^*(\omega)\right\} \tag{12.5}$$

Operating shapes, i.e. the spatial distribution of the complex displacement amplitudes at a given frequency, depend on the magnitude and location of an applied point force. This may be expressed locally in terms of reactive power by the expression [1]:

$$\vec{\nabla} \cdot \vec{R} = P_r - 2\omega(\langle U \rangle - \langle T \rangle) \tag{12.6}$$

where $P_r = \frac{1}{2}\Im\{F_i^* V_i\}$ is the reactive power input by the external forces, $\langle U \rangle$ is the time-averaged strain energy density, $\langle T \rangle$ is the time-averaged kinetic energy density, ω is the angular frequency and $\vec{\nabla} \cdot$ denotes the divergence of a vector field.

12.2.1 *Structural Intensity (Active Power) Expressions*

Intensity is power flowing per unit area in a given direction within a continuum. In the case of two-dimensional waveguides, such as thin plates, the intensity is given in units of power per unit length (orthogonal to the flow direction) and, in the case of one-dimensional waveguides such as bars and beams, in units of power directly. While acoustical intensity has a widespread use in the industry, structural intensity is still basically restricted to research applications. This may be explained by the fact that while acoustical waves propagate in the form of primary waves (P-waves) only, in solids the two kinds of wave, primary and secondary (S-waves) interact, so that different types of wave are generated in different waveguides: longitudinal in bars, longitudinal, flexural, and torsional in beams, Rayleigh surface waves in semi-infinite media, Lamb waves in doubly bounded media, and so forth [2]. Hence, deriving expressions for structural intensity and computing it from measurements is much more elaborate than for acoustical intensity.

Nevertheless, neglecting near-field effects and assuming there is no interaction between different types of waves propagating, simple expressions for structural intensity may be obtained for many types of waveguides.

Since the original work of Noiseaux [3], many papers have been published where expressions for the structural intensity in different types of structural elements are derived. Using these expressions, it is possible to compute analytically or numerically, as well as estimate experimentally, the intensity vector from two or more spatially distributed measurements. The simplest case is the linear acoustics field, where two pressure measurements are sufficient to estimate a plane-wave intensity. There is a strong similarity between intensity expressions for different media [4]. Using equations (12.1) and (12.4) it is possible to generalize an intensity estimator using two measurements, E_1 and E_2, spaced by a distance Δx by an expression such as:

$$I(\omega) = F(\omega)\Im\{E_1{}^*E_2\} \qquad (12.7)$$

For acoustical plane waves:

$$F(\omega) = \frac{k_a}{\rho\omega\sin(k_a\Delta x)} \quad \text{where} \quad k_a = \omega\sqrt{\frac{\rho}{B}} \qquad (12.7a)$$

For rods:

$$F(\omega) = \frac{EA\omega k_l}{2\sin(k_l\Delta x)} \quad \text{where} \quad k_l = \omega\sqrt{\frac{\rho}{E}} \qquad (12.7b)$$

For beams:

$$F(\omega) = \frac{EJk_b^3}{\omega\sin(k_b\Delta x)} \quad \text{where} \quad k_b^2 = \omega\sqrt{\frac{\rho A}{EJ}} \qquad (12.7c)$$

For shafts:

$$F(\omega) = \frac{GJ_p k_t}{2\omega\sin(k_t\Delta x)} \quad \text{where} \quad k_t = \omega\sqrt{\frac{\rho}{G}} \qquad (12.7d)$$

For plates in bending:

$$F(\omega) = \frac{Dk^2}{\omega\Delta x} \text{ where } D = \frac{Eh^2}{12(1-\nu^2)} \text{ and } k_x^2 + k_y^2 = k^2 = \omega\sqrt{\frac{\rho h}{D}} \tag{12.7e}$$

where E, G,ν,B are the usual linear elastic material constants, ρ is the mass density, the ks are wavenumbers, the Js are area moments, A is the area of the cross-section of a rod or beam, and h is the plate thickness.

Structural intensity analysis may be used to show energy flow paths in structures [5], clearly indicating energy sources and sinks, as will be seen in the next section.

More recently, using equation (12.6), Alves and Arruda [6] have shown that it is also possible to localize an external point force in beams and plates using the divergence of the reactive power field. The above equation shows that the sum of the divergence of the reactive power field and two times the frequency times the Lagrangian density $(\langle T \rangle - \langle U \rangle)$ gives the spatial distribution of the external reactive power acting upon the structure, P_r. An example will be shown in the next section.

12.3   EXPERIMENTAL POWER FLOW TECHNIQUES

12.3.1   Measurement Techniques

Structural intensity maps are derived from out-of-plane vibrations measured with accelerometers, strain gages [7], laser holography [8,9] and laser vibrometry [10]. The frequency domain formulation makes it straightforward to transform velocities into accelerations and vice versa.

Structural intensity is usually computed from closely spaced measurements using expressions such as equation (12.7e) or more elaborate expressions using higher order derivatives. However, it has been shown that, due to measurement uncertainty, it is preferable to use simplified expressions with first-order derivatives only [5]. Computing the first-order spatial derivative from measured data is a delicate operation, as measurement noise can easily affect the results. Therefore, techniques that curve fit the spatial data with smooth functions before differentiation are recommended.

Two-dimensional velocity fields measured over the surface of a structure at a given frequency, given by amplitudes and phases measured relative to a reference signal (which can be the velocity at a given location) are usually referred to as operational deflection shapes (ODS) or, when the reference signal is the input force, operational mobility shapes (OMS). ODS and OMS are frequently measured over rectangular grids using scanning laser Doppler vibrometers.

Different two-dimensional models can be used to interpolate or approximate the measured data. Once the data is fitted, the partial spatial derivatives required to compute the structural power flow can be easily computed in the wavenumber domain.

12.3.2 Post-Processing Two-Dimensional Data Using a Regressive Discrete Fourier Series

Two-dimensional ODS's measured over equally-spaced rectangular grids, say $H_{mn}(\omega)$, can be interpolated using the two-dimensional discrete Fourier transform (DFT). The difficulty with using the DFT is that its implicit *periodization* introduces high-frequency components that account for the sharp edges present in the *wrapped-around* data. This phenomenon is known as *leakage*. In the data smoothing process, leakage is prejudicial, as it causes distortion of the low-pass filtered data. When dealing with plate vibrations, this problem does not exist for clamped boundaries, is not too serious in the case of simply supported boundaries, but becomes critical in the case of free boundaries. It should be noted that the boundaries can be actual boundary conditions or just measurement field boundaries.

The usual way to reduce leakage is windowing, but this technique is not suitable in the case of finite-length, spatial-domain, data. To overcome the leakage problem, the proposed technique consists of representing the data by a two-dimensional regressive discrete Fourier series (RDFS) proposed by Arruda [11], which will be briefly reviewed here. Unlike the DFT, in the RDFS the original length of the data is not assumed to be equal to the signal period, nor is the number of frequency lines assumed to be equal to the number of data points. With the two-dimensional, equally spaced RDFS model, the mobility shape H_{mn} (where the frequency dependency is omitted for simplicity) is expressed as:

$$H_{mn} = \sum_{k=-p}^{p} \sum_{l=-q}^{q} Z_{kl} W_{\mathrm{M}}^{mk} W_{\mathrm{N}}^{ln} + \varepsilon_{mn} \quad m = 0,\ldots,M-1; \; n = 0,\ldots,N-1; \quad (12.8)$$

where H_{mn} represents the discretized data with constant spatial resolutions Δx and Δy, $W_{\mathrm{M}} = \exp(i2\pi/\mathrm{M})$, $W_{\mathrm{N}} = \exp(i2\pi/\mathrm{N})$, and ε_{mn} accounts for the noise and higher frequency contents of H. Note that $\mathrm{M} \neq M$ and $\mathrm{N} \neq N$. The length of the data in x is $M\Delta x$, but the period of the RDFS is $\mathrm{M}\Delta x > M\Delta x$. Data reduction is achieved because $p \ll M$ due to the expected low wavenumber of the mobility shape surface. In the y direction $\mathrm{N}\Delta y > N\Delta y$ and $q \ll N$. The $M \times N$ data in H are represented by a $(2p+1) \times (2q+1)$ complex matrix Z of elements Z_{kl}.

The RDFS is an approximation instead of an interpolation of H_{mn}. Thus, the Euler–Fourier coefficients cannot be calculated by the DFT. Rewriting equation (12.8) in matrix form:

$$H = W_{\mathrm{M}} Z W_{\mathrm{N}} + \varepsilon, \quad (12.9)$$

The least-squares solution is given by:

$$Z = (W_{\mathrm{M}}^{H} W_{\mathrm{M}})^{-1} W_{\mathrm{M}}^{H} H W_{\mathrm{N}}^{H} (W_{\mathrm{N}} W_{\mathrm{N}}^{H})^{-1} \quad (12.10)$$

where the matrices to be inverted have a very small size, $(2p+1) \times (2p+1)$ and $(2q+1) \times (2q+1)$, respectively, and ^{H} denotes the matrix complex conjugate. The smoothed data $H^{(s)}$ may be obtained from:

$$H^{(s)} = W_{\mathrm{M}} Z W_{\mathrm{N}}, \quad (12.11)$$

where W_M and W_N can be calculated for the desired spatial resolution. The reduction of the data is achieved as Z represents the data using only $(2p+1)(2q+1)$ values, instead of the original MN values. The formulation of the RDFS for nonequally spaced data given by Arruda and Mas [12] can be used in the place of the formulation above when the mobility shapes are mapped over nonregular, arbitrary grids, while the DFT is only applicable to equally spaced, rectangular grids. The spatial derivatives are obtained simply by multiplying each coefficient Z_{kl} by the product of the corresponding wave-number by $\sqrt{-1}$ to the appropriate power.

12.3.3 Applications of Active Intensity Maps

Many examples of the use of active intensity maps to localize sources and sinks can be found in the literature. One of them will be shortly described here.

Arruda and Mas [12] have treated an example consisting of a flat aluminum plate bolted through rubber mounts to a massive steel block. A shaker was attached to the back of the plate through a piezoelectric force transducer. In this example, the shaker is the power source and the rubber mounts are the energy sinks.

The responses over the surface of the plate were measured with a scanning laser Doppler vibrometer (LDV). An equally spaced measurement grid of 46×40 measurement points was used. The paper shows results for a single frequency. The measured OMS were filtered with a median filter to remove noise 'spikes' which are typical in LDV measurements. The filtered OMS was curve-fit using a two-dimensional RDFS with $p = q = 4$, $M/M = N/N = 1.5$.

Using the Fourier series model, spatial derivatives were computed and the active intensity map (arrows of length proportional to the intensity and direction according to the flow in the two orthogonal directions) was plotted. The active intensity maps indicate the path of the power flow in the plate, such that the locations from which the arrows diverge indicate the input of external power and the location to which the arrows converge indicate an energy sink. To facilitate the localization of sources and sinks, divergence plots can be computed from the vector intensity field. However, it was shown that the divergence plots are more sensitive to measurement errors, so that only the divergence plots computed using the RDFS allowed the localization of the source and the sinks.

Pascal and Li recently proposed a technique which consists of separating the intensity vector field into a rotational and an irrotational component, the latter allowing the localization of sources and sinks much more easily [13].

The rationale behind the idea of using structural intensity maps and divergence maps to localize faults is that some types of fault generate local friction and, therefore, local energy sinks, which could be detected with such techniques. Nevertheless, this potential has not yet been fully demonstrated, although some attempts have been made [14].

Some of the major drawbacks of this technique for fault detection are the necessity of spatially dense measurements over a free surface, which is not always available, its sensitivity with regard to measurement errors, and the present limitation to beam and plate-like structures. The technique can be extended to shell structures, but this may lead to the necessity for also measuring in-plane vibrations. New optical measurement techniques are becoming available for this purpose, such as speckle interferometry.

Commercial electronic speckle pattern interferometry (ESPI [15]) systems allow both in-plane and out-of-plane measurements. Optical full-field measurement systems such as ESPI provide spatially dense data for spatially based post-processing techniques such as intensity mapping.

12.3.4 Reactive Power Example

In many cases, structures under test exhibit very low damping, and the active part of the measured power is much lower than the reactive part. Therefore, the active intensity computed from experimental data is contaminated with the reactive part, and the reactive intensity maps become meaningless. Recently, Alves and Arruda [6] proposed the use of the reactive intensity map to localize energy sources. Figure 12.2 shows a scheme the simple structure used to investigate this possibility. A homogeneous straight cantilever beam with dissipation in the support is excited at mid span. Figure 12.3(a) shows the plot of the divergence of the reactive field and twice the Lagrangian density times the frequency. Figure 12.3(b), shows the difference between these two curves, which corresponds to spatial distribution of the input power, clearly indicating the location of the external force. Results were obtained using a spectral element model with two elements.

Figures 12.4(a) and 12.4(b) show experimental results for a beam with one free end and one end plunged into a sand box excited by a shaker at mid span. The computed divergence of the reactive power and the Lagrangian term multiplied by 2ω are shown in Figure 12.4(a) and their difference, which is the reactive input power, in Figure 12.4(b). The latter clearly indicates the force location. Measurements were performed using a laser Doppler vibrometer and a piezoelectric force transducer.

The same technique was used on plates [16], but in this case the experimental results were not so conclusive, and the technique needs further investigation.

The information about the location of the force is contained in the operating mode, as this depends on the location of the force, which dictates the relative contribution of the different modes to the operating shape. Therefore, one should not try to use equation (12.6) to localize a source at a frequency close to resonance, where only one mode is dominant and the operating shape will not depend on the location of the excitation force. This technique may be used to localize dynamic forces in structures and, therefore, it may also be useful in detecting damage that causes shifts in the location of dynamic forces in structural parts. The notion of active and reactive power can also be used in other approaches to fault detection, as the existence of faults can change the balance between kinetic and potential energy in the structure.

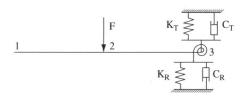

Figure 12.2 Scheme of a cantilever beam with flexible, damped support

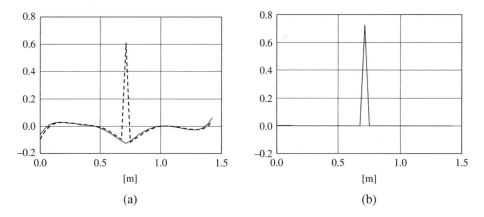

(a)   (b)

Figure 12.3 Spatial distribution of (a) the divergence of the reactive power (——) and the Lagrangian term (---), and (b) the difference between them, indicating the force location (numerical prediction)

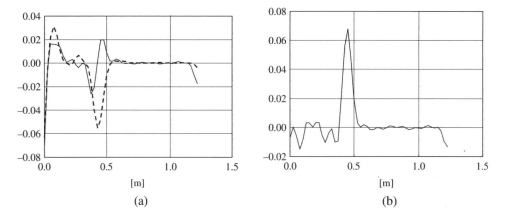

(a)   (b)

Figure 12.4 Spatial distribution of the (a) divergence of the reactive power (——) and the Lagrangian term (---), and (b) the difference between them, indicating the force location (experimental)

12.4 SPATIAL FILTERING FOR FAULT DETECTION

Once a spatially dense vibration field has been measured over a vibrating surface, for each degree of freedom (e.g., out-of-plane velocity or in-plane velocity in two orthogonal directions), a two-dimensional data matrix is generated. This data can be processed directly or curve-fitted for post-processing.

One of the simplest spatial processing techniques in wavenumber-domain filtering. In one dimensional structures (e.g., frames) a one-dimensional low-pass, high-pass, or band-pass, filter may be applied. As commented before, structural faults tend to be very localized and, thus, appear more clearly at higher wavenumbers. Therefore, high-pass spatial filtering the data will tend to reveal the fault locations. The more usual situation is the two-dimensional structure (e.g., plates and shells) where two-dimensional filters apply. The simplest way to implement a filter is to use the two-dimensional discrete Fourier transform (DFT) and eliminate some of the wavenumber lines, usually in terms of $k = \sqrt{k_x^2 + k_y^2}$, where k_x and k_y are the wavenumbers at two orthogonal directions. For high-pass filtering, the lower wavenumbers are discarded. Usually, one also discards the very high wavenumbers, as they are frequently related to noise. Sharp edges caused, for instance, by cracks, will also cause high wavenumber components, and this low-pass filtering aimed at filtering part of the noise can blur the sharp edges of the spatial data. A compromise must be made in such cases. Faults such as delaminations in composite plates normally cause local modes at high frequencies that can be exposed with this high-pass filtering technique.

There is great potential for new techniques of fault detection based upon spatial filtering of spatially dense vibration data obtained with optical techniques. An investigation effort is necessary to establish the potential and limitations of such experimental techniques.

12.5 ACOUSTICAL MEASUREMENTS AS A TOOL FOR FAULT DETECTION

Most of the time, structures are surrounded by a fluid, generally air. Therefore, vibrations at their surfaces cause waves in the fluid that can be measured with pressure sensors such as microphones. This offers an inexpensive, noncontact, measurement technique for full-field surface vibration measurement. There are numerous microphone array techniques [17] that are aimed at noise source localization, which can be used for structural fault location. It is not necessary that faults generate noise, though this may happen, but just the fact that a different vibration pattern causes changes to the pressure field in the surrounding fluid is sufficient to allow fault detection.

Noise-source location methods usually aim at reconstructing a sound source, which often is a vibrating surface. One of the first experimental microphone array techniques was nearfield acoustic holography (NAH) [18]. Based upon Fourier acoustics, it consists of obtaining the three dimensional sound field from pressure measurements near the surface that radiates the sound. The classical NAH technique presents good results for sources with simple, or separated, geometry (e.g., plane or cylindrical). However, it has limitations in the case of arbitrarily shaped sources.

The boundary element method (BEM) is a numerical technique employed for radiation problems and, by itself or associated to structural models based on the finite-element method (FEM), allows obtaining solutions for complex vibroacoustic problems. Inverse BEM models can be used to characterize acoustic sources [19]. Although numerical methods are well accepted, the computational time can still be prohibitively high, mainly for higher frequencies.

Alternative modeling methods can be used in an inverse scheme for the characterization of acoustic fields. The acoustic holography via the elementary source method (ESM) [20] can be situated in this context. The ESM consists of replacing the vibrating structure by a set of elementary acoustical sources, usually lying in the interior of the structure, which, together, give rise to an acoustic field matching the field outside the structure. Differently from standard NAH, the acoustic holography technique by ESM allows to estimate the pressure field anywhere, independently of the noise source geometry, the measurement grid or the coordinate system utilized. The inverse radiation model by ESM frequently results in a discrete, ill-conditioned problem. Therefore, regularization tools, such as Tikhonov's, are used in order to improve the reconstruction of the acoustic source strength.

Acoustic array techniques also present a great potential as a fault detection and localization tool, as it yields spatially dense vibration data over a structure surface. Most techniques used with optical measurements can be used here too.

12.6 DETECTING NONLINEARITY WITH SPECIAL EXCITATION SIGNALS

Faults may cause nonlinearity in an otherwise linearly behaving structure. Therefore, techniques aimed at detecting nonlinearity may be used to detect some kinds of structural fault. The basic property of any linear, time-invariant, system is responding to a sinusoidal excitation with a sinusoidal response of that same frequency. This is straightforward to prove using the superposition and time-invariance properties of the system and properties of the complex exponential. Therefore, a direct way to detect nonlinearity is to excite a system with a sine wave and analyze the frequency spectrum of the response. However, this is time consuming if a large frequency range must be tested.

A research group from the Electrical Engineering Department of the Free University of Brussels has been working for many years on tailoring specialized signals suitable for broadband testing to detect certain types of nonlinearity. These signals use the basic principle of arranging the phases of a series of sine waves added together to synthesize a signal with the desired spectrum, so that the crest factor is minimized [21]. Such signals are usually called multisines. It has been shown that it is possible to use multisine signals in multishaker modal tests [22].

With the basic algorithms for phase optimization of an arbitrary spectrum multisine, one can design a signal in order to extract linearized models under either constant force or constant response levels, for instance. Otherwise, by not exciting some frequencies it is possible to investigate the harmonic distortion caused by nonlinearity. A no-interharmonic-distortion (NID) multisine can be designed to contain energy only at carefully selected frequencies, so that quadratic and cubic nonlinearities can be detected and Volterra kernels estimated [23]. The Brussels group proposed an odd–odd multisine where energy is put only in frequencies $f_k = (4k + 1) f_0$. Using the odd–odd multisine [24] it is possible to evaluate both quadratic and cubic distortion, although it is not possible to determine the Volterra kernels.

A recent PhD dissertation of the Vrije Universiteit Brussel, by Vanlanduit [25], explores the possibilities of nonlinear detection tools for fault detection. He was able to detect cracks in a plexiglass plate using odd–odd multisine excitation and looking at the spatial distribution of the response at nonexcited harmonics. However, the results are very dependent on the frequencies chosen for analysis. The multisine approach for detection of faults via nonlinear distortion analysis seems very promising.

12.7 FREQUENCY LIMITS OF NUMERICAL MODELING TECHNIQUES – THE MIDFREQUENCY PROBLEM

Many structural fault detection and localization techniques rely upon a numerical model of the structure under test. As discussed before, faults are most visible at higher frequencies. However, at higher frequencies, the numerical methods currently used in structural dynamics, namely the finite-element method (FEM) and the boundary element method (BEM), are not effective due to the refined mesh that is necessary and to the uncertainty in the physical parameters of the models caused by manufacturing tolerances and environmental conditions.

Deterministic predictions of one single system may vary considerably across a group of samples of nearly identical systems (e.g., automobiles, airplanes, home appliances, and machinery in general). Many examples of the scattering of the frequency response functions for structures with randomly varying parameters can be found in the literature [26]. They show that the essential features of the dynamic behavior of the system cannot be grasped from a single deterministic response. In such cases, it is better to compute asymptotic values obtained by averaging over frequency (using third octave bands, for instance) and over the spatial distribution. However, computing average responses from deterministic predictions may have a prohibitive computational cost. Therefore, intensive research is currently under way to try to come up with methods for efficiently predicting the average response of structural systems at higher frequencies.

Statistical Energy Analysis (SEA) was the first method to deal with this problem [27]. Originally proposed by Richard Lyon in the 1960s, the method is based upon the assumption that, in a network of weakly coupled subsystems, energy flows from the subsystem with higher energy to the subsystem with lower energy. This flow of energy is proportional to the energy difference and the coefficient of proportionality is the coupling loss factor. SEA models subsystems as a group of modes that store energy and, therefore, may be characterized as a lumped, modal-based dynamical model [28]. Analytical estimates of modal density and coupling-loss factors are usually employed. It is assumed that complex local geometrical details are not important at higher frequencies in terms of global, average behavior, in a sort of 'dynamical Saint Venant's principle'. SEA is applicable to distributed parameter systems when the modal density is sufficiently high.

However, a difficulty is frequently encountered when substructures with high modal density such as thin shells – for which SEA is applicable – are coupled to substructures with low modal density such as beams, where FEM or some other deterministic method should be used. This is the case with aircraft fuselages and car bodies, for instance. To overcome this difficulty a new class of methods is currently being developed. They are referred to as local–global (or hybrid) methods [29, 30]. In these methods, a Ritz approach is developed where the structure is partitioned in long wavelength and short wavelength shape functions. The low modal density, long-wavelength modes are treated in a deterministic way, while the short wavelength, local modes, are treated statistically using a SEA formulation. According to Shorter and Langley [31], the interaction between the global and the local modes can be described in terms of spatial correlation fields associated with each local subsystem. Figure 12.5 shows the effect of a soft rod attached to a more rigid clamped-free rod (200 times more rigid). It can be observed that the effect of the local behavior (high modal density soft rod) is basically to damp the global (low modal density rigid bar) modes. The rods have the same length. These responses were computed using a deterministic method called spectral element method (SEM) [32].

In the SEM, the continuous dynamical equilibrium equations for a given structural element is solved in the Fourier domain by using a direct stiffness approach similar to the finite-element method. It is useful in the context of higher frequency dynamics because it yields exact solutions within the framework of the structural element theory used, without the need to increase the mesh refinement (only one element is needed for each homogeneous span, free of discontinuities). Its major drawback is that it is limited to one-dimensional coupling of the elements. Using spectral elements, it is straightforward to model one-dimensional waveguides such as acoustical ducts, rods, shafts, and beams. Thus, three-dimensional frame structures can be modeled using the spectral Timoshenko frame elements [33]. Two-dimensional elements have been proposed for plates, but in such cases, as the coupling can be made in only one direction, the boundary conditions in the other direction must be defined in the element. Rectangular plate elements have been developed for infinite-periodic and simply supported [34] boundary conditions. There are ongoing research efforts to formulate a spectral element for bounded plates with arbitrary boundary conditions [35].

The idea behind hybrid methods is to capture, in the same dynamical model, the low frequency behavior in an exact way, while an asymptotic response is predicted for the higher frequencies. The degrees of freedom related to the large number of local shape functions are reduced to one degree of freedom per subsystem, which is the subsystem energy.

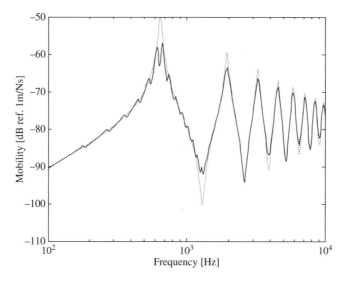

Figure 12.5 Influence of a soft rod on the response of a more rigid clamped-free rod. (---) Mobility of the clamped-free rod; (—) mobility of the rod coupled to a soft rod

Other approaches, inspired by SEA, but seeking to obtain the spatial distribution of vibration energy, are based upon energy conservation, dissipation mechanism equations, and the relation between intensity and energy density, the latter constituting the main approximation. One of these methods is the so-called power flow finite-element method [36], later formalized as thermal analogy methods by Bernhard and collaborators [37] and generalized by Jezequel and collaborators [38–40].

Looking at structural dynamics and vibroacoustics in terms of energy balance and energy flow gives new insights to old linear structural dynamic problems. The basic issue of subsystem interaction, which is usually overlooked, becomes central. A good input/output dynamical model is of little value if the inputs are unknown. In mechanics the input, or right-hand side of the dynamical equation, usually consists of dynamical forces, but these forces are a result of a complex interaction between subsystems, usually cannot be predicted and is awkward to measure. Therefore, looking at subsystem interaction using energy and intensity concepts may be very useful, particularly when it is necessary to average or smooth the solutions.

It is expected that energy flow methods and global/local methods will converge into a family of hybrid methods capable of representing both the low frequency deterministic behavior and the statistically averaged high frequency behavior in a practical way with the same convenience as that of the present finite-element analysis tools. Research groups worldwide are developing numerical methods capable of conducting such tasks with enough efficiency and reliability.

Other techniques, which are gaining importance in recent years for dynamic systems are the so-called meshless methods [41]. Meshless methods are more flexible and easy to implement than are finite elements, and it may be easier to integrate SEA with a meshless Galerkin-type model than with FEM. This is an interesting topic for further research.

Prediction methods that are more adapted to higher frequencies, particularly in the frequency range between what can be treated by deterministic numerical methods such as FEM and BEM and what can be modeled with SEA, are necessary for model-based fault detection and localization. Perhaps even more important, methods that properly take into account the variability due to fabrication tolerances and environment changes are of key importance for the applicability of model-based fault detection techniques to real-life problems.

REFERENCES

[1] Alfredsson, K.S. (1997) 'Active and reactive structural energy flow', *ASME J. of Vibrations and Acoustics*, **119**, 70–79.

[2] Doyle, J.F. (1997) *Wave Propagation in Structures*, Springer, New York.

[3] Noiseux, D.U. (1970) 'Measurement of power flow in uniform beams and plates', *J. Acoust. Soc. Am.*, **47**, 238–247.

[4] Arruda, J.R.F. and Pereira, A.K.A. (1999) 'Intensity Methods for Noise and Vibration Analysis and Control'. *Proceedings of the 1999 International Symposium on Dynamic Problems in Mechanics and Mechatronics (EURODINAME'99)*, Ulm Research Conferences, University of Ulm, Schloss Reisensburg, Germany, July 11–16, pp. 277–284.

[5] Arruda, J.R.F. and Mas, P. (1996) 'Predicting and Measuring Flexural Power Flow in Plates', *Proceedings of the 2nd International Conference on Vibration Measurements by Laser Techniques*, Ancona, Italy, September 23–25, International Society of Optical Engineering – *SPIE* Vol. 2868, pp. 149–163.

[6] Alves, P.S.L. and Arruda, J.R.F. (2001) 'Energy Source Localization Using the Reactive Structural Intensity', *Proceedings of the 17th International Congress on Acoustics*, University of Rome 'La Sapienza', Rome, Italy, September 2–7, 2 p.

[7] Meyer, B. and Thomasson, D. (1989) 'L'intensimétrie vibratoire: description, mise en oeuvre, résultats. *EDF Bulletin de la direction des études et recherches -Série A*, **3**, Suppl. 4, 69–86.

[8] Pascal, J.-C., Loyau, T. and Mann III, J.A. (1990) 'Structural Intensity from Spatial Fourier Transformation and Bahim Acoustical Holography Method', *Proceedings of the 3rd International Congress on Intensity Techniques*, Centre technique des Industries Mécaniques, CETIM, Senlis, France, pp. 197–204.

[9] Pascal, J.-C., Loyau, T. and Carniel, X. (1993) 'Complete determination of structural intensity in plates using laser vibrometers', *Journal of Sound and Vibration*, **161**, 527–531.

[10] Blotter, J.D. and West, R.L. (1994) 'Experimental and Analytical Energy and Power Flow Using a Scanning Laser Doppler Vibrometer', *Proceedings of the 1st International Conference on Vibration Measurements by Laser Techniques*, Ancona, Italy, The International Society of Optical Engineering – SPIE, *Proceedings Series* Vol. 2358, pp. 266–275.

[11] Arruda, J.R.F. (1992) 'Surface smoothing and partial derivatives computation using a regressive discrete Fourier series', *Mechanical Systems and Signal Processing* **6**, 41–50.

[12] Arruda, J.R.F. and Mas, P. (1996) 'Predicting and Measuring Flexural Power Flow in Plates', *Proceedings of 2nd International Conference on Vibration Measurements by Laser Techniques*, Ancona, Italy, September 23–25, *SPIE*, **2868**, pp. 149–163.

[13] Pascal, J.-C. and Li, J.F. (2001) 'Irrotational Acoustic Intensity and Boundary Values', *Proceedings 17th International Congress on Acoustics*, University of Rome 'La Sapienza', Rome, Italy, September.

[14] Santos, J.M.C., Arruda, J.R.F. and Cruz, R.M. (1999) 'Localizing Structural Damage Using Structural Intensity Divergence Plots', *Proceedings of the 17th. IMAC International Conference on Modal Analysis*, Society for Experimental Mechanics (SEM), Orlando, FL, February, pp. 664–669.

[15] Pedrini, G. and Tiziani, H.J. (1994) 'Double-pulse electronic speckle interferometry for vibration analysis', *Applied Optics*, **33**, 7857–7863.

[16] Alves, P.S.L. and Arruda, J.R.F. (2001) 'Active and Reactive Power Flow Estimation Using Mindlin Plate Theory', *Proceedings of the IX International Symposium on Dynamic Problems of Mechanics (DINAME)*, Brazilian Society of Mechanical Sciences, Florianopolis, SC, 5–9 March, pp. 459–464.

[17] Batel, M., Marroquin, M., Hald, J., Christensen, J.J., Schuhmacher, A.P. and Nielsen, T.G. (2003) 'Noise source location techniques – Simple to advanced applications', *Sound and Vibration*, March, 24–36.

[18] Maynard, J.D., Williams, E.G. and Lee, Y. (1985) 'Nearfiled acoustic holography: I Theory of generalized holography and the development of NAH', *J. Acoust. Soc. Am.* **78**, 1395–1413.

[19] Schuhmacher, A. (2000) *Sound source reconstruction using inverse sound field calculations*, PhD dissertation, Department of Acoustic Technology, Technical University of Denmark, report No. 77, ISSN 1397–0547.

[20] Nelson, P.A. and Yoon, S.H. (2000) 'Estimation of acoustic source strength by inverse methods, Part I: Conditioning of the inverse problem', *Journal of Sound and Vibration*, **233**, 643–668.

[21] Schroeder, M.R. (1970) 'Synthesis of low-peak-factor signals and binary sequences with low autocorrelation', *IEEE Transactions on Information Theory*, January, 85–89.

[22] Arruda, J.R.F. (1993) 'Multisine multiexcitation in frequency response function estimation', *AIAA Journal*, **31**, 215–216.

[23] Evans, C., Rees, D., Jones, L. and Weiss, M. (1996) 'Periodic signals for measuring nonlinear Volterra kernels', *IEEE Transactions on Instrumentation and Measurement*, **45**, 362–371.

[24] Schoukens, J., Dobrowiecki, T. and Pintelon, R. (1998) 'Parametric and non-parametric identification of linear systems in the presence of nonlinear distortions', *IEEE Transactions on Automatic Control*, **43**, 176–191.

[25] Vanlanduit, S. (2001) *High Spatial Resolution Experimental modal Analysis*, PhD dissertation, Vreij Universiteit Leuven, Belgium.

[26] Mace, B.R. and Shorter, P.J. (2001) 'A local modal perturbation method for estimating frequency response statistics of built-up structures with uncertain properties', *Journal of Sound and Vibration*, **242**, 793–811.

[27] Lyon, R.H. and DeJong, R.G. (1995) *Theory and Application of Statistical Energy Analysis*. Second Edition, Butterworth-Heinemann, Burlington, MA.

[28] Bernhard, R.J. (2000) 'The Family of EFA Equations and Their Relationship to SEA', *Proceedings of the 1st Conference on Noise and Vibration Pre-Design and Characterization Using Energy Methods (NOVEM'2000)*, INSA, Lyon, France, August 31–September 2, 12 pp.

[29] Langley, R.S. and Bremner, P. (1999) 'A hybrid method for the vibration analysis of complex structural-acoustic systems', *J. Acoust. Soc. Am.*, **105**, 1657–1671.

[30] Bliss, D.B. and Franzoni, L.P. (2001) 'The Method of Local-Global Homogenization (LGH) for Structural Acoustics'. *Proceedings of the 17th International Congress on Acoustics*, University of Rome, 'La Sapienza', Rome, Italy, September 2–7, 2 p.

[31] Shorter, P. and Langley, R. (2000) 'The Spatial Correlation of Vibrational Wavefields and Their Application to Mid-Frequency Structural-Acoustics', *Proceedings of the 1st Conference on Noise and Vibration Pre-Design and Characterization Using Energy Methods (NOVEM'2000)*, INSA, Lyon, France, August 31–September 2, 12 pp.

[32] Doyle, J.F. (1997) *Wave Propagation in Structures*, Springer, New York.

[33] Ahmida, K.M. and Arruda, J.R.F. (2002) 'On the relation between complex modes and wave propagation phenomena', *Journal of Sound and Vibration*, **255**, 663–684.

[34] Lee, U. and Lee, J. (2000) 'The spectral-element method in structural dynamics', *The Shock and Vibration Digest*, **32**, 451–465.

[35] Alves, P.S.L. (2001) *Numerical and Experimental Analysis of the Energy Flow in Structures*, PhD dissertation, Universidade Estadual de Campinas, Brazil (in Portuguese).

[36] Nefske, D.J. and Sung, S.H. (1989) 'Power flow finite element analysis of dynamic systems: basic theory and application to beams', *J. Vib., Acoust., Stress, and Reliability in Design*, **111**, 94–100.

[37] Wohlever, J.C. and Bernhard, R.J. (1992) 'Mechanical energy flow models of rods and beams', *Journal of Sound and Vibration*, **153**, 1–19.

[38] Lase, Y., Ichchou, M.N. and Jezequel, L. (1996) 'Energy flow analysis of bars and beams', *Journal of Sound and Vibration*, **192**, 281–305.

[39] Ichchou, M.N., Le Bot, A. and Jezequel, L. (1997) 'Energy models of one-dimensional, multipropagative systems', *Journal of Sound and Vibration*, **201**, 53–554.

[40] Le Bot, A. (1994) *Energy Equations in Vibration Mechanics – Application to Mid and High Frequencies*, PhD dissertation, École Centrale de Lyon (in French).

[41] Cheng, A.H.-D., Golberg, M.A., Kansa, E.J. and Zammito, G. (2003) 'Exponential convergence and h-c multiquadric collocation method for partial differential equations', *Numerical Methods for Partial Differential Equations*, **19**, 571–594.

13

Impedance-Based Structural Health Monitoring

Gyuhae Park[1] and Daniel J. Inman[2]

[1]*Los Alamos National Laboratory, Los Alamos, New Mexico, USA*
[2]*Center for Intelligent Material Systems and Structures Virginia Polytechnic Institute and State University Blacksburg, Virginia, USA*

13.1 INTRODUCTION

The development of a real-time, in-service, structural health monitoring and damage detection techniques has recently attracted a large number of academic and industrial researchers. The goal of this research is to allow systems and structures to monitor their own integrity while in operation and throughout their lives, in order to prevent catastrophic failures and to reduce the costs by minimizing explicit preemptory maintenance and inspection tasks.

Impedance-based structural health monitoring techniques have been developed by utilizing the electromechanical coupling property of piezoelectric materials (Sun *et al.*, 1995) and form a new nondestructive evaluation (NDE) method. The basic concept of this approach is to monitor the variations in structural mechanical impedance caused by the presence of damage. Since structural mechanical impedance measurements are difficult to obtain, impedance methods utilize the electrical impedance of piezoelectric materials, which is directly related to the mechanical impedance of the host structure, and will be affected by the presence of structural damage. Through monitoring the measured electrical impedance and comparing it to a baseline measurement, one can

Damage Prognosis – For Aerospace, Civil and Mechanical Systems   Edited by D.J. Inman, C.R. Farrar, V. Lopes Junior and V. Steffen Junior   © 2005 John Wiley & Sons, Ltd

qualitatively determine that structural damage has occurred or is imminent. In order to ensure high sensitivity to incipient damage, the electrical impedance is measured at high frequencies (typically greater than 30 kHz). At such high frequencies, the wavelength of the excitation is small and is sensitive enough to detect minor changes in the structural integrity. More importantly, high-frequency (kilohertz) signals require very low voltage (less than 1 volt at microwatts) to produce a useful impedance excitation in the host structure. By integrating the impedance technique with self-sensing smart materials (Dosch *et al.*, 1992), it has been demonstrated that the impedance-based method is suitable for use in a wide variety of structural health monitoring applications.

13.2 ELECTRO-MECHANICAL PRINCIPLE

The health monitoring method utilizes impedance sensors to monitor changes in structural stiffness, damping and mass. The impedance sensors consist of small piezoelectric patches, usually smaller than $25 \times 25 \times 0.1$ mm, which are used to measure directly the local dynamic response.

Piezoceramic transducers acting in the 'direct' manner produce an electrical charge when stressed mechanically. Conversely, a mechanical strain is produced when an electrical field is applied. The process to be used with the impedance-based monitoring method utilizes both the direct and converse versions of the piezoelectric effect simultaneously to obtain an impedance signature. When a PZT patch is attached to a structure, and is driven by a fixed alternating electric field, a small deformation is produced in the PZT wafer and the attached structure. Since the frequency of the excitation is very high, the dynamic response of the structure reflects only a very local area to the sensor. The response of that local area to the mechanical vibration is transferred back to the PZT wafer in the form of an electrical response. When a crack or damage causes the mechanical dynamic response to change (a frequency phase shift or magnitude change in the mechanical dynamic response), it is manifested in the electrical response of the PZT wafer.

The electromechanical modeling which quantitatively describes the process is presented in Figure 13.1. The PZT is normally bonded directly to the surface of the structure by a high-strength adhesive to ensure a better electromechanical coupling. The surface-bonded PZT is considered to be a thin bar in axial vibration due to an applied alternating voltage. One end of the bar is considered fixed, whereas the other end is connected to the external structure. This assumption regarding the interaction at

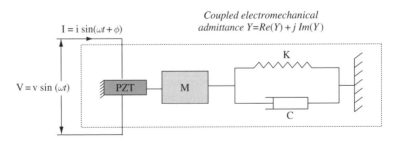

Figure 13.1   1-D model used to represent a PZT-driven dynamic structural system

two discrete points is consistent with the mechanism of force transfer from the bonded PZT transducer to the structure.

The solution of the wave equation for the PZT bar connected to the structure leads to the following equation for a frequency-dependent electrical admittance (Liang *et al.*, 1994):

$$Y(\omega) = i\omega\, a\left(\bar{\varepsilon}_{33}{}^{T}(1 - i\delta) - \frac{Z_s(\omega)}{Z_s(\omega) + Z_a(\omega)} d_{3x}{}^{2}\, \hat{Y}_{xx}^{E}\right) \qquad (13.1)$$

In equation (13.1), Y is the electrical admittance (inverse of impedance), Z_a and Z_s are the PZT material's and the structure's mechanical impedances, respectively, $\hat{Y}_{xx}^{E}$ is the complex Young's modulus of the PZT with zero electric field, d_{3x} is the piezoelectric coupling constant in the arbitrary x direction at zero stress, $\varepsilon_{33}{}^{T}$ is the dielectric constant at zero stress, δ is the dielectric loss tangent of the PZT, and a is a geometric constant of the PZT. This equation indicates that the electrical impedance of the PZT bonded onto the structure is directly related to the mechanical impedance of a host structure. The variation in the PZT electrical impedance over a range of frequencies is analogous to that of the frequency response functions (FRF) of a structure, which contains vital information regarding the health of the structure.

Damage to a structure causes direct changes in the structural stiffness and/or damping and alters the local dynamic characteristics. In other words, the mechanical impedance is modified by structural damage. Since all other PZT properties remain constant, it is Z_s, the external structure's impedance, that uniquely determines the overall admittance. Therefore, any change in the electrical impedance signature is considered an indication of a change in the structural integrity.

An experimental modal testing using the electrical impedance of PZT patches (as colocated actuators and sensors) is presented by Sun *et al.* (1996). In this paper, the authors discuss that both the point frequency response functions of a single location and the transfer frequency response function between two locations on a structure can be obtained by measured electrical impedance. This work provides a critical insight into the impedance-based structural health monitoring technique, in which the electrical impedance of piezo-ceramic materials constitutes a unique signature of the dynamic behavior of the structures.

Experimental implementation of the impedance-based structural health monitoring technique has been successfully conducted on several complex structures; a four-bay space truss (Sun *et al.*, 1995), an aircraft structure (Chaudhry *et al.*, 1995), complex precision parts (Lalande *et al.*, 1996), temperature varying applications (Park *et al.*, 1999), a spot-welded structural joints (Giurgiutiu *et al.*, 1999), civil structural compon-ents (Park *et al.*, 2000), a reinforced concrete bridge (Soh *et al.*, 2000), and civil pipelines (Park *et al.*, 2001). A complete review summarizing both hardware and software issues of the impedance methods can be found in Park *et al.* (2003b).

13.3 PARAMETERS OF THE TECHNIQUE

13.3.1 Frequency Range

The sensitivity of NDE techniques in detecting damage is closely related to the fre-quency band selected. To sense incipient-type damage that does not result in any measurable change in the structure's global stiffness properties, it is necessary for the wavelength of excitation to be smaller than the characteristic length of the damage to be detected (Stokes and Cloud, 1993). Hence, the frequency range typically used in the impedance methods is in the range of 30 kHz to 250 kHz. The range for a given structure is determined by a trial and error method. There is little analytical work done about the vibration modes of complex structures at these ultrasonic frequencies. It has been found that a frequency range with a high mode density exhibits a higher sensitivity since it generally covers more structural dynamic information (Sun *et al.*, 1995). In the impedance-based method, multiple numbers (usually two or three) of a frequency range containing 20–30 numbers of peaks are usually chosen, since a number of peaks implies that there is a greater dynamic interaction over that frequency range. A higher frequency range (higher than 150 kHz) is found to be favorable in localizing the sensing, while a lower frequency range (lower than 70 kHz) covers more sensing areas. This is due to the fact that damping becomes more dominant at high frequency. It must be noted that there are two different kinds of peaks on measured electrical impedance. One reflects the structural resonant frequencies and the other is the PZT's resonant frequencies. For lightweight structures, it is advisable to avoid PZT's resonances when selecting frequency bands because they are much greater in magnitude compared with structural resonances.

13.3.2 Sensing Region

Under the high frequency ranges used in this impedance-based method, the sensing region of the PZT is localized to a region close to the sensor/actuator. Extensive theoretical modeling efforts based on the wave propagation approach have been performed to identify the sensing region of the impedance-based method (Esteban, 1996). Esteban's work also included a parametric study on the sensing region of a PZT sensor/actuator by considering the various factors, such as mass loading effect, discontinuities in cross-section, multimember junctions, bolted structures, and energy absorbent interlayers. At such high frequency ranges, however, exact measurements and quantification of energy losses became very difficult and very little additional information was obtained. Based on the knowledge acquired through various case studies, it has been estimated that (depending on the material and density of the structure) the sensing area of a single PZT can vary anywhere from 0.4 m (sensing radius) on composite reinforced concrete structures, to 2 m on simple metal beams. Castanien and Liang (1996), and Kabeya (1998) used transfer impedance or transfer admittance to interrogate the structure in order to extend the sensing region of the impedance-based health monitoring technique.

13.3.3 Damage Assessment

While the impedance response plots provide a qualitative approach for damage identification, the quantitative assessment of damage is traditionally made by the use of a scalar damage metric. In the earlier work (Sun *et al.*, 1995), a simple statistical algorithm was used, based on frequency-by-frequency comparisons, referred to as 'root mean square deviation' (RMSD),

$$M = \sum_{i=1}^{n} \sqrt{\frac{[\mathrm{Re}(Z_{i,1}) - \mathrm{Re}(Z_{i,2})]^2}{[\mathrm{Re}(Z_{i,1})]^2}} \qquad (13.2)$$

where M represents the damage metric, $Z_{i,1}$ is the impedance of the PZT measured at healthy conditions, and $Z_{i,2}$ is the impedance for the comparison with the baseline measurement at frequency interval i. In a RMSD damage metric chart, the greater numerical value of the metric, the larger the difference between the baseline reading and the subsequent reading indicates the presence of damage in a structure. Raju (1998) adopted another scalar damage metric, referred to as the 'correlation' metric, which can be used to interpret and quantify the information from different data sets. The correlation coefficient between two data sets determines the degree of linear relationship between two impedance signatures, and provides an aesthetic metric chart. In most cases, the results with the correlating metric are consistent with those of RMSD, in which the metric values increase when there is an increase in the severity of damage.

The damage metric simplifies the interpretation of impedance variations and provides a summary of the information obtained from the impedance response curves. Using this damage metric in conjunction with a damage threshold value, this technique can warn inspectors in a green/red light form, whether or not the threshold value has been reached.

Temperature changes, among all other ambient conditions, significantly affect the electric impedance signatures measured by a PZT. Some of PZT material parameters, such as the dielectric constant, are strongly dependent on temperature. Generally speaking, an increase in temperature causes a decrease in the magnitude, and leftward shifting of the real part of the electric impedances. The RMSD and correlation based damage metrics do not account for these variations. Park et al. (1999) used a modified RMSD metric, which compensates for horizontal and vertical shifts of the impedance in order to minimize the impedance signature drifts caused by the temperature or normal variations.

Lopes et al. (2000) incorporated neural network features with the impedance method for somewhat quantitative damage analysis. The authors proposed a two-step damage identification scheme. In the first step, the impedance-based method detects and locates structural damage and provides damage indication in a green/red light form with the use of the modified RMSD. When damage is identified, the neural networks, which are trained for each specific damage, are used to estimate the severity of damage. Zagrai and Giurgiutiu (2001) investigated several statistics-based damage metrics, including RMSD, mean absolute percentage deviation (MAPD), covariance change, and correlation coefficient deviation. It has been found that the third power of the correlation coefficient deviation, $(1-R^2)^3$, is the most successful damage indicator, which tends to decrease linearly as the crack in a thin plate moves away from the sensor. Tseng et al. (2002) also investigated the performance of RMSD, MAPD, covariance and correlation coefficients as indicators of damage. The RMSD and the MAPD were found to be suitable for characterizing the growth and the location of damage, whereas the covariance and the correlation coefficient are efficient in quantifying the increase in damage size at a fixed location.

The main limitation of the use of the aforementioned damage metrics in impedance methods is how to set appropriate decision limits or thresholds values. The decision is typically made based on arbitrary values, i.e. 'small variations' for undamaged cases and 'large variations' for damaged cases. In order to diagnose damage with levels of statistical confidence, the impedance-based monitoring is cast in the context of an outlier detection framework (Park *et al.*, 2003a; Fasel *et al.*, 2003). An auto-regressive model with exogenous inputs (AR-ARX) in the frequency domain (Adams, 2001) is incorporated into the impedance methods for nonlinear damage discrimination (Fasel *et al.*, 2003). Because nonlinear feature identification requires separate input and output measurement, which is not possible with the traditional impedance analyzers, a modified frequency AR-ARX model is proposed (Park *et al.*, 2003a). The damage sensitive feature is computed by differentiating the measured impedance and the output of the ARX model. Furthermore, because of the non-Gaussian nature of the feature distribution tails, extreme value statistics (EVS) is employed to develop a robust damage classifier (Sohn *et al.*, 2003).

13.4 COMPARISONS WITH OTHER DAMAGE IDENTIFICATION APPROACHES

Traditional NDE techniques include ultrasonic technology, acoustic emission, magnetic field analysis, penetrant testing, eddy current techniques, X-ray analysis, impact-echo testing, global structural response analysis, and visual inspections. Each of these various techniques has their positive and negative virtues. For instance, the ultrasonic method is useful in providing details of damage in a structure, however, this method requires the knowledge of damage location *a priori* and renders the structure unavailable throughout the duration of the test. Many traditional NDE methods require out of service periods, or can be applied only a certain intervals, while the impedance-based method provides continuous, on-line monitoring with the potential for autonomous use.

13.4.1 Comparison to Global Structural Vibration-Based Methods

Like the global structural methods, the impedance-based approach involves the comparison of vibratory patterns ('signatures') taken at various times during the life of the structure. The major difference, however, deals with the frequency range used to detect the changes in structural integrity. Relying on the lower-order global modes, the low-frequency global techniques are not sensitive to damage that has occurred at a very early stage. It has been shown (Banks *et al.*, 1996) that high frequency responses are very sensitive to changes in the structural integrity. By employing a high frequency range, the impedance-based method provides an alternative procedure that can identify local, minor changes in structural integrity.

13.4.2 Impedance Signature vs Ultrasonic Testing

In ultrasonic testing of structural components, a piezo-transducer is used to produce an acoustic wave in the component. Based on the time delay of the wave transmission, the change in length (strain) and/or density of the component is determined. Usually the mechanical nature of the component must be fairly well known before testing so that the frequency of the ultrasonic signal can be chosen to correlate with the mechanical response of the component. Typically, a single frequency wave or only a few different frequencies are used in ultrasonic methods. A broadband signal is not obtained as in the impedance signature method. The ultrasonic method is useful in some structures for obtaining a picture of various embedded components or material anomalies. This method however does not lend itself to autonomous use as does the impedance method and experienced technicians are required to review the ultrasonic data to discern detail.

13.4.3   Impedance Signature vs Acoustic Emission

The acoustic emission (AE) method uses the elastic waves generated by crack initiation, moving dislocations, and disbonds for detection and analysis of structures. The AE method is suitable for long-term, in-service, monitoring like the impedance method. Both methods are ideal for monitoring critical sections where high structural integrity should be maintained. However, the AE method requires stress or chemical activity to generate the acoustic emission, while the impedance method can easily solve the problems associating with 'how to excite structures' by using the concept of self-sensing actuation (Dosch *et al.*, 1992). The advantage of the self-sensing actuator is more obvious in the sense that, in the AE method, the existence of multiple numbers of travel paths from the source to the sensor can make signal identification difficult (Bray and McBride, 1992). In addition, the AE method needs to filter out the electrical interference and ambient noise from the emission signals, whereas the limited sensing area of the impedance method helps in isolating changes in the impedance signature due to other far-field changes such as mass loading and normal operational vibrations.

13.4.4   Impedance Signature vs Impact-Echo Testing

For impact-echo (IE) testing, a stress pulse is introduced into the structure from an impact source and resulting stress waves are measured and analyzed by a transducer. The pulse propagates into the structure and is reflected by cracks or disbonds of the structures. IE testing has been used to assess the conditions of various civil structures, including concrete, wood, and masonry materials. However, IE testing requires an external source to excite a pulse and does not lend itself to autonomous use like the impedance method. The IE testing technique has been shown to be fairly effective for detecting and locating large scale voids and delaminations, but is not sensitive to the presence of small cracks and discontinuities due to the relatively low frequencies involved.

The principal advantages of the impedance approach compared to other techniques are as follows:

- The technique is not based on any model, and thus can be easily applied to complex structures.
- The technique uses small nonintrusive actuators to monitor inaccessible locations.

- The sensor (PZT) exhibits excellent features under normal working conditions, has a large range of linearity, fast response, light weight, high conversion efficiency, and long term stability.
- The technique, because of high frequency, is very sensitive to local minor changes.
- The measured data can be easily interpreted.
- The technique can be implemented for on-line health monitoring.
- The continuous monitoring provides a better assessment of the current status of the structure, which can eliminate scheduled base inspections.

Impedance-based structural health monitoring could provide a compromise interface between global structural methods and the traditional high frequency NDE techniques. With a limited number of sensors and actuators, critical areas of a structure can be monitored, which is one of the advantages of global structural methods. Damage in an incipient stage can be accurately identified, which only local inspection techniques, such as ultrasonics, can possibly detect.

13.5   PROOF-OF-CONCEPT APPLICATIONS

The impedance-based health monitoring technique has been successfully applied to several structural components. Damage detection on two structures, including a pipe-line structure and a quarter-scale bridge section, is presented to illustrate the potential of the impedance-based method for locating local damage in civil applications.

13.6   HEALTH ASSESSMENT OF PIPELINE STRUCTURES

Pipelines convey natural gas, oil, and water, and some pipelines contain communication and power cables, all of which are very important for maintaining functional residential and industrial facilities. Pipelines are also required for economic and community recovery after natural disasters. However, pipelines are severely damaged by shaking, liquefaction, and landslides during earthquakes (O'Rourke, and Palmer, 1996; Koseki *et al.*, 1998) and the immediate assessment of pipeline facilities is critical in preventing fires, explosions, and pollution from broken gas or sewage lines. Although extensive research efforts have been focused on assessing the conditions of pipelines after earth-quakes (Hwang *et al.*, 1998), the condition monitoring of these structures is still based on limited information. Therefore, the possibility of implementing the impedance-based health monitoring technique for pipeline structural damage assessment has been inves-tigated and its ability immediately to detect and locate damage has been demonstrated.

13.6.1   Experimental Setup

Bolted joints are frequently used to connect segmented pipelines in building piping systems. This interface can be the most critical source of failure of the pipelines, since significant seismic loadings can stress the joint beyond its yield or buckling capacity,

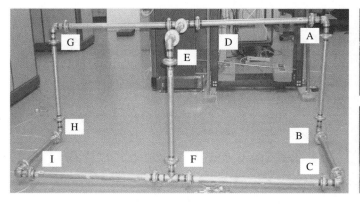

Figure 13.2   A pipeline used in the experiment

while the main body of the pipe remains elastic (Eidinger, 1999). Therefore, the conditions of these joints need to be monitored to ensure the integrity of entire pipelines.

A model of a pipeline with bolted joints is shown in Figure 13.2. This model consists of segmented pipes (d.40 mm), flanges, elbows, and joints connected by more than 100 bolts. The size of this structure is 2 m wide and 1.3 m tall. One PZT sensor/actuator (15 × 15 × 0.2 mm) is bonded on each joint to monitor the conditions of this structure. The HP4194 electrical impedance analyzer was used for the measurement of PZT's electrical impedance in the frequency range of 80–100 kHz. The total impedance of each junction (two or three joints), labeled A to I in Figure 13.2, was utilized to track the damage. The total impedance refers to 'a single impedance signal acquiring from distributed PZTs'; the leads from the several distributed PZTs are physically connected together and this single lead is then connected to the terminal on the impedance analyzer. This procedure may reduce the sensitivity to structural damage due to the multiplexing nature of measurements; however it drastically reduces the interrogation time as compared with that of analyzing each PZT separately. After measuring the baseline impedance signature, damage was introduced by slightly loosening the bolts over several joints on this structure.

13.6.2   Observations and Analysis

The impedance measurements (real part) of PZTs at Junction D with four levels of local damage are shown in Figure 13.3. Only the real portion of the electrical impedance is analyzed to predict damage because it is more sensitive to change than the imaginary part or magnitude. The damage was simulated by completely loosening bolts at junction D, as shown in the labels. It can be seen from the figure that, with increasing damage, the impedance signature shows a relatively large change in shape and is clearly indicative of imminent damage.

For the first level of damage (loosening two bolts), only a small variation along the original signal (undamaged curve) was observed. This is because the first level of damage can be categorized as the incipient stage. When four bolts had been loosened, the impedance showed more pronounced variations as compared with previous

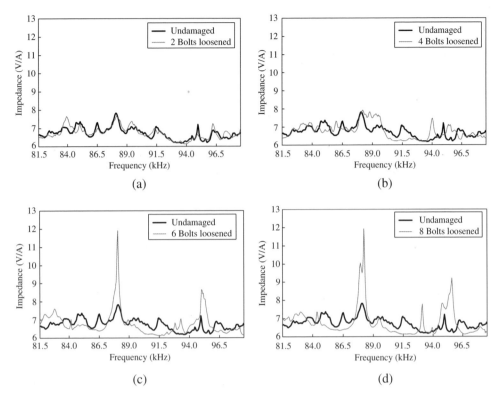

Figure 13.3 The electrical impedance measurements of PZTs at Junction D. The variation in the impedance is increased as the level of damage is increased. (a) Two bolts loosened; (b) Four bolts loosened; (c) Six bolts loosened; (d) Eight bolts loosened

readings, and finally, when six and eight bolts had been loosened, it showed a distinct change in the signature pattern, i.e. new peaks and valleys appear in the entire frequency range. This change occurs because the damage modifies the apparent stiffness and damping of the joint. This variation shows the extreme sensitivity of the impedance-based method to the presence of damage in the sensing area.

A damage metric chart is illustrated in Figure 13.4. The damage metric chart based on RMSD was constructed after each measurement had been taken. As can be seen in the figure, with an increase in extent of damage, there was a corresponding increase in the damage metric values. Although the impedance method cannot precisely predict the exact nature and size of the damage, the method provides somewhat quantitative information on the conditions of a structure by showing an increasing damage metric with increased severity of damage.

Another experiment was performed on the global scale. Three conditions were imposed on this structure in sequence, as shown below:

- Damage 1: loosening three bolts at Junctions A and B, respectively;
- Damage 2: loosening two bolts at Junctions E and G, respectively;
- Damage 3: loosening four bolts at Junctions F, G, and H, respectively.

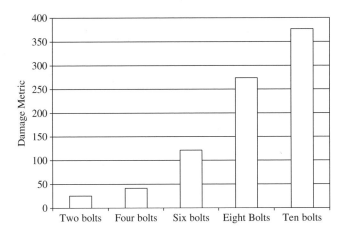

Figure 13.4 Damage metric chart. Comparison of metric values with induced damage

The impedance measurements (real part) of PZTs located at junction A are shown in Figure 13.5. For junction A, when damage 1 was introduced, the measurement was significantly different from the baseline measurement, which is indicative of damage. However, when the other two damage conditions were imposed, the remaining curves followed the same pattern as that of the second reading, since those are well out of sensing range of PZTs at Junction A. Other impedance measurements of junction G are shown in Figure 13.6. The location of damage 1 was out of the sensing range of PZTs, hence almost no change in impedance curve was observed. However, when damage 2 was introduced, the impedance measurement was significantly different from the previous readings and was affected by the presence of damage. When damage 3 was introduced, the measurements indicated another complete change in the signature pattern.

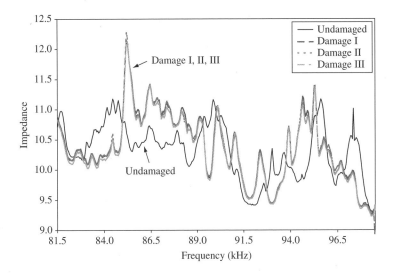

Figure 13.5 The electrical impedance measurements of PZTs at junction A

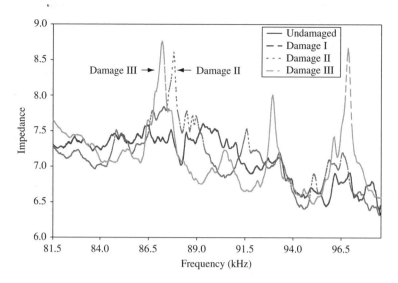

Figure 13.6   The electrical impedance measurements of PZTs at junction G

The damage metric chart demonstrates the results more clearly, as can be seen in Figure 13.7. It can be seen that at damage 1, there is a large increase in the damage metric value for PZTs at A and B. The other PZTs show a very small change in the damage metric, because they are distant from the damage. Similar results are obtained when damages 2 and 3 were induced. Each PZT shows an increase in the damage metric value, if damage is induced close to the sensors. By looking for variations in the impedance measurement and in the damage metric value, structural damage can be detected and the integrity of the structure can be monitored throughout its service life or immediately after the natural disaster, as shown in this example.

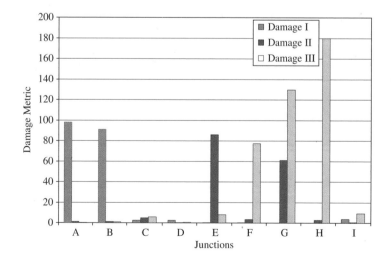

Figure 13.7   Damage metric chart over the different locations

The use of the damage metric charts in providing a quick, accurate summary of the health of the structure became obvious during the testing. The time necessary to take the impedance measurements and to construct the damage metric is less than 5 minutes, which is quick enough for an on-line implementation of this technique.

13.7 ANALYSIS OF A QUARTER SCALE BRIDGE SECTION

In almost all practical field applications, the structure being monitored is constantly undergoing changes due to external boundary conditions, such as loading, low-frequency vibrations of the structure, and changes in the ambient temperature. These effects make it difficult to detect and locate structural damage because such ambient changes also modify the response of a structure. However, a compensation technique was developed to minimize the effects of temperature and normal variations, making the impedance-based structural health monitoring technique stable under all types of environmental condition (Park *et al.*, 1999).

13.7.1 Experimental Setup

An investigation on a massive quarter scale model of a steel truss bridge joint is presented. A model of a steel bridge joint is shown in Figure 13.8. The bridge model consists of steel angles, channels, plates, and joints connected by over 200 bolts. The size of this structure is 1.8 m tall and has a mass of over 250 kg. Four PZT sensor/actuators

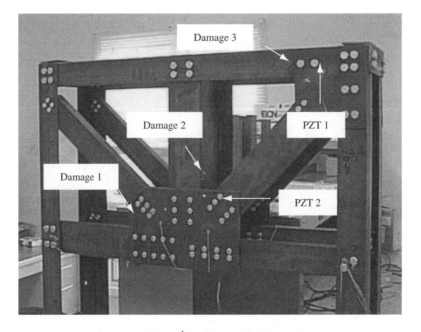

Figure 13.8 A $\frac{1}{4}$-scale steel bridge section

are bonded on the critical sections to monitor actively the conditions of this typical high-strength civil structure. The purpose of the experiment presented here is to examine the effect of external boundary conditions on the impedance signature, and to obtain a better understanding of practical issues that are a result of monitoring the health of structures in an uncontrolled field environment.

The following three ambient boundary conditions were imposed on the structure in an attempt to simulate real-life variation;

- repeatability – variations of the signal over a given time period is monitored;
- vibrations – structure is manually hammered while the measurements are being taken;
- loading – a load of 15-kg mass is added to the structure. The weight is placed in the vicinity of PZT sensors, so that it induces the stresses on bolted connections within the sensing range of PZT sensor/actuators.

These sets of readings from four PZTs are repeated over a period of 3 weeks. After identifying the range of the impedance signature variations due to the boundary condition changes, damage was induced by loosening the bolts over several locations on the structure. The HP 4194 impedance analyzer is used to interrogate each PZT. Throughout the analysis, the compensation technique (Park *et al.*, 1999) to minimize the effects of any boundary condition changes was applied.

13.7.2   *Test Observations and Analysis*

The impedance measurements of two PZTs are presented in Figure 13.9, with two different frequency ranges. Each plot shows the variations of the impedance signature with three ambient condition changes imposed on the structure. Only fourteen measurements, which show the largest variations, are shown without the labels. As can be seen, the variations remained relatively small and would be considered as minor changes. The vibration produced the largest variations, to be was expected as the structures were being hammered while measurements were being taken. As compared with the modal analysis experiments, where a small orientation change results in marked changes in

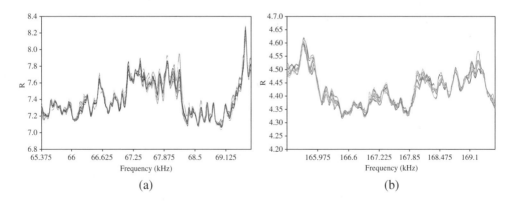

Figure 13.9   Impedance (real) vs frequency plots for (a) PZT 1, and (b) PZT 2

resonant frequencies, mode shapes, and modal damping, the impedance signature patterns showed relatively small variations. The measurements were found to be repeatable and no noticeable degradation with time was observed.

The damage metric chart is presented in Figure 13.10. The first fourteen variations are those from the changes in boundary conditions. These pronounced ones are the damage metric due to the vibration. As depicted, the values are very small and hence, negligible. The location of damage 1, 2 and 3 are shown in Figure 13.8. Damage is simulated by loosening a bolt (1/8 turn) in that location. The exact sensing range of each PZT sensor was difficult to predict, since a number of bolts were presented in this structure. Consistent with the results of others (Esteban, 1996), the bolted joints were the major contributors of energy dissipation in the structures.

Damage 1 is believed to be well out of sensing range of both PZT 1 and PZT 2. Hence, only a small increase in damage metric is shown for both PZTs. However, PZT 2 shows an increase in metric value (due to damage 1) over that of any of the increase caused by the normal variations. This small increase cannot be used to signal the presence of damage, however, it does provide evidence of the sensitivity of this method in relatively large ranges. Damage 2 is located close to the PZT 2 and damage 3 is within sensing region of the PZT 1, hence the increase in damage metric values is the highest for both PZTs. Note that the effect of loosening a single bolt on the entire structure is minor, thus damage can be detected in its early stage. The impedance measurements of PZT 2 for the cases of both damage 2 and damage 3 are shown in Figure 13.11, for visual comparison. It can be seen that, when damage 2 was introduced, the impedance measurement was significantly different from the previous readings. However, in the case of damage 3, only a small variation in the impedance measurement was observed, since damage is distant from this PZT sensor/actuator.

An important problem associated with monitoring large-scale civil structures is that they possess very low natural frequencies and are difficult to excite, leading to difficulties in picking up very small frequency changes for the damage detection technique

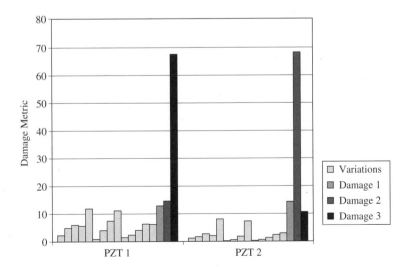

Figure 13.10 Damage metric chart for PZTs. Comparison of metric values with induced damage

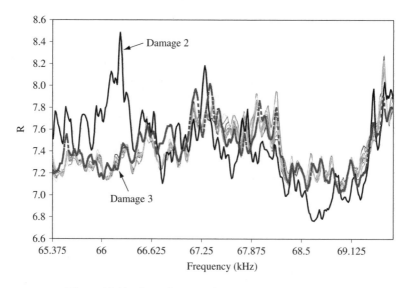

Figure 13.11   Impedance vs frequency plots for PZT 2

using low frequency vibration data (Friswell and Penny, 1997). However, as demonstrated in this example, the impedance-based structural health-monitoring technique can be easily applied to relatively large, massive civil structures by employing a very-high-frequency local excitation.

Bolted and riveted connections are commonly found in civil structures. These connections invariably promote damage growth due to the nature of the geometry and local stress concentrations. It has been estimated that approximately 70 % of all mechanical failure occurs due to fastener failure (Park *et al.*, 2003c). However, these connections are often difficult to inspect due to the geometry and/or the location in the structure. Various types of bolt failure that occur include tensile overload, shear overload, hydrogen embrittlement and fatigue failure. Although only the changes in joint stiffness are used to simulate real-time damage, the results of this experiment support the effectiveness of the impedance-based technique in monitoring the condition of various civil applications that are subjected either to adverse environments that can degrade the connections, or to strenuous loading cycles causing bolt cracking and fatigue damage. It should be noted that the damage considered in this article contains bolted joint failures only; this is mainly because they are easy to simulate, control, and enable repeatable tests. There is ample evidence in the references (Park *et al.*, 2003b) that the impedance method can successfully detect and locate any possible types of damage in structures.

13.8 SUMMARY

The experimental investigations of impedance-based health monitoring techniques on various components typical of civil infrastructures were presented. The basic concept of this health monitoring technique is to monitor the variations in the structural mechanical impedance caused by the presence of damage at high frequency range (typically higher than 30 kHz), utilizing the electromechanical coupling properties of piezoelectric materials.

Impedance-based structural health monitoring is slowly coming into full view of the structural NDE community. With continual advances in sensor/actuator technology, signal processing techniques, and damage prognosis algorithms, the methods will continue to attract the attentions of researchers and field engineers for monitoring of various structural applications.

REFERENCES

Adams, D.E. (2001) 'Frequency domain ARX models and multi-harmonic FRF estimators for nonlinear dynamic systems', *Journal of Sound and Vibration*, **250**, 935–950.

Banks, H.T., Inman, D.J., Leo, D.J. and Wang, Y. (1996) 'An experimentally validated damage detection theory in smart structures', *Journal of Sound and Vibration*, **191**, 859–880.

Bray, D. and McBride, D. (1992) *Nondestructive Testing Techniques*, John Wiley & Sons, Inc., New York.

Castanien, K.E. and Liang, C. (1996) 'Application of Active Structural Health Monitoring Technique to Aircraft Fuselage Structures', *Proceeding of SPIE Smart Structure Conference*, **2721**, pp. 38–50.

Chaudhry, Z., Lalande, F., Ganino, F., Rogers, C.A. and Chung, J. (1995) 'Monitoring the Integrity of Composite Patch Structural Repair Via Piezoelectric Actuators/Sensors', *Proceedings of AIAA/ASME/ASCE/AHS/ASC 36th Structures, Structural Dynamics and Materials Conference, Adaptive Structures Forum*, AIAA Publishing, pp. 2243–2248.

Dosch, J.J., Inman, D.J. and Garcia, E. (1992) 'A Self sensing piezoelectric actuator for collocated control', *Journal of Intelligent Material Systems and Structures*, **3**, 166–185.

Esteban, J. (1996) *Modeling of the Sensing Region of a Piezoelectric Actuator/Sensor*, PhD Dissertation, Virginia Polytechnic Institute and State University, Blacksburg, Virginia.

Eidinger, J.M. (1999) 'Girth Joints in Steel Pipelines subjected to Wrinkling and Ovalling', *Proceedings, Fifth US Conference on Lifeline Earthquake Engineering*, ASCE Technical Council on Lifeline Earthquake Engineering, pp. 100–109.

Fasel, T.R., Sohn, H., Park, G. and Farrar, C.R. (2003) 'Active sensing using impedance-based ARX models and extreme value statistics to damage detection', *Earthquake Engineering & Structural Dynamics Journal*, submitted for publication.

Friswell, M.I. and Penny, J.E. (1997) 'The Practical Limits of Damage Detection and Location using Vibration Data', *Procdings, Eleventh VPI&SU Symposium on Structural Dynamics and Control*, Blacksburg, Virginia, pp. 1–10.

Giurgiutiu, V., Reynolds, A. and Rogers, C.A. (1999) 'Experimental investigation of E/M impedance health monitoring of spot-welded structural joints', *Journal of Intelligent Material Systems and Structures*, **10**, 802–812.

Hwang, H., Lin, H. and Shinozuka, M. (1998) 'Seismic performance assessment of water delivery systems', *Journal of Infrastructure Systems*, **4**, 118–125.

Kabeya, K. (1998) *Structural Health Monitoring Using Multiple Piezoelectric Sensors and Actuators*, Master's thesis, Virginia Polytechnic Institute and State University, Blacksburg, Virginia.

Kitaura, M., Miyajima, M. and Namatame, N. (1998) 'Damage to Water Supply Pipelines Druing the 1995 Hyogeken Nambu Earthquake and Its Seismic Response Analysis', *Proceedings, Third China-Japan-US Trilateral Symposium on Lifeline Earthquake Engineering*, pp. 81–88.

Koseki, J. Matsuo, O. and Tanaka, S. (1998) 'Uplift of sewer pipes caused by earthquake-induced liquefaction of surrounding soil', *Soils and Foundations*, **38**, 75–87.

Lalande, F., Childs, B., Chaudhry, Z. and Rogers, C.A. (1996) 'High-Frequency Impedance Analysis for NDE of Complex Precision Parts', *Proceedings of SPIE Conference on Smart Structures and Materials*, SPIE Publishing, Vol. 2717, pp. 237–245.

Liang, C., Sun, F.P. and Rogers, C.A. (1994) 'Coupled electromechanical analysis of adaptive material system – determination of actuator power consumption and system energy transfer', *Journal of Intelligent Material Systems and Structures*, **5**, 21–20.

Lopes, V., Park, G., Cudney, H. and Inman, D.J. (2000) 'A structural health monitoring technique using artificial neural network and structural impedance sensors', *Journal of Intelligent Material Systems and Structures*, **11**, 206–214.

O'Rourke T.D. and Palmer, M.C. (1996) 'Earthquake performance of gas transmission pipelines', *Earthquake Spectra*, **20**, 493–527.

Park, G., Kabeya, K., Cudney, H. and Inman, D.J. (1999) 'Impedance-based structural health monitoring for temperature varying applications', *JSME International Journal*, **42**, 249–258.

Park, G., Cudney, H. and Inman, D.J. (2000) 'Impedance-based health monitoring of civil structural components', *ASCE Journal of Infrastructure Systems*, **6**, 153–160.

Park, G., Cudney, H. and Inman, D.J. (2001) 'Feasibility of using impedance-based damage assessment for pipeline systems', *Earthquake Engineering & Structural Dynamics Journal*, **30**, 1463–1474.

Park, G., Rutherford, A.C., Sohn, H. and Farrar, C.R. (2003a) 'An outlier analysis framework for impedance-based structural health monitoring', *Journal of Sound and Vibration*, accepted for publication.

Park, G., Sohn, H., Farrar, C.R. and Inman, D.J. (2003b) 'Overview of piezoelectric impedance-based health monitoring and path forward', *The Shock and Vibration Digest*, **35**, 451–463.

Park, G., Muntges, D.E. and Inman, D.J. (2003c) 'Self-repairing joints employing shape memory alloy actuators', *Journal of Minerals, Metals and Materials Society*, **55** (12), 33–37.

Raju, V. (1998) *Implementing Impedance-Based Health Monitoring Technique*, Master's thesis, Virginia Polytechnic Institute and State University, Blacksburg, Virginia.

Soh, C.K., Tseng, K., Bhalla, S. and Gupta, A. (2000) 'Performance of smart piezoceramic patches in health monitoring of a RC bridge', *Smart Materials and Structures*, **9**, 533–542.

Sohn, H., Allen, D.W., Worden, K. and Farrar, C.R. (2003) 'Structural damage classification using extreme value statistics', *ASME Journal of Dynamic Systems, Measurement, and Control*, submitted for publication.

Stokes, J.P. and Cloud, G.L. (1993) 'The application of interferometric techniques to the nondestructive inspection of fiber-reinforced materials', *Experimental Mechanics*, **33**, 314–319.

Sun, F., Chaudhry, Z., Liang, C. and Rogers, C.A. (1995) 'Truss structure integrity identification using PZT sensor-actuator', *Journal of Intelligent Material Systems and Structure*, **6**, 134–139.

Sun, F., Roger, C.A. and Liang, C. (1996) 'Structural Frequency Response Function Acquisition via Electric Impedance Measurement of Surface-Bonded Piezoelectric Sensor/Actuator', *Proceedings of the AIAA/ASME/ASCE/AHS/ASC Structures, Structural Dynamics, and Materials Conference*, pp. 3450–3461.

Tseng, K., Basu, P.K. and Wang, L. (2002) 'Damage Identification of Civil Infrastructures using Smart Piezoceramic Sensors', *Proceedings of the First European Workshop on Structural Health Monitoring*, D.L. Balageas (Ed.), DES tech Publications, Lancaster, PA, pp. 450–457.

Zagrai, A.N. and Giurgiutiu, V. (2001) 'Electro-mechanical impedance method for crack detection in thin plates', *Journal of Intelligent Material Systems and Structures*, **12**, 709–718.

14

Statistical Pattern Recognition Paradigm Applied to Defect Detection in Composite Plates

Hoon Sohn

Carnegie Mellon University, USA

14.1 INTRODUCTION

Many aerospace, civil, and mechanical engineering systems continue to be used despite aging and the associated potential for damage accumulation. Therefore, the ability to monitor the structural health of these systems is becoming increasingly important from both economic and life-safety viewpoints. The author has been tackling the damage detection problems based on the statistical analysis of measured test data, and has completed a recent thorough review of damage identification methods based on this statistical pattern recognition point of view [1].

This statistical pattern recognition paradigm can be described as a four-part process: (i) operational evaluation, (ii) data acquisition, (iii) feature extraction, and (iv) statistical model development. More detailed discussion of the statistical pattern recognition paradigm can be found in Reference [2]. In particular, this statistical pattern recognition paradigm has been applied to address a specific problem: delamination detection in composite materials.

Damage Prognosis – For Aerospace, Civil and Mechanical Systems   Edited by D.J. Inman, C.R. Farrar, V. Lopes Junior and V. Steffen Junior   © 2005 John Wiley & Sons, Ltd

This study contributes to damage detection of composite laminates by developing an improved, wavelet-based, signal-processing technique that enhances the visibility and interpretation of the Lamb-wave signals related to defects in a multilayer composite plate. In addition, a statistically rigorous damage classifier is developed to identify wave propagation paths affected by damage. Finally, a new damage location algorithm is proposed to locate damage based on signal attenuation rather than time-of-arrival information.

14.2 STATISTICAL PATTERN RECOGNITION PARADIGM

14.2.1 Operational Evaluation

Operational evaluation begins to set the limitations on what will be monitored and how the monitoring will be accomplished. This evaluation starts to tailor the damage detection process to features that are unique to the system being monitored, and tries to take advantage of unique features of the damage that is to be detected. In general, the following issues must be answered as part of the operational evaluation step in the SHM process.

(a) What are the life-safety and/or economic justification for performing the structural health monitoring?
(b) How is damage defined for the system being investigated and, for multiple damage possibilities, which cases are of the most concern?
(c) What are the conditions, both operational and environmental, under which the system to be monitored functions?
(d) What are the limitations on acquiring data in the operational environment?

In recent years, there have been increased economic and life-safety demands continuously to monitor the conditions and long-term deterioration of civil infrastructure and mechanical assemblies to ensure their safety and adequate performance throughout their life spans. For instance, the United States Air Force is currently interested in deploying an onboard SHM system for unmanned aerial vehicles (UAV) such as the Predator shown in Figure 14.1. The Predator was originally designed for surveillance purposes, and it was modified later on to carry Hellfire missiles during the Afghanistan war. Because of this additional payload, the United States Air Force is concerned about fatigue and delamination damage near the connections between the wings and the fuselage. The damage can further alter the flutter characteristics of the aircraft, causing its instability and unexpected failure. It is projected that the United States Military will spend over 15 billion dollars by 2009 on UAV research and development for military applications [3].

The ultimate goal of this study is to develop an embeddable SHM system for UAV. Composite materials are commonly used in UAV to improve performance by reducing weight and cost and by increasing stiffness and strength. In fact, the whole structural components and skins of the Predator are made of composite materials, and the delamination detection experiment for a composite plate presented in the following section is inspired by this Predator.

Figure 14.1 An Unmanned aerial vehicle predator: The United States Air Force is interested in developing an online monitoring system for the Predator (Courtesy of the United States Air Force)

14.2.2 Data Acquisition

The data acquisition portion of the structural health-monitoring process involves selecting the excitation methods, the sensor types, number and locations, and the data acquisition, storage, transmittal hardware. This portion of the process will be application specific. Economic considerations will play a major role in making decisions regarding the data acquisition hardware to be used for the structural health-monitoring system. The intervals at which data should be collected are another consideration that must be addressed. For earthquake applications, it may be prudent to collect data immediately before and at periodic intervals after a large event. If fatigue crack growth is the failure mode of concern, it may be necessary to collect data almost continuously at relatively short time intervals once some critical crack length has been obtained.

The overall test configuration of this example is shown in Figure 14.2(a). The test setup consists of a composite plate with a surface mounted sensor layer, a personal computer with a built-in data acquisition system, and an external signal amplifier. The dimensions of the composite plates are $60.96\,\text{cm} \times 60.96\,\text{cm} \times 0.6350\,\text{cm}$ ($24\,\text{in} \times 24\,\text{in} \times 1/4\,\text{in}$). The layup of this composite laminate contains 48 plies stacked according to the sequence $[6(0/45/-45/90)]$S, consisting of Toray T300 graphite fibers and a 934 epoxy matrix.

A commercially available thin film with embedded piezoelectric (PZT) sensors is mounted on one surface of the composite plate as shown in Figure 14.2(b) [4]. A total of 16 PZT patches are used both as sensors and actuators to form an 'active' local sensing system. Because the PZTs produce an electrical charge when deformed, the PZT patches can be used as dynamic strain gauges. Conversely, the same PZT patches can also be used as actuators, because elastic waves are produced when an electrical field is applied to the patches. In this study, one PZT patch is designated as an actuator, exerting a predefined waveform into the structure. Then, the adjacent PZTs become

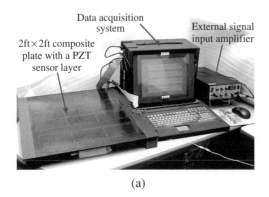
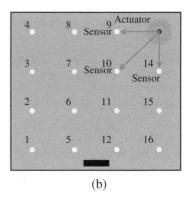

Figure 14.2 An active sensing system for detecting delamination on a composite plate: (a) testing configuration; (b) layout of the PZT sensors/actuators

strain sensors and measure the response signals. This actuator–sensor sensing scheme is graphically shown in Figure 14.2(b). This process of the Lamb-wave propagation is repeated for different combinations of actuator–sensor pairs. A total of 66 different path combinations were investigated in this study. The data acquisition and damage identification were fully automated and completed in approximately 1.5 minutes for a full scan of the plate used in this study. These PZT sensor/actuators are inexpensive, generally require low power, and are relatively nonintrusive.

The personal computer shown in Figure 14.2(a) has built-in analog-to-digital and digital-to-analog converters, controlling the input signals to the PZTs and recording the measured response signals. In this experiment the optimal input voltage was designed to be near 45 V, producing 1–5 V output voltage at the sensing PZTs. PZTs in a circular shape are used with a diameter of only 0.64 cm (1/4 in). The sensing spacing was set to 15.24 cm (6 in). A discussion on the selection of design parameters such as the dimensions of the PZT patches, sensor spacing, and a driving frequency can be found in Reference [5].

14.2.3 Feature Extraction

The area of the structural health-monitoring process that receives the most attention in the technical literature is the identification of data features that allow one to distinguish between the undamaged and damaged structure. A damage-sensitive feature is some quantity extracted from the measured system response data that indicates the presence of damage in a structure. Fundamentally, the feature extraction process is based on fitting some model, either physics-based or data-based, to the measured system response data. The parameters of these models or the predictive errors associated with these models then become the damage-sensitive features. Inherent in many feature selection processes are various forms of data normalization in an effort to separate changes in the measured response caused by varying operational and environmental conditions from changes caused by damage.

One of the unique aspects of this example insertion into a structure of a carefully designed input waveform to make the response signal more sensitive to damage. The use of a known and repeatable input makes the subsequent signal processing for damage detection much easier, and the extracted features are less sensitive to ambient variation of the system. A similar approach to noise elimination in ultrasonic signals for flaw detection can be found in Reference [6]. In addition, a multiresolution processing scheme based on wavelet analysis is developed to extract a portion of the response signal that is more amenable to signal interpretation for detecting defects. Finally, the vibration characteristics of the response signals are investigated under varying temperature and boundary conditions.

In this example, a Morlet wavelet with a narrowband driving frequency around 110 kHz is designed as the input waveform [see Figure 14.3(a)]. Figure 14.3(b) shows the time response of the PZT patch No. 1 when the Morlet input waveform is generated at the PZT patch No. 6. The numbering of the PZT patches is shown in Figure 14.2(b). The solid line represents the baseline signal, and the dashed line shows the response time signal when delamination is simulated within a direct line of the actuator and sensor path. The response signal is composed of several wave modes because of the dispersive nature of the excitation signal at the input frequency. From Figure 14.3(b), it is observed that some modes are more sensitive to the simulated delamination than are other modes. The first mode, which looks like a sine wave modulated by a cosine function in Figure 14.3(b), is the first arrival of the A_o mode associated with the direct path of the wave propagation. The second mode is another A_o mode that is reflected from the edge of the plate. [Note that the wave propagation path between the PZT Nos 1 and 6 pair is near the edge of the plate, see Figure 14.2(b).] Observation of Figure 14.3(b) clearly reveals that the first A_o mode is the most sensitive to the delamination damage. Therefore, only the A_o mode was used in this study.

As a wave propagates through a solid medium, energy is transferred back and forth between kinetic and elastic potential energy. When this transfer is not perfect because of heating, wave leaking, and reflection, attenuation occurs. In particular, the attenuation is increased by the presence of the delamination, and the energy of the input force spills over from the driving frequency to neighboring frequency values. The energy loss in a damaged area is a result of reflection and dispersion caused by microcracks within the laminate resulting in the excitation of high-frequency local modes. Based on these observations, a damage index is defined as the function of a signal's attenuation at a limited time span (a signal portion corresponding to the first A_o mode) and at a specific frequency (the input frequency of the signal). Note that the attenuation is correlated to

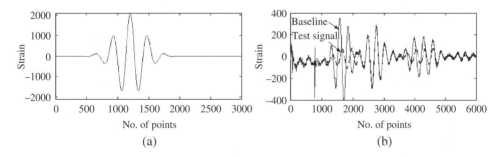

Figure 14.3 Lamb wave time signals from the tested composite plate: (a) an input Morlet waveform; (b) response strain time signals

the amount of energy dissipated by damage. In other words, the proposed damage index measures the degree of the test signal's energy dissipation compared to the baseline signal, especially at the first A_o mode and at the input frequency value.

To achieve this goal, a wavelet transform is first utilized to obtain time-frequency information for the baseline signal [7]. This procedure is schematically shown in Figure 14.4(a)–(e). In this study, the real Morlet wavelet is used for the family of basis functions. [Note that the same Morlet wavelet is used as the input waveform in the experiment.] Because the shape of the actuating input is same as the shape of the wavelet basis functions, the wavelet analysis procedure becomes more accurate and efficient [6, 8]. This wavelet, $\psi(t)$, is defined as:

$$\psi(t) = e^{-t^2/2} \cos(5t) \tag{14.1}$$

where 1500 data points are sampled between −4.2 and 4.2 of time t. Assuming the sampling rate of 20 MHz, this waveform results in a central frequency of 110 kHz. The wavelet transform, $Wf(u, s)$, is obtained by convolving the signal $f(t)$ with the translations (u) and dilations (s) of the mother wavelet:

$$Wf(u, s) = \int_{-\infty}^{\infty} f(t) \frac{1}{\sqrt{s}} \psi_{u,s}^*(t) dt \tag{14.2}$$

where

$$\psi_{u,s}^*(t) = \frac{1}{\sqrt{s}} \psi\left(\frac{t-u}{s}\right) \tag{14.3}$$

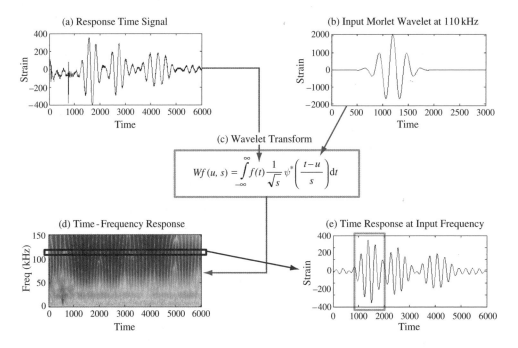

Figure 14.4   A wavelet analysis procedure to extract a damage sensitive feature

Note that each value of the wavelet transform $Wf(u, s)$ is normalized by the factor $1/\sqrt{s}$ to ensure that the integral energy given by each wavelet is independent of the dilation s.

Once the time-frequency information is obtained [Figure 14.4(d)], the signal component corresponding only to the input frequency is retained for additional signal processing. By looking at this filtered view of the transmitted energy from the actuator to the sensor at the input frequency, one could gain insight as to how the intensities of the input energy have been shed into sideband frequencies as a result of damage. Then, the energy content of this baseline signal component is computed only at the first A_o mode of the signal [Figure 14.4(e)]. These procedures [Figure 14.4(a)–(e)] are then repeated for a test signal. Finally, the damage index (DI) is related to the ratio of the test signal's kinetic energy to that of the baseline signal:

$$DI = 1 - \int_{u_0}^{u_1} Wf_t(u, s_o)du \bigg/ \int_{u_0}^{u_1} Wf_b(u, s_o)du \tag{14.4}$$

where the subscripts b and t denote the baseline and test signals, u_o and u_1 represent the starting and ending time points of the baseline signal's first A_o mode, and $0 \le DI \le 1$. Note that the value of DI becomes zero when there is no attenuation of the test signal compared with the baseline signal, and its value becomes close to one as the test signal attenuates more as a result of damage.

14.2.4 Statistical Inference

The portion of the structural health monitoring process that has received the least attention in the technical literature is the development of statistical models for discrimination between features from the undamaged and damaged structures. Statistical model development is concerned with the implementation of the algorithms that operate on the extracted features to quantify the damage state of the structure. The algorithms used in statistical model development usually fall into three categories. When data are available from both the undamaged and damaged structure, the statistical pattern recognition algorithms fall into the general classification referred to as *supervised learning*. Group classification and regression analysis are categories of supervised learning algorithms. *Unsupervised learning* refers to algorithms that are applied to data not containing examples from the damaged structure. *Outlier* or *novelty detection* is the primary class of algorithms applied in unsupervised learning applications. All of the algorithms analyze statistical distributions of the measured or derived features to enhance the damage detection process.

In the current example, outlier analysis is applied to the damage index values. Once the damage index value is computed, a method must be employed for determining a statistically rigorous threshold, whose exceedance indicates the presence of damage in a path. This establishment of the decision boundary (or the threshold) is critical to minimize false-positive and false-negative indications of damage. Although data are often assumed to have a normal distribution for building a statistical model for

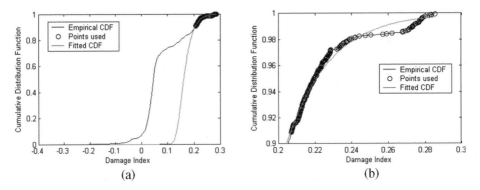

Figure 14.5   Fitting of a Gumbel maximum distribution to top 10 % of damage index values. (a) Gumbel distribution curve-fitting (only top 10 % of data is used for curve fitting); (b) Gumbel distribution curve-fitting (zoomed for top 10 % of the damage index values)

damage classification, it should be noted that a normal distribution weighs the central portion of data rather than the tails of the distribution. Therefore, for damage detection applications, we are mainly concerned with extreme (minimum or maximum) values of the data because the threshold values will reside near the tails of the distribution. The solution to this problem is to use an approach called extreme value statistics (EVS) [9], which is designed accurately to model behavior in the tails of a distribution.

The pivotal theorem of EVS states that in the limit as the number of vector samples tends to infinity, the induced distribution of the maxima can only take one of three forms, Gumbel, Weibull, or Frechet, regardless the distribution types of the parent data [10]. Once the parametric model is obtained, it can be used to compute an effective threshold for damage detection based on the true statistics of the data, as opposed to statistics based on a blanket assumption of a Gaussian distribution. Details on parameter estimation of extreme value distributions can be found in [10].

In this study, the damage index value is computed from various normal conditions of the composite plate. Then, the statistical distribution of the damage index is characterized by using a Gumbel distribution, which is one of three types of extreme value distributions. Note that the Gumbel distribution is fit only to the maximum 10 % of the damage index values. From this fitted cumulative density function (CDF), a one-sided 99.9 % confidence interval was determined, resulting in a threshold value of 0.29 (Figure 14.5).

It should be noted that the computation of the threshold value based on EVS requires training data only from the undamaged conditions of a structure, classifying the proposed statistical approach as one of the unsupervised learning methods. Another class of statistical modeling is supervised learning, where training data from both undamaged and damaged conditions are required. When a SHM system is deployed to real-world applications, it is often difficult to collect training data from various damage cases. Therefore, an unsupervised learning method, such as the one presented here, will be more practical in field applications. Furthermore, by properly modeling the maximum distribution of the damage index, false alarms have been minimized.

14.3 EXPERIMENTAL RESULTS

In the experiment, delamination on the composite plate was initially simulated by attaching industrial putties in various locations of the composite plate. Because the putty attenuates the Lamb waves in a similar manner to the delamination, this material closely models the change in the Lamb wave characteristics that may be expected from the delamination. The sizes of the industrial putty patches were about 2.54 cm × 2.54 cm (1 in × 1 in) and 2.54 cm × 5.08 cm (1 in × 2 in), respectively [See Figure 14.6(a)]. Although it is not presented here, actual delamination is also seeded by shooting a steel projectile to the composite plate, and the resulting delamination is successfully detected by the active sensing system.

First, the baseline signals corresponding to 66 different actuator-and-sensor paths were recorded at a known intact condition of the plate. Once delamination was simulated by placing the industrial putty at an arbitrary location on the composite plate, the test signals were recorded for the actuator and sensor combinations identical to the baseline case. Then, the damage index value defined in Equation (14.4) was computed for each path. A given actuator–sensor path is classified as damaged when the value of the damage index becomes larger than 0.29. This threshold value for the damage index is established by using the previously described EVS. Figure 14.6(b) displays the actuator–sensor paths (damaged paths) affected by the simulated damage shown in Figure 14.6(a). These damage paths have *DI* values greater than 0.29.

Also, a damage localization algorithm is developed to identify the location and size of the delamination. The delamination shown in Figure 14.6(a) is identified from the damaged paths previously detected in Figure 14.6(b). The detail for damage localization is provided in Reference [11]. It is worthwhile to note that our damage localization method is solely based on the estimation of signal attenuation unlike other methods that use the time-of-arrival information [7, 12]. Therefore, the proposed damage location approach can be easily applied to anisotropic composite plates because the wave-speed variation in regard to the fiber orientation of the plate is irrelevant. In real-world applications, it is also important to demonstrate that a SHM system is robust when

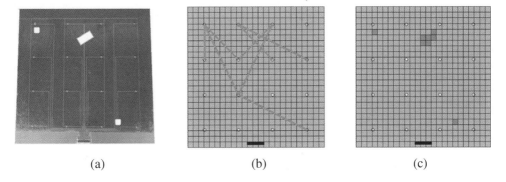

(a)   (b)   (c)

Figure 14.6 Detection of different sizes of damage using the wavelet-based approach. (a) Actual damage simulated by placing a industrial putty; (b) actuator–sensor paths affected by the damage; (c) the damage identified by the proposed method

used under varying environmental and operational conditions. The effect of varying temperature and boundary conditions on the proposed damage detection algorithm is also investigated in Reference [11].

14.4 SUMMARY AND DISCUSSION

In this paper, a SHM system is developed following a statistical pattern recognition paradigm to identify the location and area of flaws on a graphite/epoxy composite plate. The uniqueness of this study lies in that: (i) a specific input waveform to drive PZT actuators and a signal processing analysis technique are developed based on wavelet transform to extract response features sensitive to damage and insensitive to changing boundary and operational conditions; and (ii) a statistically rigorous damage classifier based on extreme value statistics is employed to minimize false indications of damage. By using the proposed data interrogation techniques and statistical tools, the damage detection algorithm presented was able properly to detect simulated flaws on an anisotropic composite plate, even in the presence of varying temperature and boundary conditions. Finally, the approach presented herein is very attractive for a continuous monitoring of composite structures because the whole monitoring process is fully automated and the sensor layer can be readily mounted or embedded in composite structures.

ACKNOWLEDGMENTS

Funding for this project was provided by the Department of Energy through the internal funding program at Los Alamos National Laboratory known as Laboratory Directed Research and Development (Damage Prognosis Solutions). The author is grateful for the support and invaluable feedback given by the damage prognosis team at Los Alamos including Charles R. Farrar, Gyuhae Park, Amy N. Robertson, Jeannette R. Wait, and Nathan P. Limback.

REFERENCES

[1] Sohn, H., Farrar, C.R., Hemez, F.M., Czarnecki, J.J., Shunk, D.D., Stinemates, D.W. and Nadler, B.R. (1994) 'A Review of Structural Health Monitoring Literature: 1996–2001', *Los Alamos National Laboratory Report*, LA-13976-MS.
[2] Farrar, C.R., Duffey, T.A., Doebling, S.W. and Nix, D.A. (2000) 'A Statistical Pattern Recognition Paradigm for Vibration-Based Structural Health Monitoring', *Proceedings of the Second International Workshop on Structural Health Monitoring*, Stanford, CA, pp. 764–773.
[3] Kosmatka, J. (2003) 'Composite Wing Applications', *presented at the Pan American Advanced Studies Institute on Damage Prognosis*, Florianopolis, Brazil, October 19–30.
[4] Acellent Technologies, Inc. (2003) <http://www.acellent.com/>
[5] Kessler, S.S. (2002) *Piezoelectric-based In-situ Damage Detection of Composite Materials for Structural Health Monitoring Systems*, PhD Dissertation, MIT, Massachusetts.
[6] Abbate, A., Koay, J., Frankel, J., Schroeder, S.C. and Das, P. (1997) 'Signal detection and noise suppression using a wavelet transform signal processor: Application to ultrasonic flaw detection', *IEEE Transactions on Ultrasonics, Ferroelectrics, and Frequency Control*, **44**, 14–26.

[7] Wang, C.S. and Chang, F.K. (2000) 'Diagnosis of impact damage in composite structures with built-in piezoelectrics network', *Proceedings of SPIE*, **3990**, 13–19.

[8] Lind, R., Kyle, S. and Brenner, M. (2001) 'Wavelet analysis to characterize non-linearities and predict limit cycles of an aeroelastic system', *Mechanical Systems and Signal Processing*, **15**, 337–356.

[9] Sohn, H., Allen, D.W., Worden, K. and Farrar, C.R. (2003) 'Structural damage classification using extreme value statistics', *ASME Journal of Dynamic Systems, Measurement, and Control*, in press.

[10] Castillo, E. (1998) *Extreme Value Theory in Engineering*, Academic Press Series in Statistical Modeling and Decision Science, San Diego, California.

[11] Sohn, H., Park, G., Wait, J.R., Limback, N.P. and Farrar, C.R. (2003) 'Wavelet-based active sensing for delamination detection in composite structures', *Smart Materials and Structures*, **13**, 153–160.

[12] Lemistre, M. and Balageas, D. (2001) 'Structural health monitoring system based on diffracted Lamb wave analysis by multiresolution processing', *Smart Materials and Structures*, **10**, 504–511.

Part III
Hardware

15

Sensing and Data Acquisition Issues for Damage Prognosis

Charles R. Farrar,[1] Phillip J. Cornwell[2], Norman F. Hunter,[1] and Nick A.J. Lieven[3]

[1] *Los Alamos National Laboratory, Los Alamos, New Mexico, USA*
[2] *Rose-Hulman Institute of Technology, Terre Haute, Indiana, USA*
[3] *University of Bristol, UK*

15.1 INTRODUCTION

Instrumentation and data acquisition issues are major concerns that must be addressed when developing damage-prognosis solutions. This chapter will address these concerns by presenting three different instrumentation strategies for damage detection and prognosis. These strategies are presented in increasing order of sophistication. The primary difference in these strategies is the up-front effort to integrate testing and numerical simulations, interdisciplinary communication, and to design instrumentation specifically for the system to be monitored. These strategies are then discussed in terms of conceptual issues associated with the development of a sensing and data acquisition system that are fundamental to making significant progress in damage prognosis. Next, more general challenges such as the technological issues of sensors, data acquisition and storage, data processing using feature extraction, and the combined role of testing/modeling in damage prognosis are addressed. This section concludes with a summary of the authors' opinions on the properties of a future sensing and instrumentation system for damage prognosis.

As previously mentioned, this report emphasizes structural dynamics while accepting that structural failures are not necessarily the dominant factor in many system failures.

Damage Prognosis – For Aerospace, Civil and Mechanical Systems   Edited by D.J. Inman, C.R. Farrar, V. Lopes Junior and V. Steffen Junior   © 2005 John Wiley & Sons, Ltd

Therefore, the sensing issues that will be addressed are primarily related to detecting structural damage through some measure of kinematic quantities. Additional sensors will be needed to assess operational and environmental conditions in an effort to separate their effects from those caused by damage. The most fundamental issue that must be addressed when developing a sensing system for damage prognosis is the need to capture the structural response on widely varying length and time scales.

15.2 SENSING AND DATA ACQUISITION STRATEGIES FOR DAMAGE PROGNOSIS

15.2.1 Strategy I

A less effective, but often used, approach to deploying a sensing system for damage detection and prognosis consists of a sparse sensor array, installed on the structure after fabrication, possibly following an extended period of service. Sensors are typically chosen on the basis previous experience and availability. The selected sensing systems have often been commercially available for some time, and the technology may be 20 years old. Excitation is limited to that provided by the ambient operational environment. The physical quantities that are measured are often selected in an *ad hoc* manner without an *a priori* quantified definition of the damage that is to be detected, or any *a priori* analysis that would indicate that these measured quantities are sensitive to the damage of interest. This approach dictates that damage-sensitive data features are selected 'after the fact' using archived sensor data and *ad hoc* algorithms. This scenario represents many real-world systems, particularly those deployed on civil engineering infrastructure.

The common damage-detection approach associated with Strategy I is that a set of undamaged and damaged structures are subjected to nominally similar excitations and their responses measured. Features determined using trial and error or physical intuition are extracted from the measured response and correlated to damage using a variety of methods that vary in their level of mathematical sophistication. A comparison of power spectral density functions illustrates the type of basic data analysis associated with this sensing system strategy. When data are not available from the damaged structure, the damage detection process reverts to some form of outlier detection. In this case, trial and error or physical intuition is again used to define the damage-sensitive features that will be classified as outliers because of damage, and to set thresholds that define when these features can be considered outliers. Despite the *ad hoc* nature of this process, Strategy I is sometimes effective for damage detection but typically shows limited success for damage location and quantification. This approach is often enhanced by a comprehensive historical database. For example, the availability of information regarding the damage state and corresponding measurement results for large numbers of nominally identical units will significantly improve the ability to detect damage in subsequent units when Strategy I is employed. A major drawback of this approach is that, typically, the sensing system is not designed to measure the parameters necessary to allow one to separate operationally and environmentally induced changes from changes caused by damage.

15.2.2 Strategy II

Strategy II is a more coupled analytical/experimental approach to defining the sensor-system definition that incorporates some significant improvements over Strategy I. First, damage is well defined and to some extent quantified before the sensing system is designed. Next, the sensing system properties, including actuator properties, are defined on the basis the results of numerical simulations or physical experiments, and are based on the data analysis procedures (e.g., feature extraction and statistical discrimination) that will be employed in the damage-detection application. This process of defining the sensor system properties will often be iterative. Sensor types and locations are chosen because the numerical simulations or physical tests show that the expected type of damage produces known, observable and statistically significant effects in features derived from the measurements. Additional sensing requirements are then defined based on how changing operational and environmental conditions affect the damage detection process. However, all sensors are still chosen from commercially available sensors that best match the defined sensing system requirements. Finally, as a result of this coupled approach to designing the sensing system, there is the possibility that the extent of the damage can be directly correlated to the sensor measurements through the numerical or physical models that were used to define the sensing system properties.

Strategy II incorporates several enhancements that will typically improve the probability of damage detection:

(1) Well-defined and quantified damage information that is based on initial system design information, numerical simulation of the postulated damage process, qualification test results, maintenance records, and system autopsies.
(2) Sensors that are shown to be sensitive enough to identify the predefined damage when the measured data are coupled with the data-analysis procedures.
(3) Active sensing that is incorporated into the process: a known input is used to excite the structure with an input waveform tailored to the damage-detection process.
(4) Sensors that are placed at locations where responses are known from analysis, experiments, and past experience to be sensitive to damage.
(5) Features extracted from the measured data that are known to be sensitive damage indicators based on analysis, experiments, and past experience.
(6) Additional measurements that can be used to quantify changing operational and environmental conditions.
(7) Damage extent estimates that are obtained by correlating sensor readings with information from numerical or physical models of the damage and its effect on the system.

The number of studies in the technical literature that take this approach to developing a sensing system to detect damage is quite small. In actuality, most sensing systems used to detect damage take an approach somewhere in between Strategy I and Strategy II. However, Strategy II still does not directly address the 'predicting remaining life' issue of damage prognosis.

15.2.3 Strategy III

Strategy II is much more effective in damage detection than Strategy I, but does not specifically address the instrumentation issues associated with damage prognosis. Damage prognosis requires various validated models to predict future loading, damage accumulation, and remaining system life. The need to develop these models, update the models as new data become available and quantify the uncertainty in these models, dictates the enhanced sensing system requirements associated with Strategy III. These models will incorporate the measurement data and produce a structural state estimate as in Strategy II. Then the model will be used to predict the evolution of this state through time, and finally to translate these progressive estimates into an estimate of remaining useful life.

In summary, Strategy III includes additional predictive capabilities that dictate sensing system requirements beyond those of Strategies I and II. The following are additional modeling capabilities and their associated sensing system requirements:

(1) Numerical models, developed in conjunction with a series of nondestructive model validation experiments, that predict the damage evolution.
(2) A sensing system designed to provide information that can be used to develop future loading models.
(3) A procedure to validate the numerical damage prediction model through test–analysis correlation using measured system responses to a known set of physical excitations.
(4) The data processing and storage capabilities necessary to perform measurements on an ongoing basis.
(5) A procedure to extract damage-sensitive features from the measured data so the features can be used to monitor the evolution of damage.
(6) Reduced order models that can predict remaining system life.

This strategy, which is the most sophisticated, clearly provides the most robust method for viable prognosis. Several serious challenges are involved in its implementation. The following section lists the major challenges related to sensing and data acquisition for damage prognosis.

15.3 INSTRUMENTATION: CONCEPTUAL CHALLENGES

The above three strategies identify conceptual challenges to effective damage prognosis from a sensing system perspective. These challenges include the following:

(1) the ability to capture local and system-level response; that is, the need to capture response on widely varying length and time scales;
(2) the need for a sensing system design methodology;
(3) the need to integrate the predictive-modeling and data-interrogation processes with the sensing-system design process;
(4) the ability to archive data in a consistent, retrievable manner for long-term analysis.

These challenges are nontrivial because of the tendency for each technical discipline to work more or less in isolation. Therefore, an integrated systems-engineering approach to the damage-prognosis process and regular, well-defined routes of information dissemination are essential. The subsequent portions of this section will address specific sensing-system issues associated with damage prognosis.

15.3.1 What Types of Data Should be Acquired

Instrumentation, which includes sensors and data acquisition hardware, first translates the system's dynamic response into a signal, such as an analog voltage signal, that is proportional to the measured quantities of interest. Next, the analog signal is discretely sampled to produce digital data. To begin defining a sensing system for damage prognosis, one must first define the types of data to be acquired. The data types fall into three general categories of kinematic, environmental and operational quantities. There are many traditional sensors that can be used to measure these various physical quantities, and there are emerging technologies that could have tremendous impact on the future of damage prognosis. Although this chapter focuses on more traditional sensing technology, it acknowledges that sensing technology is one of the most rapidly developing fields related to damage prognosis, and therefore one must always be looking for new technologies that are applicable to the prognosis problem.

15.3.1.1 Kinematic Quantities

The 'traditional' sensors used to measure kinematic quantities include wire resistance strain gauges, mechanical displacement transducers such a linear variable differential transducers, and piezoelectric accelerometers. In general, the accelerometers provide an absolute measurement at a point on the structure while the displacement and strain sensors provide relative measurements over typically short gage lengths. These sensors are used extensively for aerospace, civil, and mechanical engineering applications. Conditioning electronics for these sensors have evolved from bulky vacuum tube systems to small, sophisticated, solid-state, devices. A wide variety of sensors that can accommodate many different applications are available off the shelf.

The principal emerging kinematic sensing technologies include microelectromechanical systems (MEMS), piezoelectric (PZT) actuator/sensors (discussed in Section 15.7), and fiber-optic strain sensors. Commercially available MEMS devices can measure strain, and rotational and linear acceleration. A MEMS accelerometer is shown in Figure 15.1. Once fully developed, MEMS sensors have the potential to impact a variety of sensing activities based on their versatility, small size, and low cost when manufactured in large numbers. These properties will allow the sensor density on a structural system to increase significantly, which is essential to improve damage-prognosis technology. MEMS can be integrated with on-board computing to make these sensors self-calibrating, and self-diagnosing. This integration of the sensor with microprocessors defines the 'smart sensor' concept. Inhibiting MEMS use today are issues such as commercial availability, traceable calibration, and a track record of stability and ruggedness when used for long-term structural monitoring activities. In

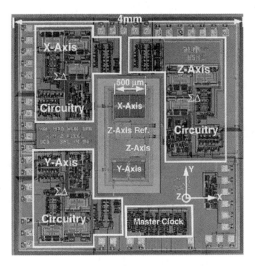

Figure 15.1 A MEMS accelerometer

contrast, current commercially available traditional accelerometers have been proven to be reliable and stable. These accelerometers incorporate on-board signal conditioning and may soon have on-board A–D conversion. However, in comparison with anecdotal reports of MEMS sensors, the traditional accelerometers typically don't measure multiple parameters such as both rotational and linear acceleration. The traditional sensors are relatively expensive (hundreds of dollars for conventional piezoelectric accelerometers versus tens of dollars for MEMS accelerometers), and they are typically not integrated with microprocessors.

Fiber-optic strain gauges are mentioned here as a nearly commercially available emerging technology. In the most sophisticated type, a selectable gage length of a single long fiber is queried to obtain the strain (with picostrain accuracy). This technology could allow a single fiber with the length of a bridge girder to monitor strain at any location along the girder, or to embed sensors in manufactured parts that could measure strain later in selected regions using Bragg grating technology [1]. Multiple Bragg gratings can be placed in a single long fiber to obtain numerous discrete strain readings. These readings can be obtained with greater accuracy than with electrical-resistance strain gauges, the readings are immune to electromagnetic and RF interference, and the sensors are not a spark source, which is a key issue if monitoring is to be done near combustible materials. Also, these sensors are nonintrusive, extremely lightweight, and have proven to be very rugged [2].

The previous discussion of measuring kinematic quantities has focused on local measurements. In addition, there are some global sensing technologies that are commercially available. More mature global sensing technologies include laser Doppler velocometers and acoustic field detectors. Global sensors can scan a surface of a structure, and in some cases with proper signal processing they can identify damaged areas. The disadvantages associated with global sensors include fairly high procurement cost, the need for visual access to the measured part, and the need to remove the structure from service to carry out the test. An emerging technology in this area of global sensing is chemical coatings that emit a particular signature when cracked. This technology has already been demonstrated through the application of pressure-sensitive paints for wind tunnel testing [3].

15.3.1.2   Environmental Quantities

If changes in environmental quantities produce changes in the damage-sensitive features similar to those produced by damage, a measure of the environmental quantity will be necessary to separate these effects. Such a case will necessitate that environmental quantities such as temperature, pressure, and moisture content be measured. Note that if damage changes the features that are in someway orthogonal to the changes produced by environmental effects, then a measure of the environmental parameters may not be necessary.

15.3.1.3   Operational Quantities

Similar to environmental quantities, operational quantities may also produce changes in damage-sensitive features and may therefore need to be measured. Operational quantities include such things as traffic volume for a bridge, mass loading of an off-shore oil platform, or amount of fuel in an airplane wing.

15.3.2   Define the Sensor Properties

One of the major challenges in defining sensor properties is that these properties need to be defined *a priori* and typically cannot be changed easily once a sensor system is in place. These properties of sensors include bandwidth, sensitivity (dynamic range), number, location, stability, reliability, power requirements, cost, telemetry, etc. To address this challenge, a significantly coupled analytical and experimental approach to the sensor system deployment should be used in contrast to the current *ad hoc* procedures used for most current damage-detection studies. This strategy should yield considerable improvements. First, critical failure modes of the system can be well defined and, to some extent, quantified using high-fidelity numerical simulations before the sensing system is designed. These high-fidelity numerical simulations can be used to define the required bandwidth, sensitivity, sensor location, and sensor number. Additional sensing requirements can also be ascertained if changing operational and environmental conditions are included in the models, so as to determine how these conditions affect the damage detection process.

15.3.2.1   Required Bandwidth

Local response characteristics are required to identify the onset of damage, which tends to manifest itself in the higher-frequency portions of the response spectrum. Global response characteristics are required to capture the influence of damage on the system level performance and to predict future performance. The global system response is typically characterized by the lower-frequency portion of the response spectrum. Therefore sensors with a high-frequency range tend to be more sensitive to local response and therefore can detect the onset of damage. This sensitivity requires a sensor with a large bandwidth. Typically, as the bandwidth goes up, the sensitivity goes down. Also, it is harder to excite higher-frequency response of a structural system. This difficulty dictates that the excitation needs to be very local, as is possible with PZT actuators (see Section 15.7). Both local and global response characteristics are required for damage prognosis.

15.3.2.2 Required Sensitivity

Adequate sensitivity and dynamic range is required to separate ambient vibration or low-level local excitation caused by damage (e.g., cracks opening and closing) from large-amplitude excitation such as that caused by impact or earthquake loading. Thirty-two bit sensors are able to resolve this sort of dynamic range, but issues remain concerning the calibration of the sensors over the entire range of possible inputs.

15.3.2.3 Number of Sensors and Sensor Locations

Two primary considerations when deciding on the number and location of sensors are whether or not the sensing system should be optimal and how much redundancy is desired. It is critical that the expected type of damage produces known, observable, and statistically significant effects in features derived from the measured quantities at the chosen transducer locations. For this reason, numerical simulations can be used to choose the number of sensors and sensor locations. It is well known from control theory that the observability of a system depends critically on the location of the sensors and the desired feature to be extracted. For instance, if one desires to measure the second resonant frequency of a structure and use this value as a metric for damage, mounting the sensor on the node of the second mode will doom any frequency-based algorithm to failure. This problem is partially addressed by the observability theory developed for control algorithms. However, this theory does not address performance, nor does it address the metric used to determine damage. The issues associated with integrating observability calculations for local damage and global behaviors of a system into the optimal sensing design, which has not been addressed in current damage identification practice, should be examined. Such methods may incorporate genetic algorithms [4] or neural networks with the ability to model the system in detail or the ability to examine the structure systematically.

Intuitively, sensors should be near expected damaged locations. With MEMS or smart dust [5] (tiny, wireless sensors) it may be possible to saturate the part with sensors to provide sensing redundancy, a clear issue for prognostics in civil aerospace applications, and to reduce the need for an optimal sensing system. As the number of sensors increases, the cost, reliability, and perhaps power requirements may become significant issues. Traditional sensing emphasizes relatively few sophisticated sensors or scanning noncontact sensing. An alternative approach to attaching a sensor to a structure is the use of a probe. The structure is examined with a probe at numerous locations sequentially. This sensing approach is common practice in many industries, for example, condition monitoring of rotating machinery or local acoustic resonance spectroscopy. Clearly a probe requires a human operator and does not allow periodic data acquisition from a remote location or automatic data acquisition during a severe event such as an earthquake.

15.3.3   Calibration and Stability

Most sensors are calibrated at a specialized calibration facility. This type of calibration is expected to endure, but to be supplemented in the future by self-checking and self-calibrating sensors. Calibration raises several important issues. It is not clear just which forms of calibration are essential, and which are superfluous. Some measurements are acceptable with 20 % error, especially if sensor-to-sensor comparisons are accurate within a few percent. In other scenarios, absolute accuracies better than 1 % are required. The calibration community needs to address these issues, including both precision and flexibility; for example, how to calibrate a 32-bit sensor over its entire dynamic range, and how to calibrate a precise sensor versus a coarse sensor.

15.3.4   Sensor Durability

Sensor survival is probably the major issue whose resolution is unclear. Confidence and robustness in the sensors are prime considerations for prognostics. If this part of the system is compromised, then the overall confidence in the system performance is undermined. For sensors implemented for prognostics, several durability considerations emerge:

(1) the nontrivial problem of sensor selection for extreme environments; e.g. in service turbine blades.
(2) sensors being less reliable than the part. For example, reliable parts may have failure rates of 1 in 100 000 over several years time. Sensors are often small, complex assemblies, so sensors may fail more often than the part to be sensed. Loss of sensor signal then falsely indicates part failure, not sensor failure.
(3) Sensors may fail through outright sensor destruction while the part being sensed endures.
(4) False indications of damage or damage precursors are extremely undesirable. If this occurs often, the sensor is either overtly or covertly ignored. The biggest cause of aircraft delay is a failed sensor. Sensor failure might be acceptable if its demise is simultaneous with the part failure, which would in itself provide an indication of damage.

15.3.5   Define Data Sampling Parameters

Sampling issues include deciding how fast to discretize the data and when to take data. These issues will most likely change depending on the structure and the expected type of damage. If it is important to characterize the environmental or operational variability, then a lot of samples may need to be taken initially and data will need to be taken from all the expected environmental and operational conditions. Once a baseline has been established, data may be obtained either periodically or only after extreme or anomalous events such as an earthquake or new environmental conditions not previously experienced.

15.3.6   Define the Data Acquisition/Transmittal/Storage System

Multiple smart sensors produce an abundance of time history data to store and manage. Another field with this characteristic is satellite imaging, where huge volumes of image data are standard. Image-data processing may yield some insights valuable for processing smart mechanical-sensor data.

Sensing, especially with a dense array of smart sensors, interacts with data acquisition. Smart sensors theoretically make it possible to extract, transmit, and store features, as well as to transmit raw-time histories. Transmitting raw-time histories clearly will require much more storage space but also allows the most flexibility for future data analysis. Continuous data transmission with smart dust could lead to information overload during analysis and storage. With smart, dense, sensors the structure might be instrumented so densely that sensors are everywhere, so wherever damage occurs there is a nearby sensor.

Near-real-time data transmission for feature extraction and analysis has the potential to enhance real-time damage detection and to increase the value of testing. Currently, a test setup is often disassembled prior to data analysis, just to discover that some modification, like a different input type or level, or another transducer location, is required. Real-time analysis takes more upfront preparation time and better communications, but provides much more value from the test, because problems can be corrected during the test. Other issues that need to be addressed are the time synchronization of a large number of sensors, A–D conversion and onboard memory.

An example of an integrated sensing and processing system is the high-explosives radio telemetry (HERT) [6] system (see Figure 15.2) developed by a Los Alamos National Laboratory–Honeywell, Inc., team for weapons flight-test monitoring. The HERT system currently can measure, record, process, and transmit data from 64-fiber-optic sensor channels. A field-programmable gate array is used for local data processing and sensor diagnostics.

15.3.7   Active Versus Passive Sensing

Most field-deployed, structural-health monitoring strategies examine changes in kinematic quantities, such as strain or acceleration, to detect and locate damage. These methods typically rely on the ambient loading environment as an excitation source and, hence, are referred to as passive sensing systems. The difficulty with using such excitation sources is that they are often nonstationary. The nonstationary nature of these signals requires robust data normalization procedures to be employed in an effort to establish that the change in the kinematic quantity is the result of damage as opposed to changing operational and environmental conditions. Also, there is no control over the

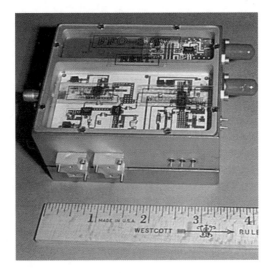

Figure 15.2   High-explosives radio telemetry system

excitation source, and it may not excite the type of system response useful for identifying damage at an early stage.

As an alternative, a sensing system can be designed to provide a local excitation tailored to the damage detection process. Piezoelectric (PZT) materials are frequently being used for such active sensing systems. Because PZT produces an electrical charge when deformed, PZT patches can be used as dynamic strain gauges. Conversely, the same PZT patches can also be used as actuators because a mechanical strain is produced when an electrical field is applied to the patch. This material can exert predefined excitation forces into the structure. The use of a known and repeatable input makes it much easier to process the signal for damage detection and prognosis. For instance, by exciting the structure in an ultrasonic frequency range, the sensing system can focus on monitoring changes of structural properties with minimum interference from global operational and environmental variations. These sensor/actuators are inexpensive (less than US$5 per PZT patch), generally require low power (less than 5 V), and are relatively nonintrusive (as shown in Figure 15.3).

Examples of documented successes in active local sensing for damage detection using PZT are the impedance-based method [7] and the Lamb-wave propagation method [8]. The impedance method monitors the variations in mechanical impedance resulting from damage, and the mechanical impedance is coupled with the electrical impedance of the PZT sensor/actuator. For this method, the PZT acts simultaneously as a discrete sensor and actuator. A schematic of the impedance method is shown in Figure 15.4. For the Lamb-wave propagation method, one PZT is activated as an actuator to launch elastic waves through the structure, and responses are measured by an array of the other PZT patches acting as sensors. The structure can be systematically surveyed by sequentially using each of the PZT patches as an actuator and the remaining PZT patches as sensors. The technique looks for possible damage by tracking changes in transmission velocity and wave attenuation/reflections. A composite plate with a PZT sensor layer is shown in Figure 15.5. Both methods operate in the high frequency range (typically above 30 kHz)

Figure 15.3   A PZT sensor/actuator being used to monitor a bolted connection

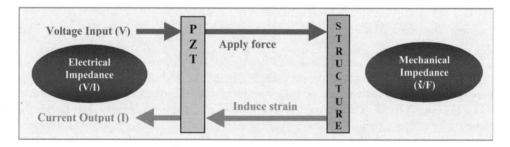

Figure 15.4   Schematic of the impedance method

Figure 15.5   A composite plate with a PZT sensor layer

where there are measurable changes in structural responses for even incipient damage associated with crack formation, debonding, delamination and loose connections.

Once structural damage has begun and is detected, numerical models can be used to capture the influence of damage on the system-level performance and to predict future performance. This procedure necessitates the measurement of global system-level response. It may be possible to use these same PZT patches in both an active and passive mode. When used in the passive mode, the sensors detect strain resulting from ambient loading conditions and can be used to monitor the global response of a system. In the active mode, the same sensors can be used to detect and locate damage on local level as described above.

15.3.8   Sensor Communication

The typical way of instrumenting a system is to run wires between the local sensors and a centralized data acquisition unit. This approach can impose serious limitations on damage prognosis because it is highly desirable to have a very large number of sensors. Recent advances in wireless communication can alleviate most of these limitations. With wireless technology, the local sensing and processing units can communicate with a centralized processing unit and each other. Potential constraints on wireless systems include the maximum range, amount of bandwidth available, energy requirement, and susceptibility to electromagnetic interference.

15.3.9   Powering the Sensing System

A major consideration in using a dense sensor array is the problem of providing power to the sensors. This demand leads to the concept of 'information as a form of energy'. Deriving information costs energy. If the only way to provide power is by direct connections, then the need for wireless protocols is eliminated, as the cabled power link can also be used for the transmission of data. Hence, the development of micropower generators is a key factor for the development of the hardware if wireless communication is to be used. A possible solution to the problem of localized power generation is technologies that enable harvesting ambient energy to power the instrumentation [9, 10]. Forms of energy that may be harvested include thermal, vibration, acoustic and solar. Although this is new technology, the overriding consideration of reliability still exists, as it does with any condition monitoring system. With two-way communication capability, the local sensing and processing units can also turn themselves off-line for energy conservation and they can be resuscitated when a 'wake-up' signal is broadcast.

15.3.10   Data Cleansing

In this section, the term 'data cleansing' refers to what is done electronically and not the data cleansing associated with software and data analysis. Data processing includes filtering, amplifying, and perhaps changing the signal source impedance. Historically, data processing prepared the signal for recording (on analog magnetic tape) or digitizing and storage (for digital acquisition). More sophisticated data processing incorporates 'smart processing' using a sensor/computer combination for amplification, A/D conversion, overload detection, transmission of time series data, and feature selection. A 'transformation matrix' preprocesses the raw sensor data into physical quantities of interest. This matrix is sensor specific and conceptually includes frequency response compensation, bounds on the sensor error, and automatic removal of out-of-band signals.

15.3.11 Sensor System Definition

Standards for definition of the sensor system are needed. A standard system defines sensor types, sensor locations, physical parameter sensed, sensor transition matrix (a matrix for translation of sensor readings to estimated physical parameters), and any other important sensor characteristics. Extensible Markup Language (XML) and Institute of Electrical and Electronics Engineers (IEEE) 1451.2 are potential templates for a standard sensor system.

Although XML was originally designed to improve the functionality of the Web by providing more flexible and adaptable information identification, it can also be used to store any kind of structured information and to enclose or encapsulate information in order to pass it between different computing systems, which would otherwise be unable to communicate.

IEEE 1451.2 standard defines smart transducer elements that can be treated as network-independent devices. The IEEE wrote 1451.2 to reduce the complexity of establishing communications between transducers in a networked environment. The specification addresses wiring, installation, and what is needed to calibrate a networked sensor. Essentially, the standard specifies a digital interface that can be used to access what might be called an electronic data sheet (nonvolatile memory as a transducer electronic data sheet, or TEDS). The specification also defines how a device reads sensor data and how it sets downstream actuators.

15.4 SUMMARY: SENSING AND DATA ACQUISITION

Fundamentally, the key issue for developing a sensor system for damage prognosis is the ability to capture system response on widely varying length and time scales. Special purpose sensing is a development of a 'smart sensor' philosophy, in which the sensor array is sensitive to and reports the presence of damage. Current developments in sensor technology indicate that MEMS devices will soon be integrated with signal processing, data interrogation, and telemetry capabilities and fabricated on the same silicon substrate. Such 'systems on a chip' may significantly improve the sensing and processing capabilities for prognostic applications by providing a dense array of sensors with low cost and low maintenance. Depending on the application, the telemetry can be accomplished either in a wired or wireless manner. For wireless telemetry, a major concern is the power source for such systems. Microparasitic generators being developed elsewhere may, when integrated with the system on a chip, provide the power that will enable a truly self-contained sensing capability. For in-service prognostics, it is possible that ambient vibration will provide both the power source and excitation for the structure. However, the authors' believe that the use of local actuation with waveforms tailored to the damage-prognosis activity will provide a more robust damage-detection and damage-monitoring capability.

Current modeling technology is not well integrated with developing sensor technology. In most cases, without the precise model of a system, it is difficult to know what exactly to measure *a priori*. Sensors that directly measure crack properties or corrosion are nonexistent. Damage prognosis requires sensors that measure those physical properties that are more directly related to the most probable damage scenarios, rather than sensors that measure only strain, strain rate, and acceleration. In addition, sensors are needed that use multiple materials or sensing elements to get increased dynamic range, and range of properties. There may be an evolution of sensor types in damage prognosis, for example, expensive, accurate sophisticated sensors initially, replaced by cheap special purpose sensors in the longer term, with the special purpose sensors targeting specific damage types.

At the 'front line' of any damage detection or prognosis system is the ability to acquire data that encapsulates any change in system properties that may affect its life or operation. Although simple sensor configurations with a limited number of sensors will provide an indicator of change to the global properties, higher-density sensor arrays are required, not only to provide localized information relating to damage, but also to provide for redundancy. Perhaps the most important aspect of the sensing system is that it must be more reliable than the system being monitored.

REFERENCES

[1] M.D. Todd, G.A. Johnson and B.L. Althouse (2001) 'A novel Bragg grating sensor interrogation system utilizing a scanning filter, a Mach–Zehnder Interferometer, and a 3×3 coupler', *Measurement Science and Technology*, **12**, 771–777.

[2] G.A. Johnson, K. Pran, G. Sagrulden, O. Farsund, G.B. Hausgard, G. Wang and A.E. Jensen (2000) 'Surface Effect Ship Vibro-Impact Monitoring with Distributed Arrays of Fiber Bragg Gratings', *Proceedings of the 18th International Modal Analysis Conference*, Society of Experimental Mechanics, San Antonio, TX, February 2000.

[3] R.H. Engler, C. Klein and O. Trinks (2000) 'Pressure sensitive paint systems for pressure distribution measurements in wind tunnels and turbomachines', *Measurement Science and Technology*, **11**, 1077–1085.

[4] W.J. Staszewski, K. Worden, R. Wardle and G.R. Tomlinson (2000) 'Fail-safe sensor distributions for impact detection in composite materials', *Smart Materials and Structures*, **9**, 298–303.

[5] http://robotics.eecs.berkeley.edu/~pister/SmartDust/

[6] R.R. Bracht, R.V. Pasquale and T.L. Petersen (2002) 'QAM Multi-Path Characterization due to Ocean Scattering', *International Telemetering Conference*, San Diego, CA, October 2002.

[7] G. Park, H. Sohn, C.R. Farrar and D.J. Inman (2003) 'Overview of piezoelectric impedance-based health monitoring and path forward', *The Shock and Vibration Digest*, accepted for publication.

[8] J.R. Wait, G. Park, H. Sohn and C.R. Farrar (2003) 'Active Sensing System Development for Damage Prognosis', *Proceedings, Fourth International Workshop on Structural Health Monitoring*, Stanford, CA, September 2003, DES tech Publications, Inc., Lancaster, PA, USA.

[9] H. Sodano, E.A. Magliula, G. Park and D.J. Inman (2002) 'Electric Power Generation using Piezoelectric Devices', *Proceedings of 13th International Conference on Adaptive Structures and Technologies*, Berlin, Germany, October 7–9, 2002, CRC Press, Boca Raton, Florida, USA.

[10] H. Sodano, G. Park, D.J. Leo and D.J. Inman (2003) 'Use of Piezoelectric Energy Harvesting Devices for Power Storage in Batteries', *Proceedings of 10th SPIE Conference on Smart Structures and Materials*, San Diego, CA, March 2–6, 2003.

16

Design of Active Structural Health Monitoring Systems for Aircraft and Spacecraft Structures

Fu-Kuo Chang, Jeong-Beom Ihn and Eric Blaise

Department of Aeronautics and Astronautics, Stanford University, California, USA

16.1 INTRODUCTION

Recent advances in smart structure technologies and material/structural damage characterization, combined with recent developments in sensors and actuators, have resulted in a significant interest in developing new diagnostic technologies for the *in-situ* characterization of material properties in manufacturing, for monitoring their integrity, and for the detection of damage to both existing and new structures in real time, with minimum human involvement. Using distributed sensors to monitor the condition of in-service structures becomes feasible if sensor signals can be interpreted accurately to reflect the *in-situ* condition of the structures through real-time data processing.

The potential benefits of such technologies are enormous, and include:

- self-warning and detection – increase *safety* and improve *reliability* and *readiness*;
- real-time monitoring and reporting – saves on maintenance;
- minimum human involvement – reduces labor and downtime.

Almost all in-service structures, such as transportation systems and civil infrastructures, would undoubtedly benefit significantly from the development of these technologies

Damage Prognosis – For Aerospace, Civil and Mechanical Systems Edited by D.J. Inman, C.R. Farrar, V. Lopes Junior and
V. Steffen Junior © 2005 John Wiley & Sons, Ltd

in maintaining the health functions of aircraft/spacecraft structures, buildings, bridges, and so on. Furthermore, the technology could fundamentally change the paradigm of the conventional philosophy of separating maintenance from design/analysis consideration. Traditional structural maintenance procedures are typically derived from maintenance manuals, which are produced mostly by manufacturers based primarily on laboratory coupon data and analytical predictions in the design and manufacturing stages. Limited or no in-service data are used in developing the manuals.

Since in-service conditions experienced by structures are typically unknown and may vary from one structure to another, traditional maintenance and inspection is required to follow strictly specific time schedules, which can be time consuming, labor intensive, and very expensive. As all structures will age, maintenance service and the associated maintenance cost is expected to increase. However, the specific procedures and time intervals derived from the manuals for performing inspections are not to be revised or changed unless a catastrophe or unexpected failure occurs between two inspection intervals. If this happens, a maintenance schedule will most likely be intensified, which, unavoidably, further increases the maintenance cost.

As a consequence, based on the current schedule-driven maintenance philosophy, the reliability of in-service structures will continue to deteriorate as structures age, regardless of the maintenance procedures and frequency of inspection. Intensifying maintenance service intervals may ease the decay of reliability, but, in return, will substantially increase the cost of maintenance as structures age. Furthermore, due to the lack of information of the actual in-service condition (loads, temperature, stresses, etc.), design data (limit load, design allowables, etc.) and analytical tools (fatigue analysis, damage analysis, finite-element simulations, etc.) used for the construction of the structures are usually excluded from consideration for appropriate updating of maintenance procedures as structures age. This results in maintenance procedures that are out of date and not appropriate in response to the actual conditions of the structures. The communication breakdown between the design/analysis and maintenance procedures is primarily due to the unknown of the structural conditions once in service. Figure 16.1 depicts the traditional design and maintenance philosophy in structural operation.

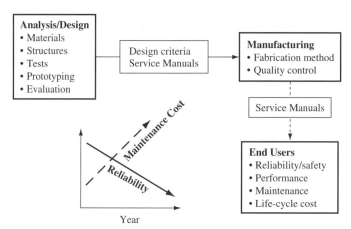

Figure 16.1 Traditional design and maintenance philosophy in structural operation. Maintenance cost continues to go up, and reliability will go down as structures age

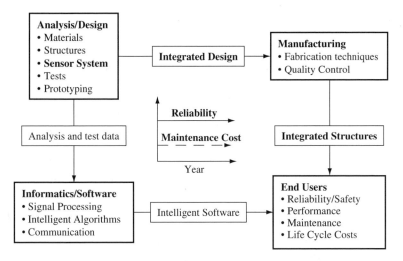

Figure 16.2 The philosophy of design/maintenance for structural operation through structural health monitoring

However, structural health monitoring (SHM) may offer a promising technology to revolutionize how structures should be maintained in the future. With a built-in sensor network on structures, SHM could provide information regarding the condition and damage state of the structures as they age. Maintenance will be executed only based on the actual performance of the structures, not on a specific time frame. Such performance-based maintenance would be able to prevent catastrophes while sustaining the reliability of the structures and reducing maintenance cost.

Furthermore, with the information about the service environments, SHM technology would enable maintenance procedures to be updated appropriately with assistance from the design and analysis tools developed originally for a specific structure. Accordingly, the maintenance procedures and time intervals will be up to date, and the structures will be maintained appropriately in response to the actual experience which the structures have exposed. Figure 16.2 describes the new philosophy of integration of design and maintenance in structural operation through structural health monitoring technology.

Since SHM technology relies on the changes in sensor measurements at two different times to diagnose the condition of a structure, this technology is applicable to both existing and new structures. However, development of such systems encompasses a broad range of technologies and requires innovative approaches to detect damage in advance of failure.

16.2 ACTIVE SENSOR NETWORK FOR STRUCTURAL-HEALTH MONITORING SYSTEMS

Based on sensor functionality, all the structural-health monitoring systems can be classified into two types: passive sensing and active sensing. For a passive sensing system, sensor measurements are constantly taken in real time, while the structures are in service, and are compared with a set of reference (healthy) data. The sensor-based system estimates the condition of the structures based on the data comparison. Because the input energy to the structure is typically random and unknown, the corresponding sensor measurements reflect the response of the structures to the unknown inputs. This type of diagnostics has been primarily applied to the determination of the unknown inputs that cause the changes in sensor measurements, such as external loads, incipient cracks, temperature, pressure, etc. For an active sensing system, known external mechanical or nonmechanical loads are input to the structures through built-in devices such as transducers or actuators. Since the inputs are known, the difference in the local sensor measurements based on the same input is strongly related to a physical change in the structural condition, such as the introduction of damage. This chapter focuses on an active sensing system based on piezoelectric materials. Typically, the active sensing network uses thin-disk type of piezoelectric material of lead zirconate titanate (PZT) ceramic as actuators and sensors. When an electric field is applied, the material produces dimensional changes. Conversely, when mechanical pressure is applied, the crystalline structure produces a voltage proportional to the pressure. The mechanical and electrical axes of operation are set during 'polling', the process that induces piezoelectric properties in the ceramic. Due to the anisotropic nature of piezoelectric materials after the polling process, its effects depend on direction.

The following electromechanical equations describe the effects of a piezoelectric material:

$$S_{ij} = s_{ijkl}^E T_{kl} + d_{kij} E_k \qquad (16.1)$$

$$D_i = d_{ikl} T_{kl} + \varepsilon_{ik}^T E_k \qquad (16.2)$$

where S_{ij} is a strain tensor, s_{ijkl}^E a compliance tensor (constant electric filed), E_k an applied electric field, ε_{ik}^T a dielectric constant (at constant stress), T_{kl} is a stress tensor, d_{kij} a piezoelectric constant, and D_i the dielectric displacement.

In order to design an active built-in diagnostic system, the following components are required:

- built-in sensor network;
- diagnostic software;
- electronic hardware.

In this presentation, the attention will focus on the built-in sensor network design and diagnostic software. The design of hardware will not be discussed.

16.2.1 Sensor Network Design

Lin and Chang [1] have invented a manufacturing technique to integrate a network of piezoelectric sensors into a dielectric film (so-called 'SMART layer') using electronic-circuit-printing techniques. A SMART layer can be surface mounted on metallic structures or embedded inside composite structures (Figure 16.3). The SMART layers

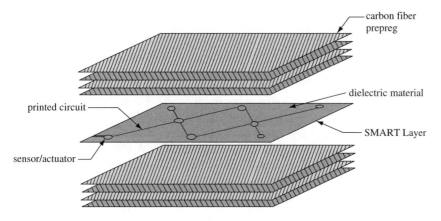

Figure 16.3   Rectangular SMART layer with embedded sensor networks

can be customized in different shapes depending on the applications or geometry of the structures.

Two sensor network designs based on the SMART-layer concept will be introduced for each selected application; fatigue crack detection in fuselage joints and disbond detection in cryogenic fuel tanks.

16.2.1.1   Design for Crack Detection at Fuselage Lap Joints

As illustrated in Figure 16.4, the SMART layer, which has a strip shape with the embedded array of piezoelectric sensor/actuators, has been designed for monitoring multisite cracks on fuselage lap joints. An actuator SMART strip will be installed between

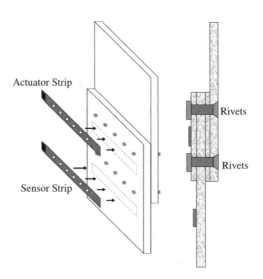

Figure 16.4   The sensor network design for riveted lap joints

rivet rows and a sensor SMART strip will be installed at a distance. This is a pitch-and-catch configuration where multiline inspection is possible for covering a large area.

16.2.1.2 Design for Disbond Detection in Cryogenic Fuel Tanks

The SMART layer concept was further extended by Blaise and Chang to apply to detecting disbond in cryogenic fuel tanks [2,3]. A particularity of cryogenic fuel tank structures is the high thermal gradient experienced by the structure. The temperature of the inner part of the tank is the boiling temperature of the fuel, which for hydrogen corresponds to −253 °C, while the temperature of the outer part of the tank varies during the flight cycle, but is warmer than the inner skin. As with any lightweight structure, sandwich structures are susceptible to damage that can compromise the safety of part or of the whole structure.

NASA [4] showed that the properties of the piezoelectric ceramics are greatly reduced once the temperature approaches the temperatures of the inner part of the tank. Therefore to minimize the temperature effect, the piezoelectric sensors have to be installed as far from the cold side as possible.

The proposed design mounts the sensors close to the outer skin, which is the warmest part of the sandwich structure. The sensors are mounted inside the honeycomb core and the space between the both skins is filled with the sensor and a connecting rod. Due to its internal location, the sensor is protected from damage due to mechanical loading or thermal stresses. The connecting rod constrains the sensor between the two skins, thereby using the axial as well as the radial properties of the piezoelectric ceramic disc. Thus, a greater portion of the strain energy generated by the actuator is transmitted to the structure. Since the sensors are located inside the structure, we can use larger actuators to produce more energy, thus further increasing the sensor spacing. Figure 16.5 illustrates how the sensors fill the space inside the honeycomb core of the structure. Honeycomb made by either the expansion process or the corrugation process doesn't have a geometric tolerance that is very precise; usually the tolerated error is ±10 % on cell dimensions. This can be problematic for accurately positioning an array of actuators and sensors within the core. To align an array of sensors with the corresponding cells, two degrees of freedom are required for each sensor. This can only be achieved when the sensor allows some flexibility in its positioning. Figure 16.5 illustrates how the sinusoidal shape of the connecting wire can be used to achieve the needed flexibility to position the sensor in the appropriate cell. The wire layer connecting the sensor to the rest of the circuit is also represented.

16.2.1.3 Temperature Effect

Before a structural health monitoring system can be implemented in a real structure, such as a cryogenic fuel tank, changes of signal due to variations of environmental variables have to be understood and efficiently modeled. The temperature of the fuel tank varies with the amount of cryogenic fuel. The temperature also varies during the flight of the reusable launch vehicle. With the varying temperature, the measured signal will change in amplitude and in time of arrival. To capture this change tests were done on a panel with embedded sensors.

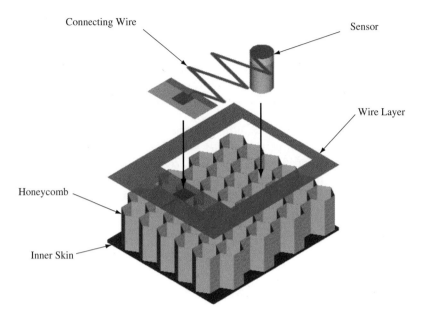

Figure 16.5 Sensor assembly showing the flexibility of the connecting wire

Test Methodology

A sandwich panel with four embedded sensors was manufactured. The piezoelectric ceramic sensors had a disc shape with a diameter of 3/8 in and thickness of 1/8 in. The sensors were bought from APC International Ltd, and the model used was APC 850. The sandwich panel was made with aluminum skins, which were 0.05-in thick, and an aluminum honeycomb core that was furnished by Hexcel. The honeycomb core had 3/8-in cell size, with a wall thickness of 0.003 in and a cell height of 5/8 in. The skins and the core were bonded using Hysol EA9268 film adhesive. The panel measured 15-in square and the four embedded piezoelectric ceramics with their matching aluminum rods were spaced 6 in apart. To test the signal propagation at different temperatures, the panel was insulated from the ambient air by placing it in a recessed Styrofoam panel. Above the insulated panel, an open bath of liquid nitrogen was used to lower the temperature of the test panel. The thermocouples were placed in the middle of the panel, one on the topside and the other on the bottom. Once the temperature reached the lowest reading, corresponding to the time when the boiling liquid nitrogen had all evaporated, the open bath was removed. At this time the test panel was covered by a Styrofoam plate, thereby encasing the test panel inside the insulation material. With the test panel fully insulated, the temperature remained constant long enough to take measurement at a given temperature.

Test Results

With the colder temperature, the amplitude of the signal decreases due to reduction of the electromechanical coupling factor of the sensors. Another behavior that can be recorded is the wave propagation speed. With colder temperatures, the stiffness of the structure increases slightly. This change in stiffness makes the elastic waves travel faster in the test panel. Piezoelectric ceramic selection is very important, since the choice of the sensor can influence its behaviour at different temperature. From the data given by Morgan Electro-Ceramics, which plots the electromechanical properties of the different ceramics as a function of temperature, the ceramics PZT-5A and PZT-5H exhibit a linear behavior over a wide range of temperatures. Both these ceramics show constant linear behavior from $-150\,°C$ to $25\,°C$. This corresponds to the temperatures that the sensors located close to the outer skin might experience in an application such as the cryogenic fuel tank. The properties of the ceramics used in this study (APC 850) correspond to the general properties of the PZT-5A. The linear behavior of the coupling factor is reflected in the measured signal amplitude. The signal amplitude varies linearly with the ambient temperature as shown in Figure 16.6. The time of arrival, which changes only a little, has a variation that can be approximated as linear. Since both amplitude and time of arrival vary linearly with temperature, it is possible to

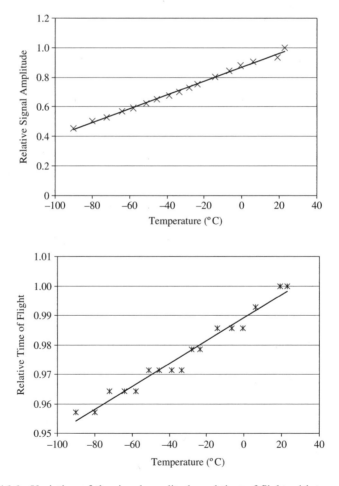

Figure 16.6   Variation of the signal amplitude and time of flight with temperature

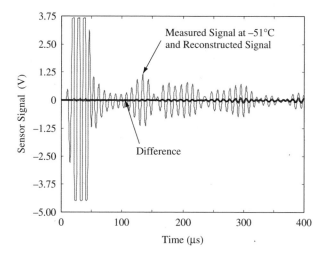

Figure 16.7 Comparison between the measured and reconstructed signal at $-51\,^{\circ}C$

generate the baseline at any temperature from a set of two measurements at different temperatures.

A simple linear interpolation or extrapolation, depending upon the case, is needed to generate the signal for a given temperature. Equation 16.3 gives the formula for calculating the reconstructed signal $y_T(t)$, which is a function of time (t), for a given temperature T, from two previously recorded signals $y_{T_1}(t)$ and $y_{T_2}(t)$ at two different temperatures T_1 and T_2. Figure 16.7 shows the reconstructed signal compared with the data taken at $-51\,^{\circ}C$.

$$y_T(t) = \frac{T - T_1}{T_2 - T_1}\left(y_{T_2}(t) - y_{T_1}(t)\right) + y_{T_1}(t) \tag{16.3}$$

16.3 DIAGNOSTIC SOFTWARE

The damage detection technique uses sensor signals generated from nearby piezoelectric actuators built into the structures to detect any structural flaws. It consists of three major components: diagnostic signal generation, signal processing, and damage diagnostics.

16.3.1 Signal Generation

A piezoelectric actuator can generate diagnostic waves that propagate along the structure for damage interrogation. The changes in the received signals can then be analyzed to reveal structural flaws. An appropriate diagnostic waveform has to be chosen for easy signal interpretation. The five-cycle windowed sine-burst signal shown in Figure 16.8(a) is widely used in the field of nondestructive evaluation for its good dispersion

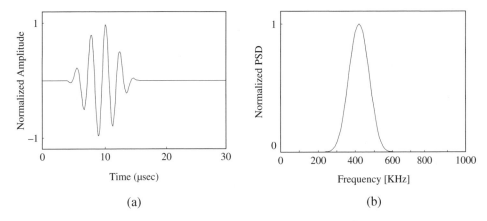

Figure 16.8 Diagnostic input waveform at driving frequency of 420 kHz. (a) Windowed burst signal with 5 cycles; (b) frequency spectrum (power spectral density)

characteristic and its sensitivity to structural flaws. A Hanning window is used to concentrate the maximum amount of energy at the desired driving frequency. As Figure 16.8(b) shows, the waveform has a narrow-band frequency spectrum with its spectral density concentrated at the driving frequency. The narrower the band width is, the less dispersion there is where the input waveform remains about the same as it travels along the medium. If the number of cycles of the input function is increased, the bandwidth will narrow but the time duration of the wave will lengthen. When the time duration of a wave is longer, the time resolution gets lower. Therefore, there is always a trade-off between a good dispersion characteristic and time resolution.

16.3.2 Signal Processing

Before sensor measurements can be interpreted and related to structural flaws (crack, delamination), the raw sensor measurements should be represented such that the interpretation can be easier and more effective. The time and frequency representation gives us the most information about the sensor measurements [5]. The plots of spectral amplitude information in time and frequency axes are called spectrograms. A Matlab^{TM} function from a signal processing tool box, *specgram.m*, was used to generate spectrograms. The *specgram.m* computes the windowed discrete-time Fourier transform of a signal using a sliding window (short-time Fourier Transform). The spectrogram is the magnitude of this function.

The STFT of a signal $s(t)$ is defined as:

$$S(\omega, t) = \frac{1}{2\pi} \int_{-\infty}^{\infty} e^{-i\omega\tau} s(\tau) h(\tau - t) d\tau \tag{16.4}$$

where $h(t)$ is a window function. Since the window function $h(t)$ has a short time duration, by moving $h(t)$ with Fourier integrals, the signal's local frequency properties evolving over time can be revealed. The Hanning window is typically used as a window function.

A joint time-frequency analysis method based on the short-time Fourier transform (STFT) can be used for generating the envelope of the time varying signal and extracting time-of-flight (TOF) information from sensor measurements. The group velocity measurements can be then obtained from:

$$Group\ velocity = \frac{Sensor\ spacing}{TOF} \tag{16.5}$$

The group velocity measurement is used as key information for the signal interpretation as it reveals the types of mode traveling along the structure. Figure 16.11 illustrates the procedure of obtaining the time of flight information and the group velocity of the sensor measurement. Roh [6] previously used this approach to obtain the time of flight information of the scattered wave from an anomaly on an aluminum plate. Plot (a) in Figure 16.9 shows the time-frequency representation of a given sensor measurement by a spectrogram with time represented by the horizontal axis, frequency represented by the vertical axis, and spectral amplitude represented by the axis coming out of the page. It shows the maximum spectral amplitude at a driving input center frequency of 340 kHz. This spectral amplitude, when plotted in time at a fixed driving frequency (340 kHz), represents the envelope of the time signal after normalization (plot (b) in Figure 16.9). Since the time signal contains lots of peaks and valleys, the unique peak of the envelope on the first arrival wave packet well defines the reference point for the time-of-flight information. The time of flight on the first arrival wave packet can be

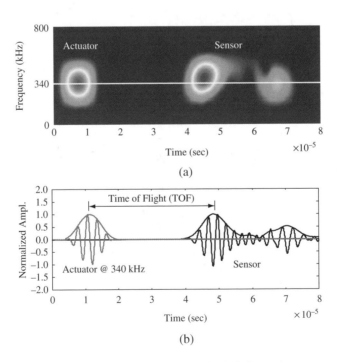

Figure 16.9   Extracting time of flight information

obtained by taking the time difference between the peak of the actuator and the peak of the first arrival envelope.

The wide-band spectrogram can be constructed from the short-time Fourier transform processes of the collected data at different center frequencies. Figures 16.10 and 16.11 show the collected sensor signals from 60 to 580 kHz input center frequencies and corresponding sensor spectrogram respectively. The wide-band sensor spectrogram shows how the amplitude of the sensor signals distributes in a wide range of frequencies and the time domain. It also helps to decide the driving center frequency where it gives the best wave propagating resolution or, the highest signal-to-noise ratio. It also shows the dispersion relation of Lamb waves in the time and frequency axes as we connect the peaks of the first and the second arrival wave packet (Figure 16.11).

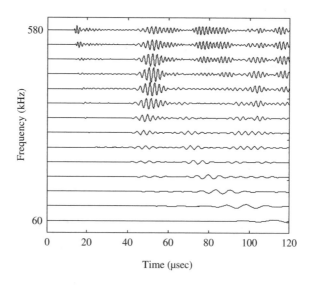

Figure 16.10   Collected sensor signal spectrogram

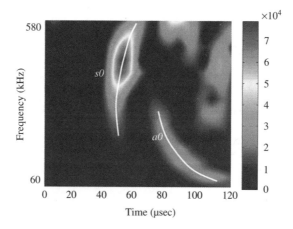

Figure 16.11   Wide-band sensor

16.3.3 Damage Diagnostics

When an elastic wave travels through a region where there is a change in material properties, scattering occurs in all directions. The directly transmitted signals are modified to the forward scattering waves, and the scattered energy provides good information about a presence of damage across the actuator and sensor path. The scatter wave in the time domain can be obtained by subtracting the baseline data recorded for the structure with the initial damage size from the sensor data for the structure with the extended damage size.

16.3.3.1 Damage Index

When a single actuator–sensor path is considered for damage interrogation, the amount of signal changes are quantified as a scattered energy and may be related to the change in local material properties. Since the scattered energy contains information of both attenuation and phase delay of the directly transmitted wave due to the presence of a flaw, the appropriate damage index can be selected to quantify the change in sensor measurements to crack size. It is well known that using piezoelectric elements as actuators, propagating Lamb waves can be generated by the actuators on thin plate-like structures [7]. The authors and others have demonstrated that Lamb waves in particular modes can be sensitive to specific types of damage [8–11]. For instance, Lamb wave in s0 mode is shown to be sensitive to crack growth, and in a0 mode is sensitive to delamination or debonding. Accordingly, by selecting appropriate Lamb wave modes from the receiving sensor measurements, the following damage indices were proposed by the authors [12,13] to quantify crack and debond size:

$$\text{Damage Index for crack detection} = \left[\frac{\text{Scattered energy of s0 wave}}{\text{Baseline energy of s0 wave}}\right]^k$$

$$\text{Damage Index for debond detection} = \left[\frac{\text{Scattered energy of a0 wave}}{\text{Baleline energy of a0 wave}}\right]^k$$

(16.6)

The damage index can also be mathematically expressed as:

$$\text{Damage Index} = \left(\frac{\int_{t_i}^{t_f} |S_{sc}(\omega_0, t)|^2 dt}{\int_{t_i}^{t_f} |S_b(\omega_0, t)|^2 dt}\right)^k$$

(16.7)

where S_{sc} denotes the time varying spectral amplitude of scatter signal, S_b denotes the time varying spectral amplitude of baseline signal, ω_0 denotes the selected driving frequency (usually the driving frequency which gives the highest signal to noise ratio), t_f and t_i denote the upper bound and lower bound of s0 or a0 mode in time domain respectively, and k is a gain factor ($0 < k \le 1$) and can be chosen empirically depending on damage type. Previous tests have shown that a damage index with a gain factor of 0.5 gives the best linearity with the actual crack length [13].

Since the damage index utilizes the information of the forward scattered (or directly transmitted) wave, it is limited to providing a line interrogation and assessing the damage only near or in the line of the actuator–sensor path. Therefore, the actuator–sensor path should be located at high stress areas, such as holes, for detecting incipient cracks.

16.3.3.2 Damage Detection Scheme

Figure 16.12 describes the damage detection scheme. Sensor measurements are taken from the sensor network or the SMART layer installed on the structure. The initial measurement is considered as a baseline that represents the initial condition of the structure. After a signal processing routine, we can measure the group velocity from the baseline, and using only a group velocity estimate from a wave model by Disperse [14,15], we can then select the s0 mode (or a0 mode) in time domain. This information is fed to the damage-index routine. The subsequent sensor measurements are subtracted from the baseline and represent the signal changes or scatter signals. The damage index is then evaluated based on the s0 mode (or a0 mode) wave packet selected initially.

16.4 VALIDATION OF THE ACTIVE SHM SYSTEM

16.4.1 Experiment Setup

Two identical aluminum single lap joints (936 × 462 mm), called 'airbus lap joint 1' and 'airbus lap joint 2', with two artificial 1.27 mm edge cracks at the center rivet, were used for a constant amplitude fatigue test under tensile loading. Three *SMART Layer*™s by Acellent Technologies were surface mounted as shown in Figure 16.13. The center strip located between the two rivet rows was used as an actuator strip, and the top and bottom strips were used as sensor strips. Between specified cyclic loading intervals,

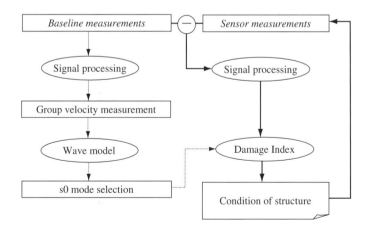

Figure 16.12 Crack detection scheme

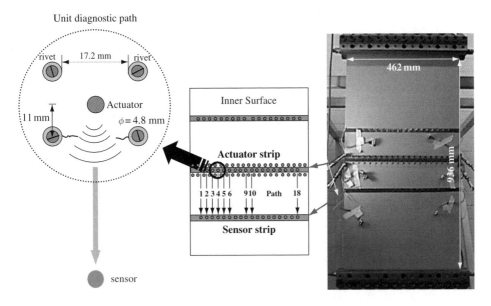

Figure 16.13 Lap joint specimen with SMART strips and diagnostic paths

ultrasonic scans and eddy current tests were performed in order to detect crack initiation and growth around 21 rivet holes. No visual inspection was possible on the lap joints. In parallel with nondestructive testing evaluations, a SMART Suitcase [16, 17] (a portable active diagnostic instrument designed to interface with piezoelectric transducers) was implemented to generate diagnostic signals from the actuator strip and record measurements from the sensor strips. Using a built-in waveform generator, a windowed sine burst wave was generated as an input signal over a wide frequency range of 100 ~900 kHz, and corresponding sensor measurements were collected at a sampling rate of 25 Msample/s.

16.4.2 SMART Strip Results

In order to evaluate the damage index for crack detection at the lap joint specimens, the s0 mode wave packet must be identified. An interactive Windows program called 'Disperse' [14,15] was utilized to generate the analytical time-of-flight (TOF) curve for the lap joints. The analytical results and experimental results of TOF are shown in Figure 16.14. The results strongly suggest that the first arrival wave packet is the s0 mode wave for the damage index evaluation for crack detection at the lap joint specimens.

Among the crack estimate results at 18 actuator–sensor paths for each lap joint specimen, Figure 16.15 shows the damage index vs the number of cycles along with the NDT estimates on the secondary axis at the selected paths. A good correlation between the damage index and NDT estimates is observed in Figure 16.15. The POD (probability of detection) is a characteristic that allows us to quantify the quality of any NDT technique [12]. To obtain the POD of the damage index based on NDT results, the different crack events monitored by NDT were ranked from the largest down to the

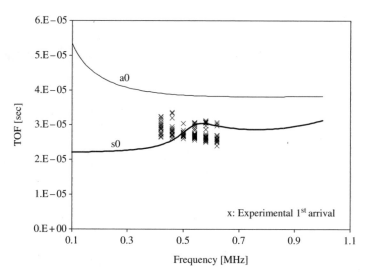

Figure 16.14   Analytical and experimental results of s0 mode time of flight

smallest crack length and then compared with the results obtained with the damage index such as:

$$POD = \frac{SC}{M + 1 - N} \tag{16.8}$$

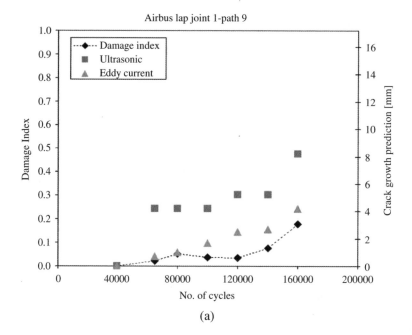

(a)

Figure 16.15   Damage Index vs number of cycles with NDT estimate in secondary axis. (a) Airbus lap joint 1, path 9; (b) Airbus lap joint 2, path 10

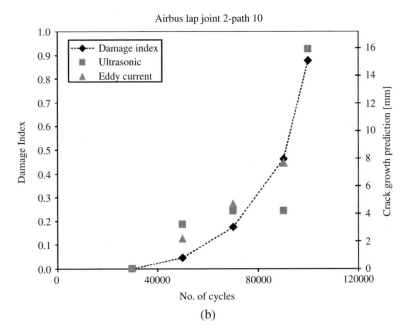

Figure 16.15 (Continued)

where *SC* is the sum of crack events recorded by the damage index, *M* is the total number of crack events recorded by NDT, and *N* is a serial event.

Using the formula above, POD was evaluated in Figure 16.16 where the damage index shows a higher probability of damage detection with the eddy current rather than the ultrasonic technique.

16.5 CONCLUSIONS

Two different sensor network designs based on the SMART layer concept were introduced for each selected application; fatigue-crack detection in fuselage joints and disbond detection in cryogenic fuel tanks. The new sensor design for cryogenic fuel tanks was tested at cryogenic temperatures to ensure their survivability, and a scheme to compensate for temperature effects on the sensor measurements was developed.

An active sensing diagnostic technique using built-in piezoelectric sensor/actuator networks was investigated and discussed. It consists of three major components: diagnostic signal generation, signal processing, and damage diagnostics. The diagnostic technique was verified with a fuselage structure where a physics-based damage index was developed for detecting cracks in riveted lap joints by extracting features in sensor signals related to crack growth. As demonstrated, the technique was used to detect crack lengths as small as 5 mm on riveted fuselage joints, with certainty equal to conventional nondestructive testing (NDT) such as the eddy current testing and ultrasonic scan methods. The damage detection ability using the proposed system can be further

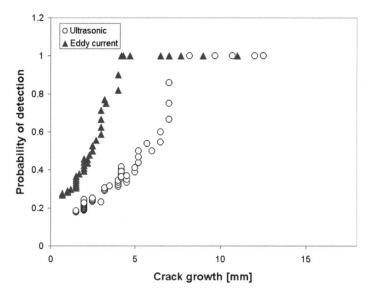

Figure 16.16 Probability of damage detection by the damage index when compared to the ultrasonic and eddy current NDT

improved by selecting the optimal actuator–sensor location, which is a function of structural geometry and initial structure conditions [13].

ACKNOWLEDGMENTS

The financial support of the Air Force Office of Scientific Research, the National Aeronautics and Space Agency, and the European Aeronautic Defence and Space Company (EADS) is gratefully appreciated. The authors would like to thank Christian Boller of the University of Sheffield (the work was done while Dr Boller was with EADS) and Holger Speckmann of EADS Airbus for the support for the single-lap tests.

REFERENCES

[1] M. Lin and F.-K. Chang (1998) 'Design and Manufacturing of Built-In Diagnostic for Composite Structures', *Proceedings of the Twelfth American Society for Composites Conference*, pp. 89–101.

[2] E. Blaise (2004) *Built-in Diagnostics for Sandwich Structures under Extreme Temperatures*, PhD Dissertation, Department of Mechanical Engineering, Stanford University, Stanford.

[3] E. Blaise and F.-K. Chang (2003) 'Active sensing monitoring of cryogenic composite structures for space transportation applications', *NASA NCC8-197-01*.

[4] Hooker, M.W. (1998) 'Properties of PZT-based piezoelectric ceramics between −150 and 250 °C', *NASA/CR-1998-208708*.

[5] Shie Qian and Dapang Chen (1996) *Joint Time-Frequency Analysis – Methods and Applications*, Prentice Hall, New York.

[6] Y.S. Roh (1999) *Built-in Diagnostics for Identifying an Anomaly in Plates Using Wave Scattering*, PhD Dissertation, Department of Aeronautics and Astronautics, Stanford University.

[7] I.A. Viktrov (1967) *Rayleigh and Lamb Waves, Physical Theory and Applications*, Plenum Press, New York.

[8] D.N. Alleyne and P. Cawley (1992) 'The interaction of Lamb waves with defects', *IEEE Trans. Ultrasonics, Ferroelectrics, and frequency control*, **39**, 381–397.

[9] Z. Chang and A. Mal (1999) 'Scattering of Lamb waves from a rivet hole with edge cracks', *Mechanics of Materials*, **21**, 197–204.

[10] C.S. Wang and F.-K. Chang (1999) 'Built-In Diagnostics for Impact Damage Identification of Composite Structures', *Proceedings of the second International Workshop on Structural Health Monitoring*, Stanford, California, Technomic Publishing, pp. 612–621.

[11] Kessler, S.S., Spearing, S.M. and C. Soutis (2002) 'Structural health monitoring in composite materials using Lamb wave methods', *Smart Materials and Structures*, **11**, 269–278.

[12] C. Boller, J.-B. Ihn, W.J. Staszewski and H. Speckmann (2001) 'Design Principles and Inspection Techniques for Long Life Endurance of Aircraft Structures', *Proceedings of the 3rd International Workshop on Structural Health Monitoring*, Stanford, California, CRC Press, pp. 275–283.

[13] J.-B. Ihn (2003) *Built-in Diagnostics for Monitoring Fatigue Crack Growth in Aircraft Structures*, PhD Dissertation, Department of Aeronautics and Astronautics, Stanford University, Stanford.

[14] B. Pavlakovic and M. Lowe (2000) *A system for Generating Dispersion Curves – User's Manual for Disperse v.2.0*.

[15] M.J.S. Lowe, B.N. Pavlakovic and P. Cawley (2001) 'Guided Wave NDT of Structures: A General Purpose Computer Model for Calculating Waveguide Properties', *Proceedings of the 3rd International Workshop on Structural Health Monitoring*, Stanford, Colifornia, CRC Press, pp. 880–888.

[16] M. Lin, X. Qing, A. Kumar and S. Beard (2001) 'SMART Layer and SMART Suitcase for Structural Health Monitoring Applications', *Proceedings of SPIE on Smart Structures and Material Systems*, March 2001.

[17] Acellent technologies http://www.acellent.com.

17
Optical-Based Sensing

Michael D. Todd

Department of Structural Engineering, University of California, San Diego, California, USA

17.1 OVERVIEW AND SCOPE OF CHAPTER

This chapter will provide the reader with an overview of primary optical sensing techniques. This topic itself is very broad, as optical sensing techniques are being used in a wide array of applications, and the scope of this chapter will be limited to techniques that are generally being used for mechanical motion detection and have made the transition from the laboratory to the field. Such techniques are limited to intensity-based modulation, phase-based modulation (interferometry), and wavelength-based modulation (e.g., Bragg gratings). This chapter is a substantially shortened version of a much more complete treatment, which may be found online at http://www.cimss.vt.edu/PASI/.

17.2 BASIC OPTICS CONCEPTS

Since light is an electromagnetic phenomenon, its propagation occurs within a continuous spectrum of all electromagnetic waves. Three regions subdivide the optical portion of the electromagnetic spectrum: the range of visible light (the range most sensitive to the human eye) is from about 380 nm to 780 nm, while on either side exists the infrared region (750 nm to 3 µm) and the ultraviolet region (100 nm to 380 nm). Because optical wavelengths are very small, the interaction of light with matter is not negligible, and the field of quantum optics, born out of the works of de Broglie, Schroedinger, Heisenberg, and others, completely describes all matter/light interactions [1]. The subset of optics called electromagnetic or wave optics describes nonquantum (classical) effects in optics such as diffraction and interference, and the effects of energy at material boundaries.

Damage Prognosis – For Aerospace, Civil and Mechanical Systems   Edited by D.J. Inman, C.R. Farrar, V. Lopes Junior and V. Steffen Junior   © 2005 John Wiley & Sons, Ltd

Finally, if one assumes that the optical wavelengths are very nearly zero, a further subset called ray or geometric optics is useful for describing certain fundamental kinematic properties of propagation such as reflection and refraction.

It is these properties of reflection and refraction that facilitate the use of *optical waveguides*, e.g. optical fiber, which are dielectric (electrically nonconductive) materials used to guide light along a specific path. One important property of a dielectric material is its *index of refraction n*, which is defined as the ratio of the speed of light in a vacuum to the speed of light in that material. When a plane optical wave impinges at angle μ_i upon a boundary between dielectric materials with different refractive indices, part of the light is known to be reflected and part is known to be transmitted into the second material. If the assumption that the waveguide length scales are large compared with optical wavelengths is made, then ray optics may be employed to depict this interaction. After incidence, the ray is reflected at an angle μ_r that is equal to the incident angle μ_i, and this result, which is experimentally observed, is derivable from Fermat's principle of least time. The transmitted part of the incident wave gets bent either towards or away from the normal line, depending on the relative magnitudes of n_1 and n_2. This bending is called *refraction*, and the amount of bending is determined by Snell's Law

$$n_1 \sin \mu_i = n_2 \sin \mu_t. \tag{17.1}$$

If $n_2 < n_1$, such as the case of light entering air from glass, and if μ_i is large enough, it would appear that no solution exists to Equation (17.1) for μ_t, since the sine function would be greater than unity. At this critical angle of μ_i, $\mu_{crit} = \mu_i = n_2/n_1$, light can not escape the initial material, resulting in *total internal reflection*. Light that grazes the boundary remains trapped inside the material. The refractive index is a property that depends on the wavelength of the light propagating inside it, and for typical dielectric materials, it decreases with increasing wavelength (known as *dispersion*). Even though a ray interpretation admits total internal reflection, a more thorough wave analysis is required to discuss the relative amplitudes (intensities) of the various reflected and/or refracted waves. Such an analysis results in the Fresnel equations, and these equations show that even under total internal reflection, the optical wave intensity is nonzero beyond the boundary. However, the intensity profile decays exponentially and may only be detected for a few wavelengths beyond the interface [2]. This wave is called an *evanescent wave*, and it has some use in optical sensing, to be discussed later.

Internal reflection is the fundamental principle behind the operation of waveguides, such as optical fibers. Optical fibers are just cylindrical waveguides constructed by concentric materials. The core is usually made of silicon dioxide-based material with a refractive index of around $n = 1.44$ at $\lambda = 1.55 \, \mu m$. Surrounding the core is the *cladding*, which is made of a glass or plastic material and a slightly higher refractive index usually 0.1 %–5 % lower than the core. Finally, an outer coating or jacketing layer of plastic (acrylate, polyimide) is added for mechanical strength and environmental protection. The theory of optical fibers, dating back to original work in 1961 [3], is well established and is discussed at length in several works [4–7]. The solution of Maxwell's equations for optical waves in a cylindrical fiber, subject to appropriate boundary conditions, results in certain characteristic modes in which the light may propagate, much as a clamped string only supports certain modes of vibration. The *guided modes* are of most importance to the present discussion, as these are the ones that propagate down the core under typical conditions. The kinds and numbers of modes that are supported by a given fiber depend on the fiber geometry, the materials, the refractive index profile, and the wavelength of the light itself. Generally, fibers may be classified as single-mode or multimode; this classification is usually quantified by the normalized frequency V of the fiber, defined as

$$V = \frac{2\pi d}{\lambda} \sqrt{n_{core}^2 - n_{cladding}^2},$$ (17.2)

where d is the core diameter, λ is the optical wavelength, and the term $\sqrt{n_{core}^2 - n_{cladding}^2}$ is called the *numerical aperture*. When $V < 2.405$, the fiber admits only a single mode, and using Equation (17.2), this establishes a given upper limit on the core diameter for a given wavelength. In the standard telecommunications range ($\lambda = 1.3 - 1.55 \, \mu m$), this limit is about $8 - 10 \, \mu m$. Typical corresponding cladding diameters for single mode fiber (SMF) are about $125 \, \mu m$ with a refractive index change of only about $0.1\% - 0.2\%$. Conversely, one may fix the value of $V = 2.405$ for single-mode operation and establish the minimum cut-off wavelength above which the fiber is single mode and below which it is multimode. It should be noted that for an accurate description of optical wave propagation in SMF, ray optics is not technically valid, as optical wavelengths are the same order of magnitude as the fiber core diameter.

In general, the sine of the numerical aperture is called the acceptance angle β_{crit}, and this angle is the maximum angle (as measured from the fiber longitudinal axis) below which all rays launched into the fiber will remain in the fiber. The geometric relationship between this angle and the critical angle for total internal reflection is shown in Figure 17.1. Again using ray diagrams, an incoming ray gets refracted as it enters the glass core from outside the fiber (e.g., air). The amount of refraction determines the critical angle μ_{crit} required for total internal reflection and guided wave propagation. In most applications, light is launched essentially parallel to the fiber longitudinal axis, so total internal reflection is guaranteed.

The most dominant material basis used to manufacture the fiber core itself is glass, and the three major types are silica, fluoride, and chalcogenide. Silica glass fibers are the most common, with very high melting points ($1000\,°F$) and low losses over a wide range ($<0.5 \, dB/km$). For applications above the $1.55 \, \mu m$ wavelength range, losses in silica glass fibers begin to accumulate rapidly, and fluorinated glasses are of interest. These fibers, often used in spectroscopy instrumentation, may transmit up to $5 \, \mu m$ with low predicted loss (e.g., $0.001 \, dB/km$ at $2.5 \, \mu m$). Chalcogenide glass fibers are based on Ge-S, As-S, etc., and are very transparent in the same infrared region as the fluoridated glasses with very

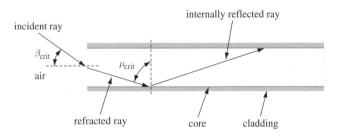

Figure 17.1 Relationship between acceptance angle β_{crit} and critical internal reflection angle μ_{crit}

long formation domains and the greatest resistance to corrosion by humidity. There are also several non-glass-based fibers such as crystalline fibers (single and polycrystalline), plastic fibers (polymethylmetacrylate), and other specialty fibers (e.g., hollow core, liquid core, and coated) in use for various selected applications [8–11].

17.3 PRIMARY FIBER OPTIC SENSING APPROACHES FOR STRUCTURAL MEASUREMENTS

Optical fiber serving as a waveguide may be integrated with other electro-optical components for sensing purposes, i.e. to detect measurands of interest through detecting some changing optical property. Very broadly, three main classes of sensing architectures may be described: sensing based on light intensity changes, sensing based on interferometry (interference of light waves), and sensing based on fiber Bragg gratings. In all three modalities, the direct physical measurand to which the fiber is sensitive is axial displacement. In other words, the fiber is stretched, compressed, or rigidly moved in a longitudinal direction such that some observable property of the light propagating in the fiber is changed. Then, if the fiber is coupled to a structure of interest (e.g., epoxied like a strain gage), the fiber is detecting the corresponding motion of the structure. Other measurands, such as acceleration or pressure, are also possible through proper coupling of the fiber to an appropriate transducer.

17.3.1 Intensity-Based Sensing

Some of the simplest and lowest-cost sensing solutions rely upon the measurand producing a detectable change in the intensity of light. Measuring optical power is easier than measuring more complicated optical properties like phase interference, wavelength shift, or polarization state. Several mechanisms can produce intensity changes, so a wide variety of architectures is possible. Typically, no specialty components or fibers are required beyond a stable optical source, a reasonable photodetector, and straightforward signal processing. Potential drawbacks for some applications include lower sensitivity than interferometric or Bragg grating-based sensing and over-sensitivity to extraneous measurands, since multiple simultaneous influences can often affect the intensity.

Three main general intensity-based sensing mechanisms will be presented: microbending in the fiber, evanescent field interaction, and fiber-to-measurand direct coupling. With microbending, utilization is made of directly coupling the measurand to bending the fiber, which induces net transmission loss over the active region. As the displacement of the microbending apparatus of Figure 17.2 (left) changes, the transmission through the fiber is affected. For multimode fibers, the transmission losses are related to a redistribution of the light resulting in an increased coupling between modes and to the cladding modes. For single mode fibers, the losses are related both to increased coupling to the highly-attenuated cladding modes and to the mismatch in the modal profile when the light encounters a bent segment of fiber [5, 12].

The most employed specific configuration is a microbender with multimode fiber, as shown in the figure. In this configuration, the modal coupling occurs when their

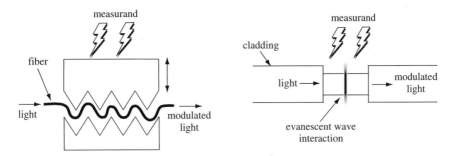

Figure 17.2 Intensity-based sensing based on fiber microbending (left) and intensity-based sensing based on interaction between the environment and the evanescent field (right)

propagation constants differ inversely proportional to the microbend period (distance between bends). For various graded-index fibers, the difference in propagation constants between adjacent modes has a general power-law relationship, and for the specific case of a parabolic graded-index fiber, an exact optimum for the microbend period may be found. For step-index fibers, sensitivity has been shown to increase for decreasing period below a threshold [12]. For single mode fibers, mode-coupling theory may also be used to relate the single propagating mode to the supported cladding modes [13].

Using the optimized microbend configuration, displacement noise floors of $0.1 \, \text{nm}/\sqrt{\text{Hz}}$, 40 dB dynamic ranges, and less than 1 % nonlinearity have been reported. The main limits tend to be thermal fluctuations, and accuracy can be strongly compromised by source intensity fluctuations or influences from extraneous measurands. Attempts to improve sensitivity have focused upon varying the spacing periodicity, using elastomeric coatings, or targeted modal propagation selection, such as a hybrid multimode/single-mode fiber structure [14].

Another intensity-based architecture involves the use of the evanescent field, which extends beyond the fiber core and was discussed in the role of optical waveguides previously. In a sensing application, if the cladding is removed or its properties directly affected by the measurand, the evanescent wave (and thus the guided mode(s)) will be also affected. A simple schematic is shown in Figure 17.2 (right): in the cladding-free sensor region, the measurand may directly affect the evanescent field. Often, biological or chemical species may directly absorb the evanescent wave, depending upon wavelength(s) chosen. More typically, the cladding is substituted by a material whose properties are directly affected by the measurand. As the interaction between the highly attenuated evanescent field and environment is weak, tapered fiber segments, made by tensioning single mode fiber at the melting point, have been proposed [15]. Up to 90 % improvement in the evanescent field strength have been reported. Finally, the evanescent field may be exploited in an architecture by which a moving medium (measurand-induced) close to the core controls reflections at the boundary, called frustrated total internal reflection. One μm changes in the medium have been shown in hydrophones and displacement sensors to produce up to 100 % change in in the reflected power [16, 17].

The final intensity-based architecture, and probably the most common for mechanical-based measurements, is based on direct coupling of fibers or fibers to physical devices. These are extrinsic devices, in which the light exits the fiber for interaction with the measurand. Two dominant configurations, transmissive and reflective modes, are shown in Figure 17.3. The dominant issue with this architecture, particularly single mode fibers, is the coupling efficiency between fibers or between fiber and transducer (in reflective mode). Single-mode fiber models successfully have employed the Gaussian approximation to the propagating mode [18], while more elaborate models such as uniform illumination [19], uniangular illumination [20], or ray optics [21] are used in multi mode fibers.

The reflective configuration executes some relationship between the position of the reflecting surface and the power of the reflected light. Once calibrated, then the distance from the fiber end to the surface is known. This configuration may involve a single fiber acting as both emitter and receiver [22], multiple receivers, or fiber bundles [20]. The reflective surface may be coupled to an inertial element to make an accelerometer or an elastic membrane to make a pressure sensor. In the transmissive mode, multiple realizations are possible. Two fibers may be fixed and aligned with a gap between, and the measurand directly affects some kind of 'valve' mechanism that impedes the transmission. If the gap is filled with absorptive material, chemical, pH, biological, and humidity sensors are realized. To improve fiber-to-fiber coupling, techniques based on lenses are proposed [23].

As mentioned, sources of error in intensity-based sensing are usually due to thermal fluctuations (affecting all components behavior), losses within connections, undesirable bending somewhere else in the fiber, or other time-dependent component drifts (either optical or electronic). For long-term stability and accuracy, compensation mechanisms have been introduced. One technique, known as optical path diversity, involves splitting the measured intensity change into two fibers. The power in each of the two paths may change in a different way with the measurand, but since both fibers follow the same path, the signal may be extracted by some judicious combination of the separate outputs, usually a ratio or sum/difference combination. A variation of this scheme employs a reference fiber that transports light from the source and follows the same path as the sensor. If it is subjected to the same environmentally induced variations, it may be compared to the sensing signal and used as a reference in a differential mode [24]; if source intensity fluctuations are the largest concern, direct measurement of its power maybe also be used comparatively [25]. Other techniques have included chromatic and wavelength referencing [26, 27], electrical AC/DC referencing [28], and Q-factor referencing [29].

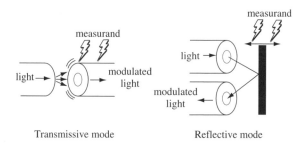

Transmissive mode   Reflective mode

Figure 17.3 Intensity-based sensing architectures based on direct coupling in the transmissive mode (left) and reflective mode (right)

17.3.2 Interferometry

The intensity or power of propagating light is not the only property that can be affected by the environment. Within the context of the full wave theory of light, one may express the propagation of the electric field from one point to a receiving point (as if passing through some generalized 'sensing' element) via a transfer matrix **T** which depends upon the properties of the sensing element, the wavelength of the light, and all the environmental variables that could affect it (stress, temperature, etc.). This transfer matrix is expressible as

$$\mathbf{T} = A\,\exp^{i\phi}\,\mathbf{C}, \tag{17.3}$$

where A is a measure of the transmission, ϕ measures the phase shift in the wave that may occur, and **C** is a matrix that accounts for birefringence, or nonisotropy that results in directional-dependent refractive properties. In 1941, R.C. Jones [30] invented a matrix algebra formalism to describe the propagation of polarized light waves through polarizing media (one of the effects of birefringence), and **C** is called the *Jones matrix*. Intensity-based sensors rely on changes to A, but *interferometric* sensors rely on changes to the phase ϕ. An interferometer is an instrument constructed by splitting an optical wave and then recombining the two split waves together coherently. Environmental affects impinged on either of the two optical paths cause the phase ϕ to be affected, and the effect is observed by measuring the intensity fluctuations at the output of the interferometer. Interferometers thus act as coordinate transformers from the phase domain to the intensity domain. The simple act of axially loading the fiber has three main influences on it: it lengthens the fiber, the refractive index of the core is changed, and the dimensions of the core are changed via a Poisson process. The first effect is the dominant one, with the second about one-fifth as large but of opposite sign, so it reduces sensitivity slightly. For the third effect, a reduction in core diameter pushes the effective core refractive index closer to that of the cladding (meaning it is being reduced), but it is a very negligible effect for most fibers.

The simplest realization of an interferometer is the splitting of an optical path into two, isolating one path, an exposing the other to the environment to be measured. The two most common versions of this are the Mach–Zehnder and Michelson configurations shown in Figure 17.4. In a Mach–Zehnder interferometer, the incoming light is split with a two-by-two coupler into two separate fibers, one the sensing arm and the other the reference arm. The reference is packaged to keep it in a relatively isolated environment, while the sensing arm is coupled to the measurand in some way. The two arms are then recombined with another two-by-two coupler, and the light is photodetected.

When the two paths are recombined at the detector, they produce an intensity output that varies periodically with the relative path difference and a period equal to the optical wavelength λ. The induced phase ϕ by the presence of an optical path difference nL is given by

$$\phi = \frac{2\pi n L}{\lambda}, \tag{17.4}$$

where n is the effective refractive index and L is the physical path mismatch. The interferometer then converts this phase shift to the intensity I at the detector (assumed a square-law processor), generally given by

$$I = I_0(1 + V\cos\phi), \tag{17.5}$$

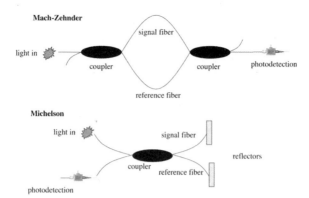

Figure 17.4 Mach–Zehnder (top) and Michelson (bottom) configurations for two-path inter-ferometers

where I_0 is the average signal level, and V is the interferometer *visibility*. Visibility, or modulation depth, depends on the relative power levels between the two fiber paths, their relative states of polarization, and their cross-coherence. Visibility may be con-sidered optimized in an interferometer if the relative power levels in the two arms are equal, the relative states of polarization are the same, and the optical path difference nL is negligible compared with the coherence length of the light. Coherence length may be generally thought of as the degree to which the light wave may continue to be described by a sine wave in space. This is usually determined by the properties of the optical source used; narrow-line lasers have very long coherence lengths, while broadband LED sources have relatively shorter coherence lengths. The relative power levels in each arm depend most upon the properties of the couplers used in the construction; coupler performance is known to vary with polarization, wavelength, and temperature effects [31]. Polarization state differences also depend on the birefringence of the fiber. Most interferometers are constructed from circular symmetric fibers, and thus some sort of birefringence control is usually needed; the simplest way is to exert controlled fiber bending [32].

The second configuration is the Michelson interferometer. The main difference between this and the Mach–Zehnder is that both the signal and reference arms termin-ate at reflectors so that the light paths are folded back on themselves and recombined in the same coupler that split them. This second pass through the system doubles the sensitivity of the device, but a disadvantage is that one of the paths is reflected into the optical source path, which can lead to instability [33], particularly if diode lasers are used. A Faraday optical isolator may be used to correct this situation.

The recombined beams in a Michelson interferometer are of equal intensity with a highest average intensity at a 50:50 coupling ratio. The influences of polarization are the same as in a Mach–Zehnder, but equality of polarization states may be obtained by using special Faraday reflectors, which are similar to the optical isolators just described but without the polarizing element. Thus, the reflected signal and reference beams are orthogonal to the forward-propagating beams, so they enter the coupler in the same state in which they left it. Birefringence in the fiber is effectively removed, and visibility is optimized.

Using an interferometer, the sensititves of fiber to strain loading, mechanical or thermal, may be presented. For example, the inherent overall strain sensitivity of a fiber at 633 nm is about 6.5 Mrad/m [34]. If one considers a thermally loaded fiber, the second effect (the refractive index change) dominates, and at 633 nm, a sensitivity of 100 rad/C has been reported for a 1 m length of fiber [35].

17.3.3   Bragg Grating-Based Sensing

So far, measurement techniques associated with direct intensity changes and direct phase changes have been presented. The final primary sensing technique involves wavelength modulation, which is achieved by using specialized in-fiber structures known as fiber Bragg gratings (FBGs). In many structural monitoring applications, more localized or shorter-gauge measurements are required, and FBGs may be thought of as the fiber-optic equivalent to conventional resistive strain gages. The telecommunications boom of the late 1990s propelled FBG technology from the laboratory into a major commercial enterprise, where uses for the technology have included add/drop filters, pulse compression, selective reflectors, and dispersion compensation.

17.3.3.1   Bragg Grating Principles

FBGs are simple intrinsic (in-fiber) periodic structures that are directly photowritten into the fiber core by ultraviolet radiation [36, 37]. Silica-based optical fibers have electronic absorption regimes in the ultraviolet range, and changes in the absorption may directly lead to changes in the local refractive index via the Kramers–Kronig effect in an essentially permanent way. Since the fiber symmetry properties inhibit nonlinear effects, defect chemistry is the prime method for inducing absorption changes. The inherent fabrication processes used to make optical fibers inevitably involve imperfect chemical reactions during the chemical vapor deposition process. The imperfect mixing ratios lead to deposits of germanium and silicon suboxides, creating a range of defects [38]. This range of Ge suboxide defects has been well correlated with photosensitivity in the ultraviolet range [39]. Such Ge-doping of fiber has led to refractive index increases by an order of magnitude after ultraviolet exposure [40].

If the ultraviolet photowriting process illuminates the fiber core in a periodic fashion, then a corresponding periodic modulation of the refractive index is achieved. The two most common photowriting processes are interferometric holography inscription [41] or, now more commonly in the commercial sector, phase mask inscription [42]. The former technique involves focusing two beams of coherent ultraviolet light transversely upon a small section of fiber (1 cm), where the two beams are interfered at some prescribed slant angle. The interference pattern established on the core provides the periodicity that results in the grating's structure. The phase mask inscription technique consists of using a one-dimensional periodic (square-wave) surface relief structure etched using photolithography. The fiber is placed in very near contact with the structure, and incident ultraviolet light normal to the structure passes through and diffracts because of the square-wave corrugations. The two first-order diffracted beams then interfere to provide the periodicity, similar to interferometric holography. The zero-order diffraction beam is kept suppressed by proper choice of the corrugation depth [43]. A more thorough discussion of these and other fabrication techniques and their associated issues may found in Reference [44].

The subsequent periodic modulation of the core refractive index acts like a narrow-band notch filter of the light propagating down the core. If light is propagated down the fiber as a guided mode, as the light interacts with each grating structure, some of the light is scattered due to the refractive index change. Figure 17.5 shows forward-propagating light interacting with each grating structure written into a fiber; the light will be diffracted by this perturbation, and the figure shows the reflected portion due to this diffraction. The various light waves reflecting from each structure will accrue constructively with the forward mode (maximal coupling), provided that an appropriate phase-matching condition is met:

$$\beta_{forward} + \beta_{backward} = \frac{2m\pi}{\Lambda},\tag{17.6}$$

where the β's reference the modal constants, m is an integer, and Λ is the physical spacing (period) between grating structures. The propagation constants are given by $\beta = 2\pi n/\lambda$, where n is the effective core index and λ is the wavelength. Substitution of these expressions into equation (17.6) yields the *Bragg wavelength* λ_r at which maximal coupling in reflection occurs

$$\lambda_r = 2n\Lambda,\tag{17.7}$$

where first-order ($m=1$) diffraction is assumed dominant [45]. The practical width of this reflected wave-length, dependent upon how the grating was manufactured and the coupling parameters, is typically $0.1-0.3$ nm. A good summary of the use of coupled mode theory to derive expressions for the reflectivity and bandwidth of FBGs is given in Reference [46].

Equation (17.7) establishes that a measurement of the reflected wavelength equates with a measurement of the grating period or the effective refractive index. Thus, any perturbation to the fiber that changes either of these two properties directly changes the reflected wavelength of that grating; for a generic physical affect χ, the change in λ_r with χ may be expressed by differentiating Equation (17.7) and dividing through by λ_r as

$$\frac{1}{\lambda_r}\frac{\partial\lambda_r}{\partial\chi} = \frac{1}{n}\frac{\partial n}{\partial\chi} + \frac{1}{\Lambda}\frac{\partial\Lambda}{\partial\chi}.\tag{17.8}$$

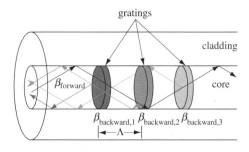

Figure 17.5   Contradirectional mode coupling in FBG-encoded fiber

The three physical parameters (χ) that affect the Bragg wavelength are strain, temperature, and hydrostatic pressure, since these directly affect either the physical dimensions of the fiber (changing Λ) or the refractive index n. The sensitivities to these parameters, $\partial n/\partial\chi$ and $\partial\Lambda/\partial\chi$, are wavelength dependent, with the first term dominating. For light at 1550 nm, typical values for these coefficients are 1.15 pm/µε, 11 pm/°C, and −3.5 pm/ MPa [43]. Fiber extension or compression at the grating location results in strain, and the relationship between applied displacement (stress) and the resulting wavelength shift (strain) may be given by

$$\Delta\lambda_r = \lambda_r\zeta\Delta\epsilon, \tag{17.9}$$

where $\zeta = 1.15$ pm/µε is the overall strain sensitivity, and $\Delta\epsilon$ is the change in average applied strain over the grating's gage length. Since the initial grating period Λ may be specified during manufacturing, a common wavelength-based multiplexing technique is to inscribe an array of gratings at unique reflection wavelengths in to the same fiber. Broadband light is launched into the fiber, and each grating will 'slice' out a 0.1–0.3 nm portion of the launched wavelength spectrum (Figure 17.6). As any of the individual fiber locations where the gratings are written are strained, a corresponding proportional shift in λ_r is observed according to equation (17.9).

Therefore, the principle requirement for developing a grating-based measurement system, regardless of what specific fields the gratings are designed to measure, is to track the various grating wavelength reflection shifts. The earliest approaches utilized some sort of wavelength-dependent optical filter (edge or bandpass) that simply had a wavelength-dependent transmittance function [47]. Since wavelength is converted to intensity, such techniques are subject to many of the influences discussed previously in

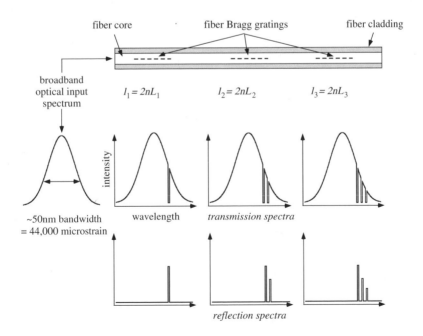

Figure 17.6 Wavelength-encoded multiplexing of Bragg gratings

intensity-based measurements. A number of approaches that directly take advantage of the wavelength encoding of the FBGs have been proposed, but most of them may be broadly classified into four groups: interferometric techniques [48], scanning Fabry–Perot filter interrogation [49], tunable acousto-optic filter interrogation [50], and prism/CCD-array techniques [51]. More recently, a new hybrid technique has been demonstrated that retains certain advantages from some of the earlier methods while improves overall performance [52, 53]. Sections 17.3.3.2–17.3.3.4 will briefly review the four primary grating interrogation methods; a good discussion of variations of some of these themes may be found in [54]. Then, in Section 17.3.3.5, the new hybrid method will be discussed, and both performance and implementation comparisons will be drawn among all the methods.

17.3.3.2   Interferometry

An unbalanced interferometer, as described previously, may be used to interrogate FBGs. Light from a broadband source is used to illuminate an FBG sensing array, and the reflections are coupled back directly into the interferometer. The interferometer converts the reflected wavelength to a phase difference according to the general relationship given by equation (17.4), where ϕ is the phase difference, λ is the peak reflection wavelength of a particular grating, n is the effective refractive index of the fiber core, and L is the path imbalance. The advantage of an interferometer over the older (but simpler) wavelength-dependent filters is the sensitivity tunability gained by modifying the path imbalance. Although larger path imbalances lead to higher sensitivity, the coherence of the FBG-reflected light drops, leading to a degraded interference pattern. An optimum path imbalance for use with a broadband source was derived by Weis *et al.* [55] to be

$$nL_{\text{opt}} = \frac{2.355\lambda^2}{2\pi\Delta\lambda_{\text{FBGband}}}, \tag{17.10}$$

where $\Delta\lambda_{\text{FBGband}} \approx 0.2$ nm is the FBG reflection bandwidth. For 1550 nm light, this optimum path imbalance is about 4.5 nm. If a high-coherence source is used, such as a fiber laser, high sensitivity may be achieved with very large path imbalances [56].

Demultiplexing may be achieved by a variety of methods, but the most common and well developed is based on a bandpass wavelength-division demultiplexer, which guides the light at each grating wavelength on to a separate detector. As with any direct interferometric signal, demodulation must be performed by one of the techniques described previously. Once the phase shift ϕ is recovered, the resulting strain ϵ on an individual grating may be calculated by

$$\Delta\epsilon = \frac{\lambda}{2\pi nL\zeta}\Delta\phi, \tag{17.11}$$

which is obtained by differentiating equation (17.4) with respect to λ and utilizing the photoelastic equation, equation (17.9). By choosing an appropriate interferometer path difference, strain resolutions as low as $1–10\ n\epsilon/\sqrt{Hz}$ with 5000 $\mu\epsilon$ dynamic ranges have been reported for frequencies above about 100 Hz [57]. However, multiplexing is burdensome due to the WDM filters. Some hybrid versions of this method incorporating time-division multiplexing (TDM), where sub-arrays of gratings may reflect at common wavelengths but are discriminated with appropriate time gating [55]. Another drawback for this method is that it requires some sort of interferometer drift compensation device for retaining accuracy at low frequencies.

17.3.3.3 Tunable Filter Methods

In the noninterferometric arena reside both the scanning Fabry–Perot and tunable acousto-optic filter methods. Both methods interrogate grating arrays on essentially the same principle but with different components for doing the individual wavelength discrimination (Figure 17.7 (left)). In the former method, light reflected from the grating array is coupled into a scanning Fabry–Perot (SFP) filter, which was described previously. A typical SFP filter has a passband of about 0.2–0.5 nm, and since FBGs have a reflection bandwidth of about 0.1–0.3 nm, only one grating reflection will pass through the SFP filter at a given time. The center wavelength of the passband may be controlled by applying a stepped voltage to a piezoelectric element controlling the spacing between the mirrors, so that all reflection wavelengths from an FBG sensor array may be passed. After passage through the SFP filter, the reflected light is sent to a single photodetector. The detector voltage trace is differentiated, and the times of the zero-crossings of the derivative (corresponding to the grating peak locations) are noted, as shown on the top part of the right side of Figure 17.7. The zero-crossing times are compared to the corresponding SFP filter drive voltage, which has been calibrated to wavelength, so that the zero-crossing times are essentially a measure of the wavelength during a given scan cycle. Strain resolution near 1 microstrain has been achieved, and the frequency response function is basically flat down to DC. However, in order to maintain accuracy and resolution, the SFP filter, since it is a mechanical device, cannot be scanned too quickly, or resonance phenomena may corrupt the measurement, as calibration between

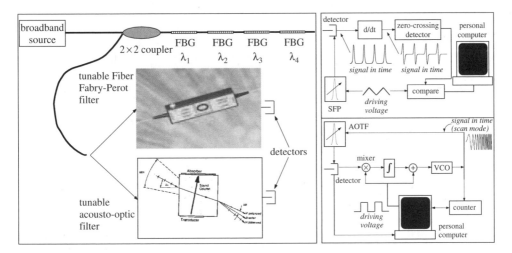

Figure 17.7 Grating interrogation architecture with tunable filters (left) such as a scanning Fabry–Perot (top, right) or a acousto-optic (bottom, right)

wavelength and SFP drive voltage is made under static conditions. SFP filters usually can not be driven past about 1 kHz when used in this capacity, and many applications of this technology have reported lower acceptable rates [58, 59]. Nevertheless, this method has become popular due to its relative simplicity and its ability to resolve static to quasistatic strains without complex componentry.

In the tunable acousto-optic method, the SFP filter is replaced with an acousto-optic filter (AOTF). An AOTF is a solid-state, electronically tunable narrow-band optical filter. In an AOTF, the broadband light interacts with a high-frequency ultrasonic sound wave inside an optically polished crystal block. At each acoustic frequency, a relatively narrow band of optical wavelengths satisfies the anisotropic Bragg diffraction condition. By sweeping the acoustic frequency, the selected wavelength band can be varied, similarly to how a stepped voltage wave was applied to the SFP filter to cover its spectral range. In other words, the wavelength tuning of the AOTF is achieved by varying the frequency of an RF signal that creates the acoustic wave via a transducer. This is often achieved with a voltage-controlled oscillator (VCO), as shown in the bottom part of the right side of Figure 17.7. AOTFs may be driven much more rapidly than SFP filters, resulting in wider bandwidth capability. However, the spectral bandwidth of typical AOTFs is in the 2–30 nm range, although a few are specified to as low as 0.4 nm. These higher bandwidths do not match grating bandwidths as well as SFP filters, and grating spacing becomes an important issue; in other words, it is not always possible to multiplex as many gratings with a single AOTF filter as it may be with an SFP filter. However, AOTF free spectral ranges cover hundreds of nanometers in most cases, so the tradeoff may be alleviated, assuming source bandwidth is available [60, 61].

17.3.3.4   CCD Linear Array Method

One potential drawback to the tunable filter methods is that they only utilize a small portion of the optical spectrum at any given time. When a wide spectral range of FBGs is interrogated at a given scanning frequency, the amount of reflected light energy from each FBG per scan is equal to the product of the FBG's reflectivity, the source spectral intensity, the FBG's spectral width, and the scan period. Using typical FBG, source, and tunable filter characteristics, the detectable light energy is of the order of 1 % of the reflected light energy. This means that very high grating reflectivities or very bright sources are needed for acceptable wavelength detection. This penalty is completely voided by the parallel whole-spectrum interrogation possible with a charge-coupled device (CCD) spectrometer. Wavelength discrimination is achieved with a fixed dispersive element such as a prism or a plane grating which converts an incipient wavelength to an image on an array of detector elements by means of a collimator, as in Figure 17.8. With a typical spectrometer and plane grating, an FBG reflection will be dispersed across more than one pixel (line) on the detector array, and a centroid calculation is performed in order to find the maximum (center) wavelength. Resolution has been demonstrated with a typical CCD spectrometer in the nanostrain range with sampling capability into the kiloHertz regime [51]. Chen *et al.* extended this technique to a planar CCD array [62]. Possible drawbacks with this method include reliance upon optical wavelengths below about 900 nm (fewer and more expensive components, including sources) and the need for bulk optics components (collimating lens).

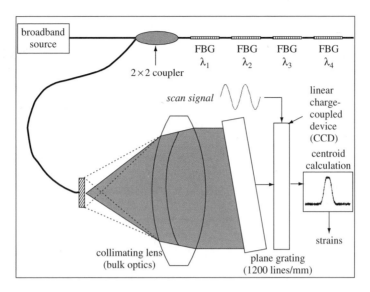

Figure 17.8   Grating interrogation architecture using a CCD spectrometer and a collimating lens

17.3.3.5   Hybrid Method Using a Tunable Filter, a Mach-Zehnder Interferometer, and a 3 × 3 Coupler

Recently, Todd, Johnson and coworkers have introduced a hybrid concept in Bragg grating interrogation that takes advantage of the high resolution of interferometry as well as the easy multiplexing capability of a tunable filter [52, 53]. The overall approach is shown in Figure 17.9. Light reflected from the FBG sensor array is coupled back through an SFP filter, which serves to pass individual grating reflections. However, in this capacity, the SFP filter is only used as an FBG 'gate' and not as a specific wavelength discriminator. In other words, in the SFP method described above, the SFP drive voltage was used via a calibration to determine specific wavelength, but in the present implementation, the filter is only used to select one grating at a time without actually determining its wavelength.

After the SFP filter passes any given FBG reflection peak, the light is transmitted to an unbalanced Mach–Zehnder interferometer (2.75 mm imbalance). The interferometer converts the reflected wavelength to a phase difference across it according to equation (17.4). Light at the interferometer output is transmitted through a 3 × 3 coupler, where an ideal phase offset of 120 ° ($2\pi/3$ rad) is induced between the three outputs, and finally the the light from the three coupler outputs is sent to photodetectors (1–3 in Figure 17.9) for conversion to voltages.

As mentioned previously, one potentially corrupting feature of interferometric measurements is the drift of the interferometer. Tracking of this drift is crucial for allowance of static to quasistatic measurements of FBG phase shift. As shown in Figure 17.9, source light is also used to illuminate an FBG reference array. The reference array consists of two gratings in a sealed package, one bonded to a glass strip (λ_G) and the other to an aluminum strip (λ_{Al}). These gratings are sensitive only to thermally induced

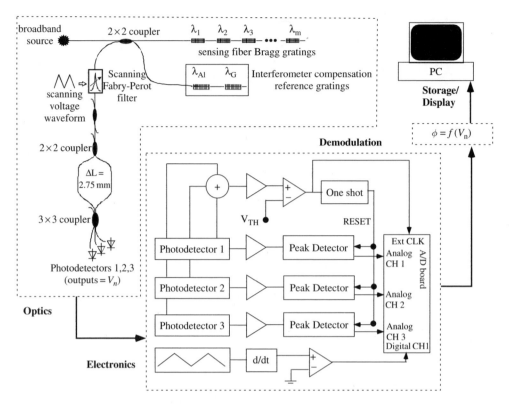

Figure 17.9 Grating interrogation architecture using a Mach–Zehnder interferometer, an SFP filter, and a 3 × 3 coupler

wavelength shifts, so the changes in phase of the FBGs in this reference device may be directly calculated:

$$
\begin{aligned}
\Delta\phi_G &= \Delta\phi_{\mathrm{env}} + \gamma_G \Delta T \\
\Delta\phi_{Al} &= \Delta\phi_{\mathrm{env}} + \gamma_{Al} \Delta T,
\end{aligned}
\tag{17.12}
$$

where γ_G and γ_{Al} are the material thermal expansion coefficients and ΔT is the temperature change. Two equations with two unknowns (ΔT and $\Delta\phi_{\mathrm{env}}$) allow for robust tracking of the interferometer and determination of temperature. Demodulation of the interferometric output is achieved using the 3 × 3 method described previously. Typical performance characteristics reported with drift compensation active are a strain resolution of about $5\,\mathrm{n\epsilon}/\sqrt{\mathrm{Hz}}$ at 0.1 Hz and a frequency bandwidth of about 20 kHz.

Table 17.1 summarizes some key performance metrics for the major Bragg grating interrogation strategies presented in this section. Each technique clearly has its strengths and weaknesses, depending on application-specific judgment of the importance of a given performance metric. For example, lower-frequency applications may benefit from an SFP filter method, but high-resolution applications would benefit from either the AOTF or hybrid method. The CCD method is very conducive to high levels of multiplexing and has the advantage of true simultaneous interrogation of all FBGs in an

Table 17.1 A summary of performance metrics for primary Bragg grating interrogation methods

Metric	System architecture				
	Scanning Fabry–Perot	Tunable acousto-optic	WDM interferometry	Hybrid method	CCD array
Dynamic resolution ($n\epsilon/\sqrt{Hz}$)	100	<200	<5	<1	50
Frequency bandwidth (Hz)	0–360	0–40 000	100–20 000	0–20 000	0–20 000
Limiting component	filter	source	filter	filter	source
Multiplexing capability (dB)	high	high	low	high	very high

array, but it relies on sub-900 nm components and bulk optics, and it hasn't been deployed in a large number of field applications to test robustness and packaging. One new item that has not made it into this table is the recent emergence of 'MEMS' SFP filters, which have large free spectral ranges (up to 100 nm) and fast scanning capability (up to the MHz range). Recent laboratory testing at the US Naval Research Laboratory on one of these components confirmed these performance specifications, but found significant nonlinearity in the voltage/wavelength relationship. Nevertheless, while still immature, such a component could expand the performance capabilities of the SFP method and the hybrid method immensely.

17.4 SUMMARY

This chapter has presented a broad overview of primary fiber optic measurement systems used in structural monitoring. Fiber-optic sensors have been developed and deployed since the late 1970s and very much came into maturity during the telecommunications revolution of the late 1990s, as component costs were driven significantly lower by technology advances and demand. In many structural monitoring applications, fiber sensors offer significant advantages over other sensing strategies, such as immunity to electromagnetic interference, high sensitivity, negligible weight penalty, self-telemetry, and embeddability inside materials such as composites. Significant research and development continues as of this writing in new kinds of grating sensors (e.g., tilted, long period, etc.), fiber laser sensors, novel fiber structures (e.g., D-section, multicore, etc.), grating fabrication during fiber drawing, micro-optical-electromechanical systems (MOEMS), nonlinear optical sensing (e.g., Raman and Brillouin effects), and new sensing strategies offering scalable gage length based on coherent Rayleigh backscatter (see [63]). Current state-of-the-art sensing schemes are already finding their way into many structural markets such as bridges, buildings, marine structures, and aerospace vehicles.

ACKNOWLEDGEMENTS

The author deeply appreciates the collaborations with many colleagues while at NRL, without whom much of the work presented in this chapter would not have occurred. Particular acknowledgement for very stimulating fiber optic sensing collaboration goes to Frank Bucholtz, Mark Seaver, and Steve Trickey (all at NRL), and to Gregg Johnson, Sandeep Vohra, Bryan Althouse, Bruce Danver, and Heather Patrick (all now with Optinel Systems, Inc.). The author also acknowledges important collaborations with the Norwegian Defence Research Establishment, SMARTEC (Switzerland), Physical Sciences Laboratory (Las Cruces, NM), and New Mexico State University. Final grateful acknowledgment is given to the Federal Highways Administration (specifically, Richard Livingston) and the Office of Naval Research for funding much of the sensor development and field deployment activities performed by NRL personnel.

REFERENCES

[1] B.E.A. Saleh and M.C. Teich (Eds) *Fundamentals of Photonics.* Wiley-Interscience, New York, 1991.

[2] M. Born and E. Wolf (1980) *Principles of Optics.* Pergamon Press, New York.

[3] E. Snitzer (1961) Cylindrical dielectric waveguide modes. *Journal of the Optical Society of America,* **51**, 491–498.

[4] D. Gloge (1971) Weakly guiding fibers, *Applied Optics,* **10**, 2252–2258.

[5] D. Marcuse (1974) *Theory of Dielectric Optical Waveguides.* Academic Press, New York.

[6] E.G. Neumann (1988) *Single-Mode Fiber,* Springer-Verlag, New York.

[7] L.B. Jeunhomme (1990) *Singe-Mode Fiber Optics,* Marcel Dekker, New York.

[8] D.A. Pinnow, A.L. Gentile, H.G. Standlee, A.J. Timper and L.M. Hobrock (1978) Polycrystalline fiber optical waveguides for infrared transmission, *Applied Physics Letters,* **33**, 28–29.

[9] T. Kaino (1992) *Polymers for Lightwave and Integrated Optics,* Marcel Dekker, New York.

[10] T.R. Wolinski, T. Nasiowski, A. Szymanska, W. Konopka, M.A. Karpierz and W. Domanski (1997) Liquid crystal core optical fibers for environmental sensing, in *Proceedings 12th International Conference on Optical Fiber Sensors,* Optical Society of America.

[11] J.M. Lopez-Higuera (Ed.) (2002) *Handbook of Optical Fibre Sensing Technology.* John Wiley & Sons, Inc., New York.

[12] J.W. Berthold III (1995) Historical review of microbend fiber-optic sensors, *Journal of Lightwave Technology,* **13**, 1193–1199.

[13] V. Arya, K.A. Murphy, A. Wang and R.O. Claus (1995) Microbend losses in single-mode optical fibers: Theoretical and experimental investigation, *Journal of Lightwave Technology,* **13**, 1998–2002.

[14] D. Donlagic and B. Culshaw (1999) Microbend sensor structure for use in distributed and quasi-distributed sensor systems based on selective launching and filtering of the modes in graded index multimode fiber, *Journal of Lightwave Technology,* **17**, 1856–1868.

[15] W. Henry (1994) Use of tapered optical fibers as evanescent field sensors, in *Chemical, Biochemical and Evironmental Fiber Sensors, SPIE 2293.*

[16] W.B. Spillman and D.H. McMahon (1980) Frustrated total internal reflection multimode fiber optic hydrophone, *Applied Optics,* **19**, 113–116.

[17] G.T. Sincerbox and J.G. Gordon (1981) Modulating light by attenuated total reflection, *Laser Focus,* **11**, 55–58.

[18] R.L. Gallawa, I.C. Goyal and A.K. Ghatak (1993) Fiber spot size: A simple method of calculation, *Journal of Lightwave Technology,* **11**, 192–197.

[19] D. Wang, M.A. Karim and Y. Li (1997) Self-referenced fiber optic sensor performance for microdisplacement measurement, *Optical Engineering,* **36**, 838–842.

[20] R.O. Cook and C.W. Hamm (1979) Fiber optic lever displacement transducer, *Applied Optics,* **18**, 3230–3241.

[21] A. Cobo, M.A. Morante, J.L. Arce, C. Jauregui and J.M. Lopez-Higuera (1999) More accurate coupling function approach for optical transducers based on power coupling between multimode fibers, in *Proceedings of the Optical Fiber Sensors Conference*, Optical Society of America.

[22] G. He and F.W. Cuomo (1991) Displacement response, detection limit, and dynamic range of fiber-optic level sensors, *Journal of Lightwave Technology*, **11**, 1618–1625.

[23] W.B. Spillman (1981) Multimode fiber-optic hydrophone based on a schleiren technique, *Applied Optics*, **20**, 465–470.

[24] E. Bois, S.J. Huard and G. Boisde (1989) Loss compensated fiber-optic displacement sensor including a lens, *Applied Optics*, **28**, 419–420.

[25] A. Shimamoto and K. Tanaka (1995) Optical fiber bundle displacement sensor using an ac-modulated light source with subnanometer resolution and low thermal drift, *Applied Optics*, **34**, 5854–5860.

[26] G.R. Jones, S. Kwan, C. Beavan, P. Henderson and E. Lewis (1987) Optical fiber based sensing using chromatic modulation, *Optics and Laser Technology*, **19**, 293–303.

[27] G.Z. Wang, A. Wang, R.G. May, A. Barnes, K.A. Murphy and R.O. Claus (1996) Stabilization for intensity-based sensors using two-wavelengths ratio technique, in *Self-Calibrated Intelligent Optical and Sensors and Systems, Proceedings SPIE 2594*.

[28] X. Fang, A. Wang and R.O. Claus (1996) A modified ac/dc compensation technique for dc measurands, in *Self-Calibrated Intelligent Optical Sensors and Systems, Proceedings SPIE 2594*.

[29] J.M. Baptista, J.L. Santos and A.S. Lage (2000) Mach-zehnder and Michelson topologies for self-referencing fiber optic intensity sensors, *Optical Engineering*, **39**, 1636–1644.

[30] R.C. Jones (1941) A new calculus for the treatment of optical systems, *Journal of the Optical Society of America*, **31**.

[31] T. Wiener and M.D. Todd (2002) The effects of thermal and polarization fluctuations on 3×3 coupler performance. In *Proc. 15th International Conference on Optical Fiber Sensors*, Portland, OR, Optical Society of America.

[32] H.C. Lefevre (1980) Single-mode fibre fractional wave devices and polarization controllers, *Electronics Letters*, **16**, 778–779.

[33] D. Anderson and J.D.C. Jones (1992) Optothermal frequency and power modulation of laser diodes, *Journal of Modern Optics*, **39**, 1837–1847.

[34] P. Akhavan-Leilabady, J.D.C. Jones and D.A. Jackson (1985) Single-mode fibre optic strain gauge with simultaneous phase- and polarization-state detection, *Optics Letters*, **10**, 576–577.

[35] M. Corke, A.D. Kersey, D.A. Jackson and J.D.C. Jones (1983) All fibre Michelson thermometer, *Electronics Letters*, **19**, 471–472.

[36] K.O. Hill, Y. Fujii, D.C. Johnson and B.S. Kawasaki (1978) Photosensitivity in optical fiber waveguides: Application to reflection filter application *Applied Physics Letters*, **32**, 647–649.

[37] J. Stone (1987) Photorefractivity in GeO_2-doped silica fibers, *Journal of Applied Physics*, **62**, 4371–4374.

[38] E.J. Friebele, D.L. Griscom and G.H. Siegel (1974) Defect centers in germanium doped silica core optical fiber, *Journal of Applied Physics*, **45**, 3424–3428.

[39] B. Poumellec, P. Niay, M. Douay and J.F. Bayon (1996). The uv induced refractive index grating in $Ge:SiO_2$ preforms: Additional cw experiments and the macroscopic origin of the index change in index, *Journal of Physics D, Applied Physics*, **29**, 1842–1856.

[40] D.L. Williams, B.J. Ainslie, J.R. Armitage, R. Kashyap and R.J. Campbell (1993) Enhanced uv photo-sensitivity in boron codoped germanosilicate optical fibers, *Electronics Letters*, **29**, 45–47.

[41] G. Meltz, W. Morey and W. Glenn (1989) Formation of Bragg gratings in optical fibers by a transverse holographic method, *Optics Letters*, **14**, 823–825.

[42] K.O. Hill, B. Malo, F. Bilodeau, D.C. Johnson and J. Albert (1993) Bragg gratings fabricated in monomode photosensitive optical fiber by uv exposure through a phase mask, *Applied Physics Letters*, **62**, 1035–1037.

[43] K.O. Hill and G. Meltz (1997) Fiber grating technology fundamentals and overview, *Journal of Lightwave Technology*, **15**, 1263–1276.

[44] A. Othonos and K. Kalli (1999) *Fiber Bragg Gratings – Fundamentals and Applications in Telecommunications and Sensing*, Artech House, Norwood, MA.

[45] A. Yariv and P. Yeh (1984) *Optical Waves in Crystals*. John Wiley & Sons, Inc., New York.

[46] T.S. Yu and S. Yin (Eds) (2002) *Fiber Optic Sensors*, Marcel Dekker, New York.

[47] S.M. Melle, K. Liu and R.M. Measures (1992) A passive wavelength demodulation system for guided-wave Bragg grating sensors, *Photonics Technology Letters*, **4**, 516–518.

[48] A.D. Kersey, T.A. Berkoff and W.W. Morey (1992) High resolution fiber Bragg grating based strain sensor with interferometric wavelength detection, *Electronics Letters*, **28**, 236–237.

[49] A.D. Kersey, T.A. Berkoff and W.W. Morey (1993) Multiplexed fiber Bragg grating strain-sensor system with a fiber Fabry-Perot wavelength filter, *Optics Letters*, **18**, 1370–1372.

[50] M.G. Xu, H. Geiger, J. Archambault, L. Reekie and J. Dakin (1993) Novel interrogation system for fibre Bragg grating sensors using an acousto-optic tunable filter, *Electronics Letters*, **29**, 1510–1512.

[51] C.G. Askins, M.A. Putnam and E.J. Friebele (1995) Instrumentation for interrogating many-element fiber Bragg grating arrays embedded in fiber/resin composites, in *Smart Structures and Materials: Smart Sensing, Processing and Instrumentation, SPIE 2444*, pp. 257–266.

[52] G.A. Johnson, M.D. Todd, B.L. Althouse and C.C. Chang (2000) Fiber Bragg grating interrogation and multiplexing with a 3 × 3 coupler and a scanning filter, *Journal of Lightwave Technology*, **18**, 1105–1105.

[53] M.D. Todd, G.A. Johnson and B.L. Althouse (2001) A novel Bragg grating sensor interrogation system utilizing a scanning filter, a Mach–Zehnder interferometer, and a 3 × 3 coupler, *Measurement Science and Technology*, **12**, 771–777.

[54] A.D. Kersey, M.A. Davis, H.J. Patrick, M. LeBlanc, K.P. Koo, C.G. Askins, M.A. Putnam and E.J. Friebele (1997) Fiber grating sensors, *Journal of Lightwave Technology*, **15**, 1442–1463.

[55] R.S. Weis, A.D. Kersey and T.A. Berkoff (1994) A four-element fiber grating sensor array with phase-sensitive detection, *Photonics Technology Letters*, **6**, 1469–1472.

[56] K.P. Koo and A.D. Kersey (1995) Bragg grating based laser sensor systems with interferometric interrogation and wavelength division demultiplexing, *Journal of Lightwave Technology*, **13**, 1243–1249.

[57] T.A Berkoff and A.D. Kersey (1996) Fiber Bragg grating array sensor system using a bandpass wavelength division multiplexer and interferometric detection, *Photonics Technology Letters*, **8**, 1522–1524.

[58] M.D. Todd, G.A. Johnson, S.T. Vohra, C.C. Chang, B. Danver and L. Malsawma (1999) Civil infrastructure monitoring with fiber Bragg grating sensor arrays, in *2nd International Workshop on Structural Health Monitoring*, Palo Alto, California, CRC Press, Lancaster, PA.

[59] S.T. Vohra, G.A. Johnson, M.D. Todd, B.A. Danver and B.L. Althouse (2000) Distributed strain monitoring with arrays of fiber Bragg grating sensors on an in-construction steel box-girder bridge, *IEICE Transactions on Electronics*, **E83-C**: 454–461.

[60] H. Geiger, M.G. Xu, N.C. Eaton and J.P. Dakin (1995) Electronic tracking system for multiplexed fiber grating sensors, *Electronics Letters*, **31**, 1006–1007.

[61] M.G. Xu, H. Geiger and J.P. Dakin (1996) Modeling and performance analysis of a fiber Bragg grating interrogation system using an acousto-optic tunable filter, *Journal of Lightwave Technology*, **14**, 391–396.

[62] Y. Hu, S. Chen, L. Zhang and I. Bennion (1997) Multiplexing Bragg gratings using combined wavelength and spatial division techniques with digital resolution enhancement, *Electronics Letters*, **33**, 1973–1975.

[63] R. Posey, G.A. Johnson and S.T. Vohra (2000) Strain sensing based on coherent Rayleigh backscattering. *Electronics Letters*, **36**, 1688–1689.

Part IV
Applications

18

Prognosis Applications and Examples

Douglas E. Adams

Purdue University, West Lafayette, Indiana, USA

18.1 INTRODUCTION

18.1.1 Background

The number and variety of engineering applications of diagnostic and prognostic (D&P) methods in structural health monitoring (SHM) for defense, commercial and consumer systems are rapidly expanding. The keys to effective D&P for structural systems are *vigilance* through near real-time nondestructive evaluation (NDE) and *foresight* through loads identification to identify and forecast damage as it accumulates. Recall that diagnostic methods aim to *assess* degradation by detecting, locating and quantifying material or structural damage *in situ* using measured data from the past and present to establish state awareness. In contrast, prognostic methods aim to *predict* the future capability of structural systems in real time with some level of statistical confidence using diagnostic information and models in addition to estimates of anticipated future loading. In other words, diagnostics involve the so-called 'inverse problem' of damage identification whereas prognostics involve the 'forward problem' of damage forecasting.

The benefits of SHM for realizing condition-based maintenance (CBM) programs to enhance safety and reduce operation and support costs have long been recognized (Figure 18.1). For example, manufacturing machine tools are now often fitted with SHM systems using D&P technologies to assess tool wear, tool-part misalignment or other sources of errors in the machining process [3]. In addition to CBM, however,

Damage Prognosis – For Aerospace, Civil and Mechanical Systems   Edited by D.J. Inman, C.R. Farrar, V. Lopes Junior and V. Steffen Junior   © 2005 John Wiley & Sons, Ltd

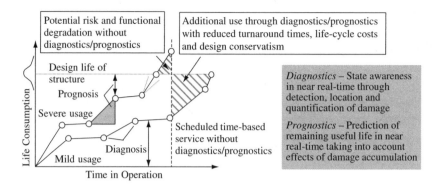

Figure 18.1 Benefits in terms of cost and safety of utilizing diagnostics and prognostics over the life-cycles of structural systems (adapted from Cronkhite and Gill, 1998 [1] and reproduced by permission of Academic Press [2]) (Reprinted from the Journal of Sound and Vibration, Vol. 262, R.L. Brown and D.E. Adams, Equilibrium point damage prognosis models for structural health monitoring, 591–611, © 2003, with permission from Elsevier.)

D&P technologies are beginning to provide commercial industries in manufacturing, transportation and other sectors with additional benefits in development and operation. For example, automotive suppliers are developing life-cycle models for their products using D&P methods, which serve to minimize the detrimental effects of manufacturing and operational variability (e.g., squeaks, buzzes and rattles). In addition, aircraft manufacturers project that the greatest benefits of SHM will be less design conservatism and lower weight/cost in new aircraft [4, 5].

D&P are also playing a major role in defense programs. For example, consider the US Air Force programmatic efforts, which are focused on developing future high-performance aircraft and spacecraft [6, 7]. In addition, the US Army is developing future combat systems (FCS) where each platform (land, air, sea) within the military system-of-systems is linked wirelessly to all other assets to determine at each instant in a military campaign the capability of the FCS as a whole. Furthermore, Army weapon systems and vehicles of the future will be manufactured using revolutionary high-strength, lightweight heterogeneous structures (e.g., laminated composites, ceramics, plastics, metal alloys and fabrics) to produce more maneuverable and lethal defense systems [8]. Multilayer armor and filament-wound, composite tube missile launchers are examples of assets for which SHM technology could serve to assess structural fitness on the battlefield and prior to deployment. Some defense applications in D&P for maximizing the utility of military assets are illustrated in Figure 18.2.

Commercial industries are also investing in D&P as a means of assessing the loads and performance of mechanical/structural systems (e.g., off-road machinery, heavy-duty trucks) over their lifetimes. For example, manufacturers of large engines are pursuing D&P technologies to develop CBM for reducing inspection times with preemptive service [9]. Even the consumer marketplace is exploring ways in which D&P systems can be implemented. For example, D&P techniques are being considered for use in assessing the effectiveness of nationwide recalls of defective products [10], and to certify the operational performance of rolling tires on consumer automobiles under a new set of stringent tire standards [11].

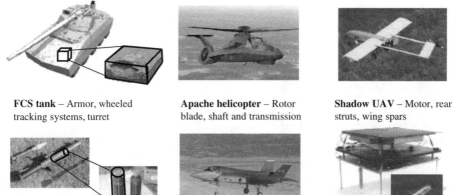

FCS tank – Armor, wheeled tracking systems, turret

Apache helicopter – Rotor blade, shaft and transmission

Shadow UAV – Motor, rear struts, wing spars

Composite weapons – Launch tube, kinetic energy missiles

JSF (F-35) – Airframe, lift fan, powerplant

Spacecraft – TPS panels, air frame, fuel tanks

Figure 18.2 Illustration of defense applications in which diagnostics and prognostics may be used to ensure the structural integrity of vehicles and weapon systems

18.1.2   Contents of Chapter

This chapter reviews various applications in D&P for mechanical and structural systems using SHM techniques to identify damage and predict remaining capability. The chapter is organized according to the three broad technical areas of loads identification, damage identification (diagnostics), and capability prediction (prognostics). Each section describes challenges and solutions in passive/active and local/global data interrogation and damage prediction that are common to all applications. Objectives, technical approaches and key results are highlighted.

18.2   APPLICATIONS

18.2.1   Technical Overview of Diagnostics and Prognostics

Consider the roll-on/roll-off (RO/RO) ramp shown in Figure 18.3 as a representative example of a D&P application in SHM. The elements of a D&P system for such a structure will now be reviewed to establish an overall technical approach for this chapter.

SHM can be viewed as a problem in statistical pattern recognition [12] involving operational evaluation, data cleansing, damage identification and life prediction. In this chapter, operational evaluation is treated in the context of loads identification, which is a precursor to D&P as illustrated in Figure 18.3. Data gathering and cleansing are considered to be part of the data interrogation process. For the RO/RO ramp, loads are first identified using structural response data and models to provide (i) guidance for further diagnostic interrogation, and (ii) future loading estimates for use in prognostic life prediction algorithms. After the structural loads are identified, diagnostic interrogation sensory systems and algorithms are implemented to detect, locate and quantify

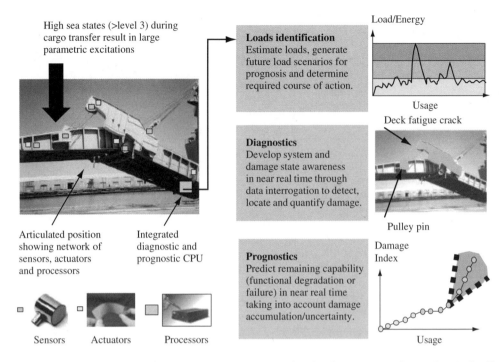

Figure 18.3 Illustration of diagnostics and prognostics in the context of a roll-on/roll-off (RO/RO) ramp that enables joint US Army–Navy expeditionary warfare

the primary damage mechanisms (e.g., matrix cracking, delamination, debonding, corrosion, preload). Some of these forms of damage are localized in nature (e.g., fasteners, bond lines) whereas others are global and can be found throughout the structure (e.g., corrosion). Interrogation methods are often classified as either local or global depending on which type of damage is being identified. These methods can then be further classified as passive or active depending on whether or not an auxiliary excitation source is used.

After structural damage is identified, prognostic algorithms are implemented using structural and damage (i.e., fault) mechanics models along with future loading estimates and current diagnostic information for state awareness to predict remaining useful life or capability. Note that it is necessary to incorporate structural system models, damage mechanics models and performance/capability models in order to render a complete prognosis because the useful life of a structure ends long before the structure catastrophically fails. The results of all three of these D&P tasks must be qualified with statistical measures that quantify user levels of confidence.

18.2.2   Loads Identification Methods

Loads identification methods aim to determine how structures are utilized in operation through sustained (i.e., static, dynamic) and transient (e.g., foreign object impact, shock) loading. As mentioned previously, this data is used to develop targeted D&P

techniques that take into account the specific kinds of loading and failure modes in a given structure. Strain gauges, accelerometers, load cells and other types of force/ motion transducers are routinely used along with data processing algorithms to estimate the temporal/spectral nature of these loads. One application of cyclic loads identification involving an air-valve assembly is described next.

The most common type of load for structural dynamic testing is operating spectral acceleration. For example, continuous cyclic loading in the form of stress reversals in structural components within an engine can cause mechanisms to degrade and eventually to fail. Figure 18.4(a) shows one such component, an air shutoff valve, and Figure 18.4(b) shows the valve instrumented for vertical dynamic shaker testing along the valve shaft in the laboratory. A 1000-lb, 1-in stroke electrodynamic shaker and 5000 Hz bandwidth accelerometers were utilized in these tests. Uniaxial testing was used because vertical motions were observed to be higher than the off-axis motions.

Although single-frequency cyclic loads are often examined in laboratory testing to identify primary failure modes, it will be demonstrated in this section that transient inputs such as impacts and nonlinear phenomena involving frictional and other such effects can conspire with stress reversals in material and mechanical systems at multiple frequencies to produce combined secondary and tertiary failure modes, which tend to disproportionally accelerate times to failure. In effect, individual components in the operating input cause failure sooner than the sum of their parts would otherwise indicate.

This particular mechanical valve was designed to close when the electromechanical solenoid [left of Figure 18.4(a)] actuates. The valve is installed on the engine between the cylinder ports and the turbocharger. The primary damage mechanism of interest in this valve was chipping of the powdered metal bushing [Figure 18.5(a)] due to repeated impacts. The measured acceleration spectrum along the axis of the valve is illustrated in Figure 18.5(b). In addition to a strong harmonic train with a fundamental 30 Hz frequency, other frequencies (166 Hz and 1333 Hz) also contribute to the shaft vibration

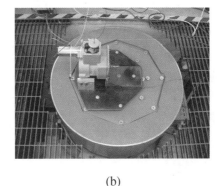

(a)   (b)

Figure 18.4 (a) Air shutoff valve assembly for large engine with solenoid and butterfly valve, and (b) valve shown instrumented for reliability testing on electro-dynamic shaker (Reproduced by permission of Kluwer Academic/Plenum Publishers from: Multiple Equilibria and Their Effects on Impact Damage in an Air-Handling Assembly, by C.G. McGee and D.E. Adams, *Nonlinear Dynamics*, 27, 2002.)

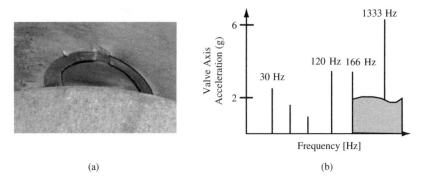

(a)   (b)

Figure 18.5 (a) Air shutoff valve with bushing damage due to impacts (Reproduced by permission of Kluwer Academic Publishers [13]) and (b) measured operational spectrum used to excite valve in the vertical direction along valve axis (Reproduced by permission of Kluwer Academic/Plenum Publishers from: Multiple Equilibria and Their Effects on Impact Damage in an Air-Handling Assembly, by C.G. McGee and D.E. Adams, *Nonlinear Dynamics*, 27, 2002.)

response. These higher frequency components and the broadband portion of the response can be attributed to the turbocharger. In summary, three types of cyclic inputs are experienced by the valve shaft: (i) lower frequency oscillations in the 30-Hz range; (ii) higher frequency cyclic motions in the 166-Hz range; and (iii) higher frequency broadband motions in the range from 166 Hz to 1333 Hz.

In the loads-identification process for this valve, it was important to note that various frequencies must be superimposed to produce the total valve excitation time history [13]. Durability testing was used to determine that the low-frequency cyclic and high-frequency excitations produced distinct failure modes. The 30-Hz component was found to cause the valve to dig into the bushing. The higher frequency motions at 166 Hz were found to accelerate impact damage by increasing the exchange of energy between the valve shaft and the bushing per unit time. The high frequency excitation from 166 Hz to 1333 Hz further accelerated the failure mode of interest due to material loss/exchange between the solenoid metal plunger and the spring-loaded valve hammer [Figure 18.6(a)], a phenomenon known as cold welding. Cold welding caused the static

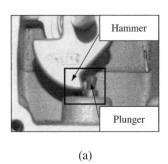

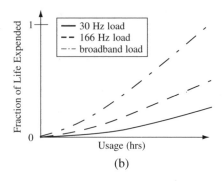

(a)   (b)

Figure 18.6 (a) Photograph of interface between spring-loaded hammer and solenoid plunger, and (b) illustration of failure trends for primary (30 Hz,— and secondary impact (30 Hz and 166 Hz, ---) and tertiary sticking (add broadband input,–·–) modes of failure in valve

frictional force between the plunger and hammer to increase beyond that provided by the solenoid and spring-loaded hammer, contributing to two failure modes. In the first mode, the static equilibrium point at which the valve was designed to operate no longer served as the operating point, resulting in a further acceleration of the impact damage. In the second failure mode, the solenoid failed to overcome the sticking forces within the mechanism due partially to thermal limitations in the solenoid. All of these failure mechanisms are used to develop a prognosis model for this valve in Section 18.2.4. Figure 18.6(b) illustrates the changes in a composite life curve for the valve as each additional damage mechanism is considered.

18.2.3 Diagnostics – Passive Local Interrogation Methods

In many operating environments (e.g., land vehicles, aircraft, spacecraft, infrastructure), it is not possible to excite structures with active interrogation sources due to power, weight or other constraints; consequently, data interrogation must be carried out using only operating responses and loads data. This limitation poses challenges in the damage identification process that can be overcome using response transmissibility. A novel nonlinear system identification technique based on transmissibility analysis is described below in the context of a three-story frame.

In some applications, it is desirable to select an *absolute* damage feature that is locally sensitive but does not require large amounts of transmissibility or other types of data to quantify damage. For example, it might be desirable to determine if a bolt is loose and to what degree using only one piece of data. Nonlinear features can be used as absolute indicators of structural health as demonstrated in this section. Figure 18.7(a) illustrates how the linear and nonlinear correlation between relative motions across a joint change as a fastener is loosened. As preload decreases, the linear correlation decreases due to torque-angle nonlinear dependencies, until a certain state at which a small gap or clearance stiffness nonlinearity is created causing a sudden (qualitative) change in the

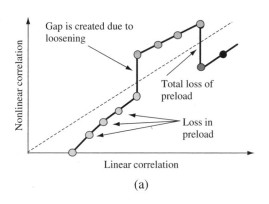

(a) (b)

Figure 18.7 (a) Illustration of variation in linear vs nonlinear correlation coefficient as damage evolves in a joint, and (b) three-story building structure with close-up view of bolted joint (Reproduced by permission of Sage Publications [14])

correlation from linear to nonlinear behavior. As the fastener is loosened further, the degree of nonlinearity changes until it gives way to linear behavior when the preload vanishes resulting in no friction or local impacts across the gap.

Adams and Farrar [14] adapted a frequency domain autoregressive exogenous input (ARX) model, which was originally developed by Adams and Allemang [15] for non-linear system identification, for the purpose of structural diagnostics using passive response data. This model relates two response spectra for DOFs i and j, $X_i(\omega)$ and $X_j(\omega)$, according to the expression

$$X_j(k) = B(k)X_i(k) + \sum_{r,s \in \Re_i} A_{r,s}(k)f_{r,s}\left(X_j\left(\frac{p_r}{q_r}k\right), X_j\left(\frac{p_s}{q_s}k\right)\right), \tag{18.1}$$

where k, r, s, $p_r k/q_r$ and $p_s k/q_s$ are contained in $\Re_1$, k is an integer frequency counter (i.e., $\omega = k\Delta\omega$) and $B(k)$ and $A_{r,s}(k)$ are complex transmissibility coefficients. The first term on the right-hand-side of equation (18.1) accounts for the nominal linear dynamics, $B(k)X_i(k)$, and the autoregressive series accounts for nonlinear dynamics in the response spectrum, $X_j(k\Delta\omega)$. The rational number arguments represent different harmonics and combinations of the excitation frequencies. The functions $f_{r,s}(\cdot)$ are so-called *describing* functions for nonlinearity in the frequency domain. Equation (18.1) is useful for damage interrogation because it relates the jth response of a structural system to the ith response and its harmonics to account for intrafrequency interactions that take place in nonlinear structures due to damage. More specifically, the $B(k)$ coefficients in equation (18.1) are nearly equivalent to transmissibility functions and are effective at detecting and locating damage; the $A_{r,s}(k)$ coefficients are quite effective at quantifying damage because they undergo relatively large changes for qualitative changes in the structure (i.e., fastener is loosened creating gap and then complete loss of preload). The fundamental advantage of utilizing features from equation (18.1) for damage assessment, and in particular damage quantification, is that both linear and nonlinear changes due to damage are taken into account. In effect, the $A_{r,s}(k)$ can be thought of as 'amplifiers' for damage that elicits nonlinear behavior.

As an example of the application of equation (18.1), consider the special case in which a first-order AR model is used [i.e., $X_j(k) = B(k)X_i(k) + A_{-1}(k)X_j(k+1) + A_1(k)X_j(k-1)$] to interrogate frequency domain data from the simulated three-story building structure shown in Figure 18.7(b). The building model consists of Unistrut columns and aluminum floor panels and was instrumented with 37 single-axis accelerometers in triaxial configurations, as shown in the close-up of Figure 18.7(b). It was excited laterally at the base with a 50-lb electrodynamic shaker to simulate a seismic input and a 500 Hz sampling frequency was used. Damage was imposed in the form of a reversible reduction in preload from 220 in·lb in the healthy state at the corner brackets to 115 in·lb and then to 5 in·lb. The most severe damage state was imposed by removing the bolts.

Figure 18.8(a) shows the estimates of the exogenous coefficients, $B(k)$, in equation (18.1), which correspond to the transmissibilities across the joint, for the healthy set of data with 220 in·lb of preload in joint 2(a) and the damaged datasets with 115 in·lb, 5 in·lb and the bolt completely removed. The relative progression of damage is clearer in the lower plot for the DOF pair 2a(x) and 2a(y) than in the top for DOF pairs 2a(z) across the joint in the three damage scenarios because the excitation is in the x–y plane. If the

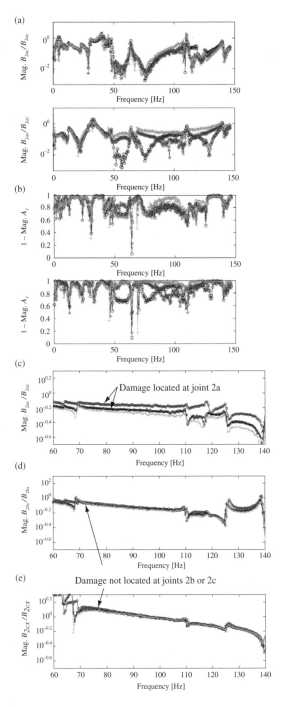

Figure 18.8 (a) Transmissibilities between vertical (z) across the damaged joint 2a (upper) and lateral (x, y) motions (lower) for (∘∘∘) 220 in·lb preload, (∗∗∗) 115 in·lb, (×××) 5 in·lb and (+++) bolt removed, (b) deviations from unity of frequency AR model coefficient, 1-Mag($A_1(\omega)$), for vertical/lateral motions (Reproduced by permission of Sage Publications [13]), (c)–(e) transmissibilities in lateral direction for third, second, first stories

transmissibilities in the x direction across the second joint on each story of the three-story building are compared, as in Figure 18.8(c–e), the damage can also be correctly localized to the third story. Many diagnostic techniques would stop at this point, having identified differences between the healthy and damaged diagnostic features, $B(k)$, and located the damage. Equation (18.1) is now used to classify linearity – nonlinearity as it pertains to damage severity to develop a more absolute measure of damage level.

To this end, Figure 18.8(b) shows the AR coefficients in the form, 1-Mag($A_l(k)$), which are indicators of nonlinearity in the joint (e.g., sliding friction, gap). Note that in the frequency range from 50 Hz to 62 Hz, all of the damage cases exhibit a much different degree of nonlinearity than do the healthy data according to the 2a(x)–2a(z) ARX model. Also note that damage does not always introduce more nonlinearity into systems; in fact, when the bolt is completely removed it is evident from the bottom of Figure 18.8(b) that the most severe damage introduces linearity into the higher frequency range but non-linearity at lower frequencies. The less-severe-damage scenarios actually introduce non-linearity into the 87–112 Hz range. As the preload is reduced from the healthy case, the joint becomes free to undergo sliding locally, which introduces more nonlinear friction. When the bolt is removed, the source of friction is removed and the vibrations are more linear across the joint. In the lower frequency range, the vacant bolt introduces more nonlinearity due to large motions that excite the stiffness–friction nonlinearities.

18.2.4 Diagnostics – Active Global Interrogation Methods

Although the data interrogation method discussed in the previous section is passive and capable of identifying local damage, it is not effective at detecting or characterizing damage throughout a relatively large structural region if dense sensor suites are not feasible. If structures can be excited with active interrogation sources, global damage can be assessed in larger structural regions. Damage in large areas can be identified using wave-propagation approaches for active data interrogation as described below.

It is often desirable to identify precisely where damage is located within a region and to characterize the geometric nature of the damage (e.g., crack length, delamination area, corrosion depth, etc.). Furthermore, noisy operational environments can distort measured FRFs, making it difficult to distinguish damage from noise sources using spectral methods. An *adaptive beamforming* technique is discussed next to locate damage and increase signal-to-noise ratios.

Wave propagation techniques for source identification are inspired by similar techniques to those employed by bats and spiders. For example, consider the predatory habits of the common orb web spider (*Araneus diadematus*), which can be observed detecting and locating prey entrapped in fibrous webs by means of elastic waves [16] sent out in a sonar-like fashion [see Figure 18.9(a)]. In wave-propagation approaches to active damage interrogation, elastic waves are transmitted through finite elastic homogeneous or heterogeneous media such as the multilayer armor shown in Figure 18.9(b), and then various scattering patterns emerge as the waves are partially reflected, refracted or transmitted by the damage. In this approach, scattering patterns are analyzed using beamformers to control the so-called 'look' direction of phased sensor arrays to detect, locate and quantify damage as illustrated in Figure 18.9(c). This technique was introduced by Sundararaman and Adams [17].

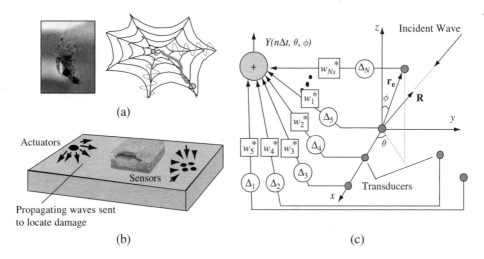

(a)

(b)

(c)

Figure 18.9   (a) Orb web spider and orb web used as medium for active interrogation by spider to detect and locate prey; (b) illustration of wave propagation approach in the context of a multi-layer armor application with pitch–catch mode of operation, and (c) phased-array beamforming arrangement with transducers undergoing weighted summation to produce array output

Beamforming is a process in which spatiotemporal filters are applied to propagating waveforms in order to enhance the amplitude of a signal with respect to the background noise and interference by combining waves from various directions in a weighted and phase-shifted summation. A beamformer array [Figure 18.9(c)] usually consists of at least two elements or arrays of elements spatially arranged to produce a directional reception/radiation pattern. In *beamforming*, a spatial filter uses tap-weights that are calculated as a function of data in real-time whereas in *beamsteering*, a spatial filter uses tap-weights that are selected *a priori*. Beamformers are adaptive and can reconfigure themselves if transducers fail.

Following the development of Dietrich [18], a simple linear narrowband beamformer can be created by weighting and summing the contributions from all the sensor readings to create the array output, $Y(n\Delta t, \theta, \phi)$, for arbitrary pairs of azimuth and elevation angles, (θ, ϕ), where $n\Delta t$ denotes the nth time sample. The summation in Figure 18.9(c) can be written as follows:

$$Y(n\Delta t, \theta, \phi) = \sum_{e=1}^{N_s} w_e^* P_o e^{j(\omega n\Delta t + \zeta_e(\theta,\phi))} = \mathbf{w}_e^H \mathbf{A_r} e^{j\omega n\Delta t}, \tag{18.2}$$

where $\mathbf{w}_e^H = [w_1^* \ w_2^* \ \dots \ w_{N_s}^*]$ is the vector of weights, which have magnitude and phase, $\zeta_e(\theta, \phi)$ is the phase delay and $\mathbf{A_r}$, the array response vector, is $A_r(\theta, \phi) = P_o[1 \ e^{-j\zeta_2(\theta,\phi)} \dots e^{-j\zeta_{N_s-1}(\theta,\phi)}]^H$.

The formation of the inner product between the weighting vector and the array response vector defines the directional sensitivity of the beamformer. When these two vectors are parallel, the output response is large but when the vectors are perpendicular, the output is small. By selecting the weighting vector appropriately, the directional sensitivity of the beamformer can be optimized. Various optimization algorithms are used to calculate these weights.

In order to verify that beamforming can be used to process propagating elastic waves for the purpose of damage interrogation, a steel plate [Figure 18.10(a)] of dimension 1.22 m × 0.915 m × 3.7 mm, was used to carry out experiments. An actuator was mounted on the surface to transmit waves along the plate. An S2-glass/epoxy composite plate [Figure 18.10(b)] of dimension 0.915 m × 0.61 m × 4.7 mm was also used in a set of experiments. It was not necessary in this composite plate to account for the anisotropic dependence of wave velocity on propagation direction. Six 10 mV/g PCB U352C22 accelerometers were affixed to the plates by means of a thin layer of wax or epoxy glue. The actuator was attached to the plate by means of high-stiffness, two-sided, sticky tape. The spacing between sensors was chosen to be at half the wavelength (2.05 cm) of the wave to obtain maximum spatial resolution (e.g., wavelength was 4.1 cm at 20 kHz). The patch actuator was placed 30 cm from the center of the sensor array along its axis. The narrow-band waveform was generated using an Agilent VXI Mainframe at a frequency of 20 kHz

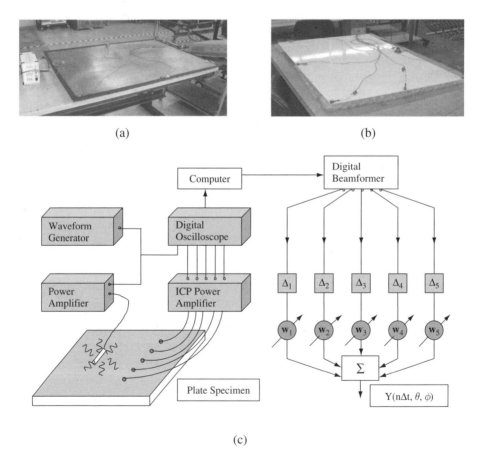

(a) (b)

(c)

Figure 18.10 (a) Steel plate specimen, (b) woven polymer matrix composite specimen used in beamforming interrogation experiments, and (c) schematic diagram of instrumentation used in experiments to generate signal for transmission, receive sensor signals and process data

and consisted of 6 cycles of a Hanning windowed sinusoidal wave. A four-channel Tektronix TDS3014B digital oscilloscope was used to acquire the acceleration response at six locations and the actuator input. Five hundred points of data were acquired at a sampling frequency of 1.25 MHz using the digital oscilloscope. The experimental setup is illustrated in Figure 18.10(c).

In one set of experiments, a surface scratch was used to simulate damage [see Figure 18.11(a)]. Figure 18.11(b) shows plots in polar (1,3), two-dimensional (2,4) and $x - y$ (5) formats when beamforming was applied with respect to the sensors closest and farthest from the actuator. The actual and estimated scratch locations are denoted in Figure 18.11(b.5) and match within one wavelength of the interrogating signal at

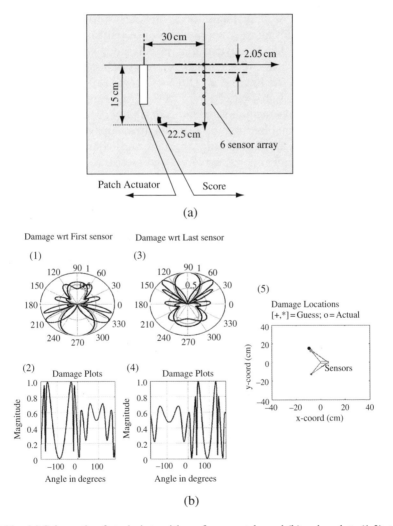

Figure 18.11 (a) Schematic of steel plate with surface scratch, and (b) polar plots (1,3), two-dim. plots (2,4) and x-y plot (5) of directional difference spectrum of array factor gain showing that the surface scratch has been correctly located

20 kHz. When damage interrogation results for traditional and optimal beamforming are compared in noise-free datasets as shown in Table 18.1, the results are nearly identical as expected; however, when a directional source is added to mimic operational noise (e.g., gear box on rotorcraft, gas turbine engine), optimal beamforming is far superior.

18.2.5 Prognostics – Signal-Based and Physics-Based Methods

As explained earlier, prognosis aims to render real-time forecasts of remaining capability by combining load and damage interrogation data with a predictive model. These predictive models can be signal based (data driven), physics based or possibly a combination of the two. Signal-based methods focus on the history of diagnostic data and involve continuous curve fitting to extrapolate on observed trends. In contrast, physics-based methods focus on the underlying, dynamically coupled structural and damage mechanics phenomena, which are functions of the structural geometry and material constitutive models. Refer to Engel *et al.* [19] for a review of prognosis requirements. The primary advantage of signal-based approaches is that trends can be analyzed independently of any structural system; however, this flexibility produces generic models that do not take into account vital information regarding the damage mechanisms and failure modes. Physics-based models provide more insight and, consequently, require greater specificity in modeling the kinematics, dynamics and mechanics of the given structure. A physics-based method is described below wherein nonlinear mechanical oscillations in a valve are modeled to predict its remaining useful life. Physics-based prognosis avoids some of the difficulties with signal-based methods just described. Note that nonlinear dynamical phenomena must be considered when developing prognostic models because nonlinear coupling between the structural and damage mechanics causes damage to grow. In this section, the tertiary failure mode described in Section 18.2.2 involving the air shut-off valve in Figure 18.4(a) is modeled in an effort to explain variability in life predictions.

To develop a prognostic model for the shut-off valve, impact forces between the valve and bushing as well as friction forces between the solenoid plunger and valve hammer

Table 18.1 Location of scratch with conventional/adaptive beamforming and triangulation

Number of sensors	Conventional beamforming			Frost constraint-based technique		
	Location (cm)	Angle wrt sensor 1 (deg)	Error (cm)	Location (cm)	Angle wrt sensor 1 (deg)	Error (cm)
Location of Score	[15, −22.5]	34				
3	[21.37, −22.12]	43	6.38	[21.36, −22.12]	44	6.37
4	[16.81, −24.0]	34	2.35	[14.79, −26.68]	29	4.19
5	[17.72, −26.28]	30	4.66	[14.83, −26.72]	29	4.26
6	[16.56, −27.57]	28	5.31	[14.8, −29.05]	27	6.55

must be modeled. The critical observation regarding the valve is that it can oscillate in the neighborhood of two different *equilibrium* points, one of which is much more susceptible to bushing impact damage than the other. Figure 18.12(a) is an illustration of the two equilibrium points that must be modeled to enable prognosis. In one equilibrium state, the design state, the solenoid plunger is retracted and the valve is at a certain angular position, θ_{o2}. In this state, impacts between the valve and bushings are less severe because the hammer on the valve axis sits in the lowest position on the plunger, which places the butterfly flap in the center of the valve. In the second equilibrium state, impacts are more severe because design tolerances between the flap and housing are reduced at the slightly larger angular position, θ_{o1}. In this position, the hammer rides up on the highest attainable point on the plunger [see schematics in Figure 18.12(a)]. The goal here is to model the valve dynamic behavior in the neighborhood of these two equilibrium points and to examine how nonlinear parameters affect the transition between these two equilibrium states.

The Newtonian equations of motion for the shut off valve system [Figure 18.13(a)] are given by:

$$M_V \ddot{x}_V + C_V(\dot{x}_V - \dot{x}_1) + F_{n1}(x_V - x_1, g(\theta)) = 0$$
$$I_V \ddot{\theta} + C_T \dot{\theta} + K_T \theta - F_N \cdot c \cos \theta = 0 \qquad (18.3a-c)$$
$$M_P \ddot{x}_P + C_P(\dot{x}_P - \dot{x}_2) + F_{n2}(x_P - x_2) + F_T = 0$$

where x_1 and x_2 are the parametric excitations along the valve and solenoid shaft axes, respectively, x_V is the absolute displacement of the valve shaft, x_P is the absolute displacement of the solenoid plunger, and θ is the rotational displacement of the valve.

The nonlinear restoring force, F_{n1}, is a function of the rotational position of the valve and describes the change in tolerance away from the design equilibrium state, θ_{o2}. This force is modeled using a bilinear (gap) stiffness function as is the force F_{n2}. The normal and tangential forces at the hammer/plunger interface are related using a Coulomb

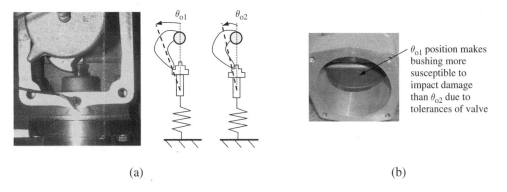

(a)   (b)

Figure 18.12 (a) Hammer and solenoid plunger and schematics illustrating two static equilibrium points for hammer above and below lip of plunger, and (b) butterfly valve in position θ_{o2} (Reproduced by permission of Kluwer Academic/Plenum Publishers from: Multiple Equilibria and Their Effects on Impact Damage in an Air-Handling Assembly, by C.G. McGee and D.E. Adams, *Nonlinear Dynamics*, 27, 2002.)

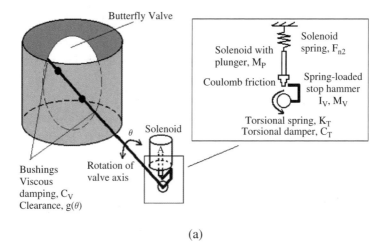

(a)

M_V, I_V, M_P	1 kg, 2.5e–5 kg·m^2, 0.2 kg
C_V, C_T, C_P	30 N·s/m, 0.001 N·m·s, 10 N·s/m
K_P, g_0	0.008 N·m/rad, 0.002 m
c	0.0254 m
Δx_{P0}	0.0015 m
F_{N1}, F_{N2}	0.033 N, 0.0055 N
θ_{10}, θ_{20}	0.105 rad, 0.018 rad
x_{10}, x_{20}	0.004 m, 0.00022 m

(b)

Figure 18.13 (a) Schematic of rigid body dynamics model and (b) parameter values for model (Reproduced by permission of Kluwer Academic/Plenum Publishers from: Multiple Equilibria and Their Effects on Impact Damage in an Air-Handling Assembly, by C.G. McGee and D.E. Adams, *Nonlinear Dynamics*, 27, 2002.)

friction model despite the rattling that is observed. When these forces are substituted into the equations of motion, along with the parameter values in Figure 18.13(b), the static equilibrium solutions obtained are $\theta_{o1} = 6°$ and $\theta_{o2} = 1°$. Note that the value of θ_{o2} determines the gap between the butterfly valve and the bushing. Oscillations in the neighborhood of the first equilibrium point, θ_{o1}, result in more energy transfer between the two components upon impact because the gap is smaller.

A fourth-order fixed time step ($\Delta t = 0.3$ msec; 3200 Hz sample rate) Runge–Kutta integrator was used to simulate the forced time responses of the valve and plunger to input motions (parametric excitations) along the valve and solenoid axes. The forced responses of the valve and plunger to excitations along the valve and solenoid axes, $x_1(t) = 0.004 \cdot \sin(2\pi \cdot 32t)$ and $x_2(t) = 0.004 \cdot \sin(2\pi \cdot 32t)$ m, respectively, and initial conditions, $x_P(0) = 0$ m and $\theta(0) = 0.105$ rad, are shown in Figure 18.14(a). The 4-mm displacement, 32 Hz excitations represent the experimental operating conditions that were imposed on the assembly to simulate field conditions. Note that the valve slowly approaches and then reaches its design equilibrium state, θ_{o2}, due to the 32 Hz harmonic excitation.

The time history in the bottom of Figure 18.14(a) shows the corresponding impacts along the x_V axis as the valve approaches its design equilibrium state. Note that the smaller gap in the initial portion of the response from 0 to 10 sec, where $\theta \approx \theta_{o1}$ and the gap is 1 mm, results in more severe impacts between the valve and housing. After a transition period from 12 to 15 s, the valve reaches its design equilibrium state and the gap increases to 2 mm, which results in less severe impacts (i.e., less exchange of energy between the valve and housing). Figure 18.14(b) shows the spectra of this time history before and after the gap closes (Hanning window; $\Delta f = 0.01$ Hz). Note that the spectrum in the first portion of the response has significantly more broadband energy than does the second portion, resulting in more severe impact damage.

In order to develop a prognostic model for the valve, the multimode failure mechanisms are taken into account using a modified Miner's rule approach [20]. First, note that the remaining life of the valve was limited by the combination of low/high frequency impacts, which damaged the bushing, and high frequency broadband rubbing, which caused the solenoid to stick due to cold welding. The solenoid actuation force also exhibited severe temperature dependencies, which exacerbated sticking. The model in equations (18.3a–c) demonstrated that the vibration levels, x_1 and x_2, controlled the rotational motion of the valve and that θ controlled the impact severity.

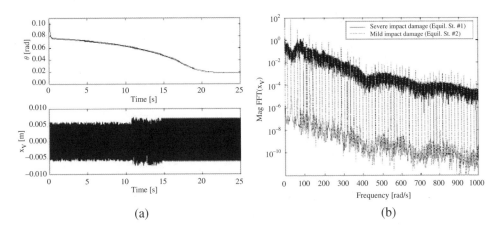

(a)   (b)

Figure 18.14   (a) Numerical time histories of θ and x_V showing transition from θ_{o1} to θ_{o2}, and (b) magnitude of valve Fourier series showing higher energy spectrum near θ_{o1} prior to transition (Reproduced by permission of Kluwer Academic/Plenum Publishers from: Multiple Equilibria and Their Effects on Impact Damage in an Air-Handling Assembly, by C.G. McGee and D.E. Adams, *Nonlinear Dynamics*, 27, 2002.)

Figure 18.15(a) summarizes these findings and Figure 18.15(b) provides a Miner's rule expression that expresses valve life in terms of the reduction in remaining life due to impacts, sticking and a temperature effect that reduced the force available in the solenoid. In this expression, n_θ and N_θ denote the number of cycles that the valve is nearer to the θ_{01} position and the average number of cycles until impact failure in this position. Likewise, n_W and N_W denote the number of cycles of high frequency broadband oscillations (due to turbocharger) and the number of cycles until sticking failures occur. T and T_o denote the current solenoid temperature and reference temperature for which the solenoid is rated in operation. This expression was used to explain the high variability in valve failure rates observed during testing and in the field.

18.3 CONCLUSIONS

The D&P methods described in this chapter included passive and active, local and global methods for damage identification in addition to signal- and physics-based techniques for prognosis. These methods addressed challenges in commercial and defense applications in loads identification, diagnostics and prognostics for mechanical and structural systems. A complete review of these methods and technologies can be found at the author's website, http://widget.ecn.purdue.edu/~FEND2.

ACKNOWLEDGMENTS

The author acknowledges the sponsorship and collaborative support of the following individuals and organizations: graduate and undergraduate research assistants (T. Johnson, C. Yang, G. McGee, S. Sundararaman, M. Nataraju, J. Hundhausen, H. Kess, R. Brown and J. Wenk), National Aeronautics and Space Administration Small Business Innovative Research program, the Department of the Army (DAAD19-

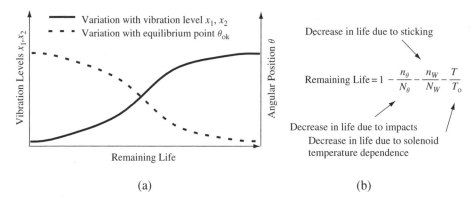

(a)   (b)

Figure 18.15   (a) Variation in remaining useful life with x_1, x_2 and angular position θ, and (b) modified Miner's rule to account for reduction in valve life due to impacts, sticking and temperature

02-1-0185, Gary Anderson), Army Research Laboratory (Elias Rigas), Aviation and Missile Command RDEC (Rob Esslinger), Air Force Research Laboratory Vehicles Directorate (Mark Derriso), Anteon Corporation, ArvinMeritor (Anthony Hicks), Caterpillar, Lord Corporation (Lane Miller), Los Alamos National Laboratory and Los Alamos Dynamics Summer School (Charles Farrar), Air Force Research Laboratory Materials and Manufacturing Directorate (Kumar Jata), PCB Piezotronics (Jim Lally), The Modal Shop (Michael Lally). The support of the US Office of Science and Technology Policy through a Presidential Early Career Award for Scientists and Engineers is also acknowledged with gratitude.

REFERENCES

[1] Cronkhite, J.D. and Gill, L. (1998) 'RTO MP-7', *RTO AVT Specialists' Meeting on Exploitation of Structural Loads/Health Data for Reduced Life Cycle Costs*, J.D. Cronkhite and L. Gill (Eds) Brussels, Belgium.

[2] Brown, R.L. and Adams, D.E. (2003) 'Equilibrium point damage prognosis models for structural dynamic systems', *Journal of Sound and Vibration*, **262**, 591–611.

[3] Schiefer, M.I., Lally, M.J. and Edie, P.C. (2001) 'A Smart Sensor Signal Conditioner', *Proceedings of the International Modal Analysis Conference*.

[4] Goggin, P., Huang, J., White, E. and Haugse, E. (2003) 'Challenges for SHM Transition to Future Aerospace Systems', *Fourth International Workshop on Structural Health Monitoring*, Stanford University, Keynote Lecture, Technomic, Lancaster, PA.

[5] Beral, B. and Speckmann, H. (2003) 'Structure Health Monitoring for Aircraft Structures: A Challenge to System Developers and Aircraft Manufacturers', *Fourth International Workshop on Structural Health Monitoring*, Stanford University Keynote Lecture, Technomic, Lancaster.

[6] Hypersonics Workshop (2001), sponsored by Air Force Office of Scientific Research and Wright Patterson Air Force Base, Dayton, Ohio, M. Camden and L. Byrd (organizers).

[7] Workshop on Multifunctional Materials (2002), sponsored by Air Force Office of Scientific Research, West Lafayette, Indiana, B. Lee (organizer).

[8] Zikry, M., Garg, D. and Anderson, G. (2000) '*Reliability of Heterogeneous Structures through Integrated Sensory Systems*', Army Research Office Memo.

[9] McGee, C.G. and Adams, D.E. (2001) 'Simulating and Detecting A Vibration-Induced Failure in An Air Handling Assembly', *Proceedings of the American Society of Mechanical Engineers Design Engineering Technical Conference on Mechanical Vibration and Noise*, Vol. 6B, pp. 1707–1711.

[10] Recall Effectiveness Meeting (2003), sponsored by US Consumer Products Safety Commission, Bethesda, Maryland, C. Kess (organizer).

[11] National Highway Transportation Safety Agency (2002) http://www.nhtsa.dot.gov/cars/rules/rulings/upgradetire/index.html.

[12] Farrar, C.R., Duffey, T.A., Doebling, S.W. and and Nix, D.A. (1999) 'A Statistical Pattern Recognition Paradigm for Vibration-Based Structural Health Monitoring', *Proceedings of the Second International Workshop on SHM*, F.-K. Chang (Ed) Lancaster, PA, Technomic, pp. 764–773.

[13] McGee, C.G. and Adams, D.E. (2002) 'Multiple equilibria and their effects on impact damage in an air-handling assembly', *Nonlinear Dynamics*, **27**, 55–68.

[14] Adams, D.E. and C.R. Farrar (2002) 'Identifying linear and nonlinear damage using frequency domain ARX models', *International Journal of Structural Health Monitoring*, **1** 185–201.

[15] Adams, D.E. and Allemang, R.J. (2001) 'Discrete frequency models: A new approach to temporal analysis', *ASME Journal of Vibration and Acoustics*, **123**, 98–103.

[16] Witt, P.N., Reed, C.F. and Peakall, D.B. (1968) *A Spider's Web: Problems in Regulatory Biology*, New York: Springer Verlag.

[17] Sundararaman, S. and Adams, D.E. (2002) 'Phased Transducer Arrays for Structural Diagnostics Through Beamforming', in *Proceedings of the American Society for Composites 17th Technical Conference*, Oct. 21–23, 2002, W. Lafayette, Indiana, Paper 177.

[18] Dietrich Jr, C.B. (2000) *Adaptive Arrays and Diversity Antenna Configurations for Handheld Wireless Communication Terminals*, PhD dissertation, Virginia Tech.

[19] Engel, S.J., Gilmartin, B.J., Bongort, K. and Hess, A. (2000) 'Prognostics, The Real Issues Involved with Predicting Life Remaining', *IEEE Aerospace Conference Proceedings*, pp. 457–469.

[20] Miner, M.A. (1945) 'Cumulative damage in fatigue', *Transactions of the ASME*, **67**, A159–A164.

19

Prognosis of Rotating Machinery Components

Michael J. Roemer[1], Gregory J. Kacpryznski[1], Rolf F. Orsagh[1] and Bruce R. Marshall[2]

[1]*Impact Technologies, Rochester, New York, USA*
[2]*NSWC Carderock Division, Philadelphia, Pennsylvania, USA*

19.1 INTRODUCTION

Prognosis is the ability to predict or forecast the future condition of a component and/or system of components, in terms of failure or degraded condition, so that it can satisfactorily perform its operational requirement. In this chapter, we will specifically focus on a few representative prognosis technologies that can enable the early detection and prediction of critical rotating machine components such as bearing and gears. In addition, it will be illustrated how this prognosis information can be used in an asset management software application developed by the US Navy.

Due to the inherent uncertainty in any prognosis system, achieving the best possible prediction of a machine's health is often implemented using various data fusion techniques that can optimally combine sensor data, empirical/physics-based models and historical information. By utilizing a combination of health monitoring data and model-based techniques, a comprehensive component prognostic capability can be achieved throughout a component's life, using model-based estimates when no diagnostic indicators are present, and monitored features such as oil debris and vibration at later stages when failure indications are detectable. Implementation of component-level, machinery prognostic modules will be illustrated in this chapter using bearing and gearbox design, test rig and failure data.

Damage Prognosis – For Aerospace, Civil and Mechanical Systems   Edited by D.J. Inman, C.R. Farrar, V. Lopes Junior and
V. Steffen Junior   © 2005 John Wiley & Sons, Ltd

19.2 BEARING PROGNOSIS FRAMEWORK

The approach described here for comprehensive prognosis of rolling element bearings builds upon existing or enabling technologies such as advanced oil-debris/condition monitoring, high/low-frequency vibration analysis, thermal-trend analysis and empirical/physics-based modeling to achieve its objectives practically. An aspect of this approach is the development and implementation of an integrated prognostic approach that is flexible enough to accept input from many different sources of diagnostic/prognostic information in order to contribute to better fault isolation and prediction on bearing remaining useful life.

The block diagram shown in Figure 19.1 illustrates an abstract representation of the integrated system architecture for the bearing prognosis module described herein. Within this architecture, measured parameters from health monitoring systems such as oil condition/debris monitor outputs and vibration signatures can be accommodated within the anomaly detection/diagnostic fusion center. Based on these outputs, specific triggering points within the prognostic module can be processed so that effective transition associated with various failure mode models (i.e. spall initiation model to a spall progression model) can be accomplished.

Also shown in Figure 19.1, various sources of diagnostic information are combined in the model-based and feature-based prognostic integration algorithms for the specific bearing under investigation, utilizing a probabilistic update process. Knowledge of how the specific bearing is being loaded, historical failure mode information and inspection data feedback can also be accommodated within this generic prognostic architecture. Finally, based on the overall component health assessment and prognosis, specific information related to remaining useful life and associated risk will be passed to operations and maintenance systems as an important input to that decision-making processes (covered in the last section of this chapter).

A more specific block diagram illustrating the implementation of this generic prognostic architecture as applied to a machinery bearing is given in Figure 19.2. In this

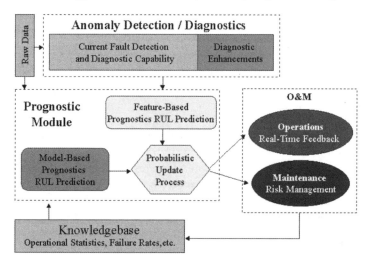

Figure 19.1 Integrated diagnostic and prognostic capability

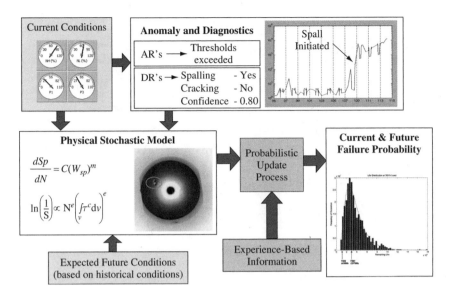

Figure 19.2   Bearing prognostic and health management module

figure, features from oil analysis monitors are combined with vibration features in the 'Anomaly and Diagnostics' module at the top. Once the appropriate feature extraction and fusion is performed, an output is passed to the 'Physical Stochastic Model' that determines the current 'state of health' of the bearing in terms of failure mode progression. This health rating or status is utilized by the model probabilistically to weight the current 'life-line location' in terms of the overall remaining useful life of the bearing. The fused features are also used as an independent measure of RUL that can be incorporated in the 'update' process. The output of the module is a continuously updated description of the components current health state, as well as a projected RUL estimate based on its individual 'signature' usage profile.

19.2.1   Fusion for Bearing Prognosis

Data, information or knowledge fusion is the process of using collaborative or competitive information to arrive at a more confident decision in terms of both diagnostics and prognosis. Fusion plays a key role in advanced prognosis processes in terms of producing useful features, combining features, and incorporating model-based information. Within a comprehensive prognostic and health-management system, fusion technologies are utilized in three main areas as shown in Figure 19.3. At the lowest level (Area 1), data fusion can be used to combine information from a multisensor data array to validate signals and create useful features. One example of this type of data fusion is combining a speed signal and a vibration signal to produce time synchronous average vibration features.

At a higher level (Area 2), fusion may be used to combine features in intelligent ways so as to obtain the best possible diagnostic information. This would be the case if a feature related to particle count and size in a bearing's lubrication oil was fused with a vibration feature, such as RMS level. The combined result would yield an improved level of

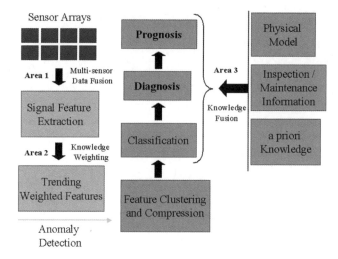

Figure 19.3 Health management fusion application areas

confidence about the bearing's health. Finally, knowledge fusion (Area 3) is used to incorporate experienced-based information such as legacy failure rates or physical model predictions with signal-based information.

Identifying the optimal fusion architectures and approaches at each level is a vital factor in assuring that the realized system truly enhances the health monitoring and prognosis capabilities. Most current fusion implementations may be categorized as belonging to one of three generic architectures. These architectures are: centralized, autonomous, and hybrid fusion, each with its own benefits specific to a given application. Fusion of raw condition-monitoring data or processed features can yield more robust diagnostic or prognostic information than can individual condition indicators. Common fusion architectures such as the centralized fusion procedure combine multisensor data while it is still in its raw form. Another common approach, called autonomous fusion, addresses some of the data management problems encountered by centralized architecture by placing feature extraction before the fusion process. In the case described herein, a hybrid fusion architecture as shown in Figure 19.4 takes the already processed features, classifies them on an individual basis and then fuses the result.

As there are many different architectures for fusion, there are also many different algorithms themselves for performing the fusion. Simple information fusion techniques include weighted average and voting combinations of evidence from various sources. More sophisticated techniques, specifically Bayesian and Dempster–Shafer combination, are briefly described to illustrate the statistical basis and information requirements of these fusion techniques. Bayesian inference can be used to determine the probability that a diagnosis is correct, given a piece of *a priori* information. Analytically, Bayes' theorem is expressed as follows:

$$P(f_1|O_n) = \frac{P(O_n|f_1) \cdot P(f_1)}{\sum\limits_{j=1}^{n} P(O_n|f_1) \cdot P(f_1)} \qquad (19.1)$$

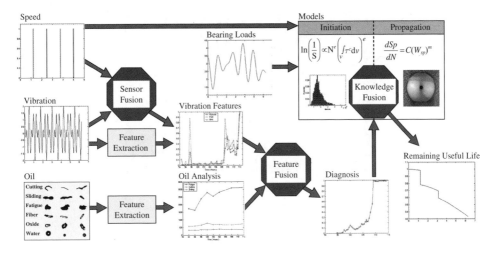

Figure 19.4 Hybrid fusion architecture

where:

$P(f|O)$ = The probability of fault (f) given a diagnostic output (O), $P(O|f)$ = the probability that a diagnostic output (O) is associated with a fault (f), and $P(f)$ = the probability of the fault (f) occurring.

In the Dempster–Shafer approach, uncertainty in the conditional probability is considered. The Dempster–Shafer methodology hinges on the construction of a set, called the frame of discernment, which contains every possible hypothesis. Every hypothesis has a belief denoted by a mass probability (m). Beliefs are combined with the following equation:

$$Belief\ (Hn) = \frac{\sum\limits_{A \cap B = Hn} m_i(A) \cdot m_j(B)}{1 - \sum\limits_{A \cap B = 0} m_i(A) \cdot m_j(B)} \tag{19.2}$$

19.2.2 Early Fault Detection for Rolling Element Bearings

Development of specialized vibration or other features that can detect the incipient formation of a spall on a bearing race or rolling element, is a critical step in the design of the integrated prognosis system mentioned above. To this end, a series of tests were designed and conducted to provide data for algorithm development, testing and validation. Vibration and oil-debris data was acquired from a ball-bearing test rig as shown in Figure 19.5. A damaged bearing was installed and used to compare the effectiveness of various diagnostic features extracted from the measured data (Table 19.1). The bearings are identical to the number 2 main shaft bearing of an Allison T63 gas turbine engine.

Although originally designed for lubrication testing, the test rig was used to generate accelerated bearing failures due to spalling. To accelerate this type of bearing failure, a fault was seeded into the inner raceway of one of the bearings by means of a small hardness

Figure 19.5 Bearing test rig setup

Table 19.1 Test rig bearing dimensions (mm)

Bearing number	Ball diameter (Bd)	Number of balls (z)	Pitch diameter (Pd)	Contact angle (β_x, deg.)
1	4.7625	8	42.0624	18.5°
2	7.9375	13	42.0624	18.5°

indentation (Brinnell mark). The bearing was then loaded to approximately 14 234 N (3200 lbf) and ran at a constant speed of 12 000 rpm (200 Hz). Vibration data was collected from a cyanoacrylate–mounted (commonly called 'super glue') accelerometer, which was sampled at over 200 kHz. Also, the quantity of debris in the oil draining from the test head was measured using a magnetic chip collector (manufactured by Eateon Tedeco). The oil data was used in determining the severity of the spall progression.

The fundamental pass frequencies of the components of a bearing can be easily calculated with standard equations. Extraction of the vibration amplitude at these frequencies from a fast Fourier transform (FFT) often enables isolation of the fault to a specific bearing in an engine. High amplitude of vibration at any of these frequencies indicates a fault in the associated component. Table 19.2 shows the calculated bearing fault frequencies.

19.2.2.1 High Frequency Enveloping Analysis

Although bearing characteristic frequencies are easily calculated, they are not always easily detected by conventional frequency domain analysis techniques. Vibration amplitudes at

Table 19.2 Test rig bearing fault frequencies (Hz)

Shaft speed	BSF	BPFI	BPFO	Cage
200	512	1530	1065	82

Note the listings are defined by: BSF, ball spin frequency; BPFI, inner raceway frequency; BPFO, outer raceway frequency; Cage, cage frequency.

these frequencies due to incipient faults (and sometimes more developed faults) are often indistinguishable from background noise or are obscured by much higher amplitude vibration from other sources, including engine rotors, blade passing, and gear mesh in a running engine. However, bearing faults produce impulsive forces that excite vibration at frequencies well above the background noise in an engine.

Impact Energy^{TM} is an enveloping-based vibration feature extraction technique that consists of first band pass filtering of the raw vibration signal. Second, the band pass filtered signal is full waved rectified to extract the envelope. Third, the rectified signal is passed through a low pass filter to remove the high frequency carrier signal. Finally, the signal's DC content is removed.

The Impact Energy^{TM} process was applied to the seeded fault test data collected using the bearing test rig. To provide the clearest identification of the fault frequencies, several band pass and low pass filters were used to analyze various regions of the vibration spectrum. Using multiple filters allowed investigation of many possible resonance's of the bearing test rig and its components. A sample Impact Energy^{TM} spectrum from early in the rig test is shown in Figure 19.6. Note that this data was collected prior to spall initiation (based on oil debris data), and the feature response is due to the indentation on the race.

For comparison, a conventional FFT (ten frequency domain averages) of vibration data was also calculated and is shown in Figure 19.7. In the conventional frequency domain plot (Figure 19.7) there is no peak at the inner race ball pass frequency (1530 Hz). However, the Impact Energy^{TM} plot (Figure 19.6) shows clearly defined peaks at this frequency and the second through fourth harmonics of the inner race ball pass frequency. These peaks were defined using a detection window of ± 5 Hz about the theoretical frequency of interest to account for bearing slippage. From the onset of the test there is an indication of a fault.

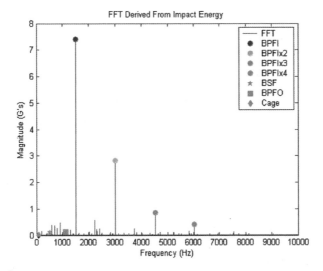

Figure 19.6   Impact Energy^{TM} FFT of test rig bearing – seeded fault

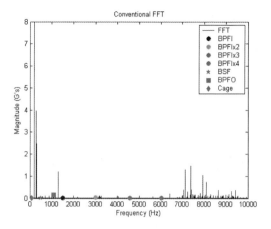

Figure 19.7   Conventional FFT of test rig bearing – seeded fault

19.3   MODEL-BASED ANALYSIS FOR PROGNOSIS

A physics-based model is a technically comprehensive modeling approach that has been traditionally used to understand component failure mode progression. Physics-based models provide a means to calculate the damage to critical components as a function of operating conditions and assess the cumulative effects in terms of component life usage. By integrating physical and stochastic modeling techniques, the model can be used to evaluate the distribution of remaining useful component life as a function of uncertainties in component strength/stress properties, loading or lubrication conditions for a particular fault. Statistical representations of historical operational profiles serve as the basis for calculating future damage accumulation. The results from such a model can then be used for real-time failure prognostic predictions with specified confidence bounds. A block diagram of this prognostic modeling approach is given in Figure 19.8. As illustrated at the core of this figure, the physics-based model utilizes the critical life-dependent uncertainties so that current health assessment and future RUL projections can be examined with respect to a risk level.

Model-based approaches to prognostics differ from feature-based approaches in that they can make RUL estimates in the absence of any measurable events, but when related diagnostic information is present (such as the feature described previously) the model can often be calibrated based on this new information. Therefore, a combination or fusion of the feature-based and model-based approaches provides full prognostic ability over the entire life of the component, thus providing valuable information for planning which components to inspect during specific overhaul periods. While failure modes may be unique from component to component, this combined model-based and feature-based methodology can remain consistent across different types of oil-wetted components in the rotating.

19.3.1   Bearing Spall Initiation and Propagation Models

Rolling-element contact fatigue is considered to be the primary cause of failure in bearings that are properly loaded, well aligned, and receive an adequate supply of uncontaminated

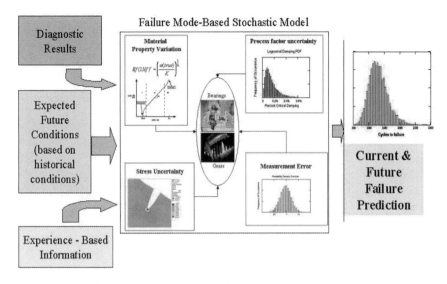

Figure 19.8   Prognostic bearing model approach

lubricant. Rolling contact fatigue causes material to flake off from the load bearing surfaces of rolling elements and raceways leaving a pit or spall as shown in Figure 19.9.

Spalling of well-lubricated bearings typically begins as a crack below the load-carrying surface that propagates to the surface. Once initiated, a spall usually grows relatively quickly, producing high vibration levels and debris in the oil. Due to the relatively short remaining life following spall initiation, the appearance of a spall typically serves as the criterion for failure of bearings in critical applications.

19.3.1.1 Spall Initiation Model

A variety of theories exist for predicting spall initiation from bearing dimensions, loads, lubricant quality, and a few empirical constants. Many modern theories are based on

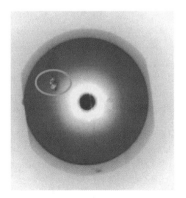

Figure 19.9   Ball bearing with spall

the Lundberg–Palmgren (L–P) model [1] that was developed in the 1940s. A model proposed by Ioannides and Harris (I–H) [2] improved on the L-P model by accounting for the evidence of fatigue limits for bearings. Yu and Harris (Y–H) [3] proposed a stress-based theory in which relatively simple equations are used to determine the fatigue life purely from the induced stress. This approach depends to lesser extent on empirical constants, and the remaining constants may be obtained from elemental testing rather than complete bearing testing as required by L–P.

The fundamental equation of the Y–H model stated in equation 19.3 relates the survival rate (S) of the bearing to a stress-weighted volume integral as shown below. The model utilizes a new material property for the stress exponent (c) to represent the material fatigue strength, and the conventional Weibull slope parameter to account for dispersion in the number of cycles (N). The fatigue initiating stress (τ) may be expressed using Sines multiaxial fatigue criterion [4] for combined alternating and mean stresses, or as a simple Hertz stress.

$$\ln\left(\frac{1}{S}\right) \propto N^e \left(\int_v \tau^c d\nu\right)^e \tag{19.3}$$

For simple Hertz stress, a power law is used to express the stress-weighted volume. In equation (19.4) below, λ is the circumference of the contact surface, and a and b are the major and minor axes of the contact surface ellipse. The exponent values were determined by Yu and Harris for $b/a \approx 0.1$ to be $x = 0.65$, $y = 0.65$, and $z = 10.61$. Yu and Harris assume that these values are independent of the bearing material.

$$\int_A \tau^c dA \cdot \lambda \propto a^x b^y \tau^z \lambda \tag{19.4}$$

According to the Y–H model, the life (L_{10}) of a bearing is a function of the basic dynamic capacity (Q_c) and the applied load as stated below. The basic dynamic capacity is also given. A lubrication effect factor may be introduced to account for variations in film thickness due to temperature, viscosity, and pressure.

$$L_{10} = \left(\frac{Q_c}{Q}\right)^{\frac{x+y+z}{3}} \tag{19.5}$$

$$Q_c = A_1 \Phi D^{\frac{(2z-x-y)}{(z+x+y)}} \tag{19.6}$$

$$\Phi = \left[\left(\frac{T}{T_1}\right)^z \frac{u(D\Sigma\rho)^{\frac{(2z-x-y)}{3}}}{(a^*)^{z-x}(b^*)^{z-y}} \frac{d}{D}\right]^{\frac{-3}{z+x+y}} \tag{19.7}$$

where: A_1 = material property; T = a function of the contact surface dimensions; T_1 = value of T when $a/b = 1$; u = number of stress cycles per revolution; D = ball diameter; ρ = curvature (inverse radii of component); d = component (race way) diameter; a^* = function of contact ellipse dimensions, and b^* = function of contact ellipse dimensions.

A stochastic bearing fatigue model based on Y–H theory has been developed to predict the probability of spall initiation in a ball/V-ring test. The stochastic model accounts for uncertainty in loading, misalignment, lubrication, and manufacturing that is not included in the Y–H model. Uncertainties in these quantities are represented using normal and log-normal probability distributions to describe the input parameters to the model. As a result, the stochastic Y–H model produces a probability distribution representing the likelihood of various lives.

19.3.1.2 Spall Progression Model

Once initiated, a spall usually grows relatively quickly, producing increased amounts of oil debris, high vibration levels, and elevated temperatures that eventually lead to bearing failure. While spall progression typically occurs more quickly than spall initiation, a study by Kotzalas and Harris [5] showed that 3 to 20 % of the useful life of a particular bearing remains after spall initiation. The study identified two spall progression regions. Stable spall progression is characterized by gradual spall growth and exhibits low broadband vibration amplitudes. The onset of unstable spall progression coincides with increasing broadband vibration amplitudes.

Kotzalas and Harris also presented a spall progression model. The model relates the spall progression rate (dS_p/dN) to the spall similitude (W_{sp}) using two constants (C and m) as shown below. The spall similitude is defined in terms of the maximum stress (σ_{max}), average shearing stress (τ_{avg}), and the spall length (S_p).

$$\frac{\mathrm{d}Sp}{\mathrm{d}N} = C(W_{sp})^m \tag{19.8}$$

$$W_{sp} = (\sigma_{max} + \tau_{avg})\sqrt{\pi Sp} \tag{19.9}$$

Spall progression data was collected during the Kotzalas/Harris study using a ball/V-ring test device. The data was collected at the Pennsylvania State University. Each of the lines shown in Figure 19.10 represents spall progression of a ball bearing. For one ball, indicated by multiple markers, the test was periodically suspended to measure the spall length. Only two spall length measurements, initial and final values, were taken for the other balls. The figure shows a large dispersion of the spall initiation life, which is indicated by the lowest point on each line. The figure also shows a large dispersion of the spall progression life, which can be inferred from the slope of the lines.

The progression model was also applied to the bearing test rig data acquired. A correlation between the recorded oil-particle quantity and the spall size was determined from the initial and final quantities and spall sizes. This allowed scaling of the oil-particle quantity to approximate spall size. The model agreed well with the spall progression during the test, as seen in Figure 19.11.

A stochastic spall progression model based on the Kotzalas/Harris theory was developed to predict the probability of spall progression in the ball/V-ring test. The stochastic model accounts for uncertainty in loading, misalignment, lubrication, and manufacturing that are not included in the Kotzalas/Harris model. As a result, the stochastic Kotzalas/Harris model produces a probability distribution representing the likelihood of various lives.

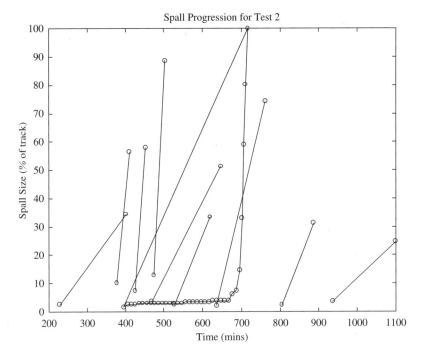

Figure 19.10   Spall progression data from phase I analysis

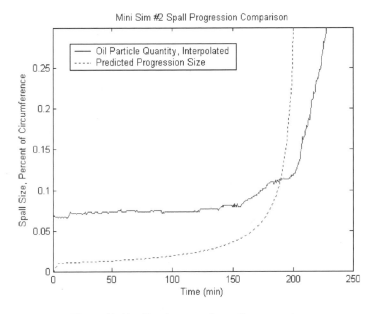

Figure 19.11   Bearing test rig spall progression

19.3.2 *Prognosis Model Validation*

Validation of the spall initiation model requires a comparison of actual fatigue life values to predicted model values. Acquiring sufficient numbers of actual values is not a trivial task. Under normal conditions it is not uncommon for a bearing life value to extend past 100 million cycles, prohibiting normal run-to-failure testing.

Accelerated life testing is one method used to generate rapidly many bearing failures. By subjecting a bearing to high speed, load, and/or temperature, rapid failure can be induced. There are many test apparatuses used for accelerated life testing including ball-and-rod type test rigs. One such test rig is operated by UES, Inc. at the Air Force Research Laboratory (AFRL) at Wright Patterson Air Force Base in Dayton, Ohio. A simple schematic of the device is shown in Figure 19.12 with dimensions given in millimeters. This rig consists of three 12.7-mm diameter balls contacting a 9.5-mm rotating central rod (see Table 19.3 for dimensions). The three radially loaded balls are pressed against the central rotating rod by two tapered bearing races that are thrust loaded by three compressive springs. A photo of the test rig is shown in Figure 19.13. Notice the accelerometers mounted on the top of the unit. The larger accelerometer is used to shutdown the test automatically when a threshold vibration level is reached, the other measures vibration data for analysis.

By design the rod is subjected to high contact stresses. Due to the geometry of the test device, the 222-N (50 lbs) load applied by the springs translates to a 942-N (211 lbs) load per ball on the center rod. Assuming Hertzian contact for balls and rod made of M50

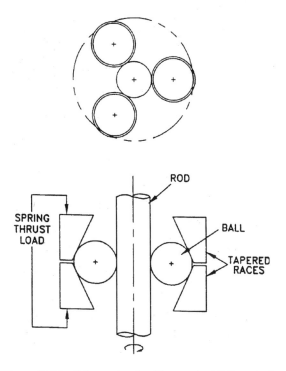

Figure 19.12   Schematic of rolling contact fatigue tester

Table 19.3   Rolling contact fatigue tester dimensions (mm)

Rod diameter (*Dr*)	9.52
Ball diameter (*Db*)	12.70
Pitch diameter (*Dm*)	22.23

Figure 19.13   Rolling contact fatigue tester

bearing steel, the 942 N radial load results in a maximum stress of approximately 4.8 GPa (696 ksi). This extremely high stress causes rapid fatigue of the bearing components and can initiate a spall in less than 100 hours, depending on test conditions including lubrication, temperature, etc. Since failures occur relatively quickly, it is possible to generate statistically significant numbers of events in a timely manner.

For validation purposes, M50 rods and balls were tested at room temperature (23 °C). The results of these tests are in Table 19.4. A summary plot is shown in Figure 19.14.

As stated above, one of the issues with empirical/physics based models is their inherent uncertainty. Assumptions and simplifications are made in all modeling and not all of the model variables are known exactly. Often stochastic techniques are used to account for the implicit uncertainty in a model's results. Statistical methods are used to generate numerous possible values for each input.

A Monte Carlo simulation was utilized in the calculation of the bearing life distribution. Inputs to the model were represented by normal or log-normal distributions to approximate

Table 19.4   RCF fatigue life results

Number of failures	Susp (#)	Susp time (cycles)
29	1 (ball failed)	83.33

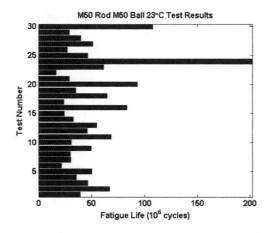

Figure 19.14   RCF fatigue life results

the uncertainty of the input values. Sample input distributions to the model are shown in Figure 19.15.

The Yu–Harris model was used to simulate the room temperature M50 RCF tests. Figure 19.16 shows the results for a series of the room-temperature RCF tests on the M50 bearing material. This test was run at 3600 rpm at room temperature with the

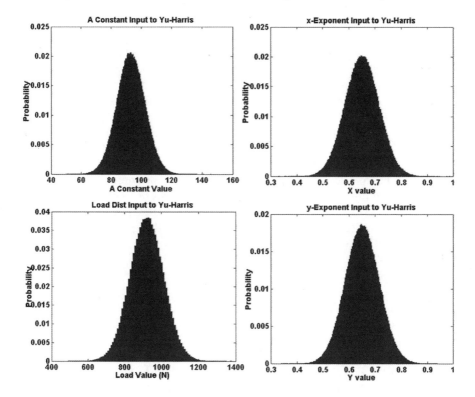

Figure 19.15   Stochastic model input distributions

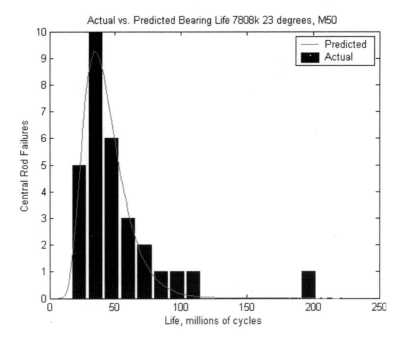

Figure 19.16   Room temperature results vs predicted

7808 K lubricant. The y-axis is the number of central rod failures, and the x-axis is the millions of cycles to failure. The predicted life from the model is also shown in Figure 19.16, superimposed on the actual test results. This predicted distribution shown was calculated from the model using one million Monte Carlo points.

In Figure 19.17, the median ranks of the actual lives (circles) are plotted against the cumulative distribution function (CDF) of the predicted lives (line). The model-predicted lives are slightly more conservative (in the sense that the predicted life is shorter than the observed life) once the cumulative probability of failure exceeds 70 %. However, since bearings are a critical component, the main interest is in the left-most region of the distribution where the first failures occur and the model correlates better. Calculation of median ranks is a standard statistical procedure for plotting failure data. During run-to-failure testing there are often tests that either are prematurely stopped before failure or a failure occurs of a component other than the test specimen. Although the data generated during these failures are the mode of interest, they provide a lower bound on the fatigue lives due to the failure mode of interest. One method for including this data is by median ranking.

The median rank was determined using Benard's median ranking method, which is stated in equation (19.10) below. This method accounts for tests that did not end in the failure mode of interest (suspensions). In the case of the ball-and-rod RCF test rig, the failure mode of interest is creation of a spall on the inner rod. The time to suspension provides a lower bound for the life of the test article (under the failure mode of interest), which can be used in reliability calculations. During the testing on the RCF test rig, significant portions of the tests were terminated without failure after reaching ten times

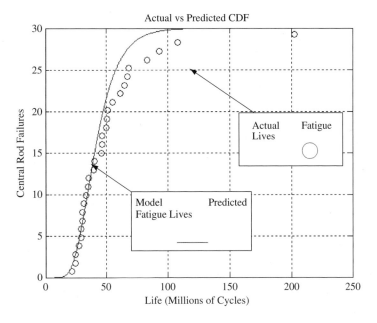

Figure 19.17   Actual life vs predicted life

the L_{10} life. There were also several tests that ended due to development of a spall on one of the balls rather than on the central rod.

$$\text{Benard's median rank} = \frac{(AR - 0.3)}{(N + 0.4)} \qquad (19.10)$$

where: AR = adjusted rank, and N = number of suspensions and failures.
The adjusted rank is calculated below.

$$AR = \frac{(\text{reverse rank}) \times (\text{previous adjusted rank}) + (N + 1)}{\text{reverse rank} + 1} \qquad (19.11)$$

Although the test does not simulate an actual bearing assembly in an engine, it does simulate similar conditions. Materials and the geometry of the bearing and the lubricants are the same for the test rig as they are in the T-63 engine. The test rig results validate the model's ability to predict the fatigue life of the material under similar conditions to an operating engine.

19.4   BEARING PROGNOSIS DISCUSSION

To achieve a comprehensive prognostic capability throughout the life of a rolling element bearing, model-based information can be used to predict the initiation of a fault before any diagnostic indicators are present. In most cases, these predictions will prompt 'just in time' maintenance actions to prevent the fault from developing. However, due to modeling uncertainties, incipient faults may occasionally develop earlier than predicted. In these situations, sensor-based diagnostics complement the model-based prediction by updating the model to reflect the fact that fault initiation has occurred. Sensor-based approaches provide direct measures of component condition that can be used to update the modeling assumptions and reduce the uncertainty in the RUL predictions. Subsequent predictions of the remaining useful component life will be based on fault progression rather than initiation models.

19.5 GEAR PROGNOSIS FRAMEWORK

As illustrated from the previous bearing-prognosis description, prognosis model types can range from purely statistical or data-driven methodologies (i.e., historical failure rates, time-series predictions) to high fidelity models that encapsulate failure mechanisms and their progression. The availability of established vibration based crack detection features, controlled test conditions, periodic inspections, and the desire for the tightest possible prognostic confidence bounds warrant the use of the most technically comprehensive and knowledge-rich prognostic model possible. For the case of gear prognosis described in this section, a high fidelity stress, fatigue and crack propagation model fits within a generic prognostic module architecture shown in Figure 19.18, configured for components that fail via fatigue mechanisms.

This comprehensive approach allows for the integration of material-level models, system-level data/information fusion algorithms, and adaptive modeling techniques for 'tuning' key failure mode variables at a local material/damage-site. The output failure rate prediction is developed within a probabilistic framework to identify directly

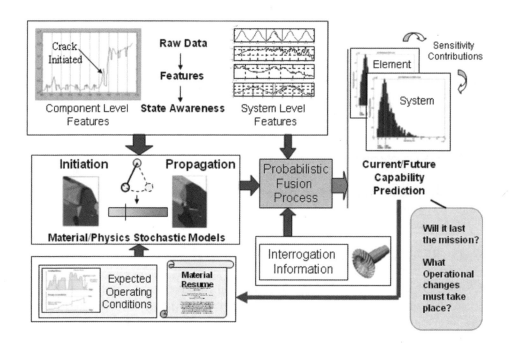

Figure 19.18   Gear prognosis model approach

confidence bounds associated with the specific component failure mode progression. By providing continuous updates/adjustments to the critical parameters used by the probabilistic fatigue/damage models, based on observing system-level measurements, more accurate failure rate predictions are possible throughout the life of the component.

The following sections will focus largely on the failure mode characterization for gears, prognosis model building, life prediction, and model calibration using vibration-based diagnostics associated with this framework.

19.5.1 Gear Failure Mode Characterization

The principal failure mode considered in this gear prognosis module development was a single fatigue crack propagating from the fillet region for a single tooth. This is an important failure mode to consider because the separation of a single tooth can lead to complete loss of power transmission under certain conditions. Common failure modes for gears involve:

(1) Tooth separation due to a fatigue crack initiating in the fillet region. Crack initiation in this area is driven by high tensile bending stresses along the fillet. A surface defect may not be required for a failure to occur.
(2) Partial tooth separation due to fatigue crack(s) initiating from surface damage sites such as one or more pits, and scoring and scuffing marks.
(3) Excessive contact surface damage due to pitting, scoring, or scuffing that can lead to high levels of vibration and noise.
(4) Propagating cracks can initiate from accidental nicks, etc., during assembly or surface damage from oil debris.

There is on-going research to identify system features that will provide reliable information on the onset and progression of failure modes 2 to 4, but the scope of this section will only cover tooth root cracking.

19.5.2 Building a Physics Model of the Gear

To determine, with a particular confidence level, the number of cycles or time until a tooth-root crack at one root location will have initiated to the point that linear fracture mechanics would apply, the following information had to be generated:

- loading profile of a pinion tooth throughout a complete load cycle with the mating gear;
- plastic zone size estimation for each notch;
- material properties associated with the material fatigue life and linear elastic fracture mechanics (LEFM).

Gleason Corporation provided Impact Technologies with the gear pinion geometry and associated load definition, which represented the pinion's load history through a

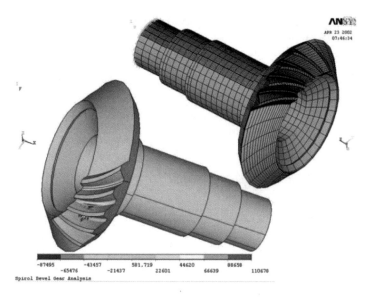

Figure 19.19   Gear finite element model

complete engagement with the mating gear. Impact Technologies used this data to perform mesh sensitivity studies and generate a full 3-D dynamic stress analysis using ANSYS (see Figure 19.19). Model data provided by Gleason Corporation was in the form of three pinion teeth defined about the global origin of the pinion and a load history, which encompassed one tooth having load applied from initially being at zero, to a maximum and then unloaded back to zero (R ratio of 0 where $R =$ Min.Stress/Max.Stress).

A typical failure scenario would involve formation and propagation of a crack in the tooth fillet region. The primary stress field in this area is nominally uniaxial and is oriented along the fillet. Therefore, the initial crack formation and propagation would be perpendicular to the base fillet. However, computational fracture mechanics models show that once the cracks grow past the field of influence of the fillet, the crack trajectory is influenced by the rim design for the gears [6].

It is important to consider that gears can be either of the through-hardened variety or a case-hardened variety. Unlike through-hardened gears, case-hardened gears can have a hardened case of up to 0.1 inch (2.5 mm) with hardness in the range 61–63 Rc. The core is significantly lower in hardness. One commonly followed convention is to define case thickness as the distance from the surface where the hardness falls to 50 Rc. This combination of hard case and tough core offers a good compromise between high surface durability/fatigue as well as reliable mechanical performance under potential dynamic gear loading. However, the presence of the hardened case leads to a continuous variation of fatigue and fracture properties within the case and therefore requires special considerations when performing life analysis. Figure 19.20 shows a representative hardness variation [7].

The compressive stresses could be as high as 40 ksi (300 MPa) and vary within the case. Further, the compressive stress variation within the case may or may not be

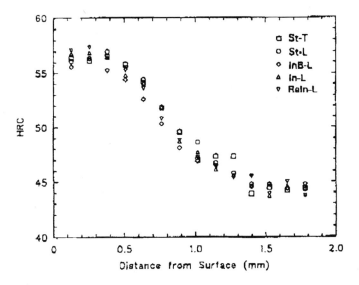

Figure 19.20   Gear material hardness variability

monotonic. To summarize, accurate prognostics for case hardened gears requires the following information:

(1) accurate definition of 3-D stress field throughout the gear;
(2) the extent of the case-hardened layer and the variation of material fatigue and fracture mechanics properties within the case-hardened layer;
(3) compressive stress distribution within the case-hardened layer and beyond;
(4) reliable small crack growth data within the case-hardened layer;
(5) material fatigue and fracture properties for the core.

Except for item (1) above, statistical distributions rather than average values are essential in order to perform probabilistic life predictions.

19.5.3 Fatigue Life Prediction

In order to allow for testing and validation of the proposed prognosis approach, an EDM notch (stress riser) was placed at the root of one of the gear teeth to accelerate the crack initiation process. For predicting the total fatigue life of a component with notches, strain-life data obtained from tests on uniaxial specimens should not be directly used because they are subject to uniform strain across the entire test cross-section. For notched components there is a severe stress gradient within the notch region and, therefore, special attention must be paid to this region by dividing it into 'sub-zones' or ligaments, and calculating fatigue life for each zone independently [8]. This approach facilitates the definition of 'crack initiation' and thereby provides an unambiguous way to transition from the crack initiation phase to the crack propagation phase and circumventing an historical dilemma. The dilemma is that if crack initiation is defined as creation of a defect (crack) of arbitrary length (an 'engineering' size crack has been defined to have a length of any where from 0.1 mm to 1 mm (4 to 40 mils)), then the associated stress intensity range (ΔK) from the Paris law [9 and Eq. 19.14] can be made arbitrarily small or large. If the assumed crack size at the end of crack initiation is too small, then the associated ΔK at the crack tip will be smaller than the $\Delta K_{threshold}$ for the material (depends on the R ratio) and therefore based on the fracture mechanics principles, the crack will not propagate. On the contrary, if crack initiation is defined as a large crack, then the predicted cycles for crack propagation could be rather small. Concisely, the prediction of total fatigue life is greatly influenced by the definition of what constitutes a crack.

19.5.4   Gear Tooth Crack Initiation

Based on the work by Socie *et al.* [10], it was demonstrated that a crack is said to have initiated when the rate of propagation of the crack due to crack propagation mechanism (using principles of fracture mechanics) exceeds that due to crack initiation or strain cycle fatigue mechanism, and is typically of the order of 0.0005–0.001 in (0.0127–0.0254 mm). The following sections follow initiation and propagation analyses for this gear application.

The EDM notch in the gear tooth acted as a stress riser leading to high elastic stresses (374 ksi at 2340 ft-lbs or 2579 MPa at 3173 N-m across the IGB). This high stress (at least in the immediate vicinity of the notch) was in excess of the yield strength (yield strength = 170 ksi or 1172 MPa) of the material, and therefore resulted in local plastic cycling. The large stress range ($0 \rightarrow 374 \rightarrow 0$ ksi) and the associated large plastic strain range experienced by the gear tooth, subjects the EDM notch to low-cycle fatigue (LCF). The total strain amplitude in the EDM notch is defined as the summation of the elastic and plastic strain amplitudes, which can be expressed as shown below [4]. Neuber's rule was employed to convert the elastically calculated stresses from the linear-elastic FE model to 'true stress' and 'true strain' values [7].

$$\frac{\Delta \varepsilon_{Total}}{2} = \underbrace{\frac{\sigma'_f}{E}(2N_f)^b}_{\text{elastic}} + \underbrace{\epsilon'_f(2N_f)^c}_{\text{plastic}} \qquad (19.12)$$

The expression above shows the number of reversals $2N_f$ (twice the number of cycles N_f) to failure as a function of elastic and plastic strain amplitudes. The strain-based approach integrated with a rain-flow counting or similar algorithm permits the treatment of 'load sequencing effects' in predicting component life. For example, in the case of helicopter gears, different amounts of material damage will accumulate at different load (torque) levels and on the sequence in which the torques were varied, i.e. the torque magnitudes as well as torque profiles both affect the amount of material damage.

In the case of the EDM notch, the plastic term is the dominant term. Using the plastic term alone, the mean number of cycles to EDM notch crack initiation is given by:

$$N_f = \frac{1}{2}\left[\frac{\Delta \epsilon_{Plastic}}{2} \cdot \frac{1}{\epsilon'_f}\right]^{\frac{1}{c}} \qquad (19.13)$$

where: N_f = number of cycles to crack initiation; $\Delta\epsilon_{Plastic}$ = plastic strain range; ε'_f = fatigue ductility coefficient; c = fatigue ductility exponent.

Because strain-life data for AISI 9130 was not available, data for a low-alloy steel with hardness similar to the assumed core hardness (38 RC/350 HB) of AISI 9130 was used [8, 9]:

Ultimate tensile strength: 179.8 ksi (1234 MPa)
Yield strength: 170.8 ksi (1178 MPa)
Elastic modulus: 27.85 Mpsi (192 GPa)
Cyclic strength coefficient (K'): 270.2 ksi (1862 MPa)
Cyclic strain hardening exponent (n'): 0.14
Fatigue strength coefficient (σ_f): 281.9 ksi (1944 MPa)
Fatigue strength exponent (b): −0.1
Fatigue ductility coefficient (ε'_f): 1.22
Fatigue ductility exponent (c): −0.73

Taking this data and applying it back to the subzone concept, as one moves away from the notch, the number of cycles to crack initiation (using the strain-life analysis procedure) at each of these sampled points will increase due to reduced stress/strain amplitude. For example, using Figure 19.21 as a reference, if sampled point 1 takes $N1$ cycles for crack initiation and the sampled point 2 takes $N2$ cycles for crack initiation, and the distance between the two points is d, then the rate of crack propagation (Length Increase/Load Cycle) over this subzone due to initiation is $d/(N2 - N1)$. If for a certain subzone defined by points I and $I+1$, the rate of propagation using the fracture mechanics data (plots of da/dN vs ΔK) is greater than that due to strain-cycle fatigue mechanism, then the distance of this subzone from the root of the notch is considered as the length of crack initiation, a_i.

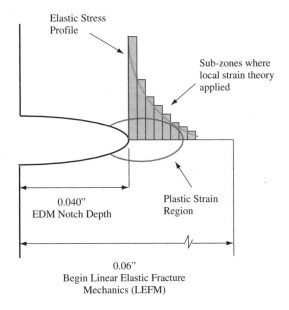

Figure 19.21 Subzone approach for crack initiation

Note that ΔK will start at a value of 0.0 when the crack length is zero and will increase with the crack size. The value a_i is taken as the initial crack size for the fracture mechanics study. Therefore, for a given component, cycles to reach final crack size, $a_f = $ [life in cycles at a distance a_i from the base of the notch based on strain-cycle fatigue mechanism] + [cycles required to propagate a crack from the crack size of a_i to a_f using fracture mechanics-based cycle estimation.]

19.5.5   Gear Tooth Crack Propagation

A 3-D linear elastic fracture mechanics (LEFM) analysis of the gear was performed. As subsequent figures in this section will illustrate, the resulting model was very accurate up to a crack length of about 0.5 inch (1.27 cm). Beyond this length, significant load sharing factors were likely present that were not modeled.

The fracture mechanics model had a total of 13 crack fronts (referred to as steps) to simulate the evolution of the 3-D crack front from the EDM notch. Each of the 13 crack fronts was subjected to a total of 18 load cases simulating instantaneous pinion loads. These corresponded to 18 discrete angular positions during the load–unload cycle for the pinion. As the crack grew, more elements were required to resolve the geometric details of the crack surface, ultimately resulting in 1 396 161 degrees-of-freedom at the twelfth step. The model was used to determine stress intensity factors and crack propagation path where mode I stress-intensity factors were primarily responsible for extent of crack growth, and the ratio between the mode II and mode I stress-intensity factors was primarily responsible for the direction of crack growth.

The predicted crack surface and fronts are shown in Figure 19.22 and along with the vector used in the crack length calculation. These can be compared to the observed fracture surface shown in Figure 19.23 and Figure 19.24 [11]. Qualitatively, the surfaces and fronts are very similar, especially near the heel end of the tooth. It should be noted that the seeded fault at the toe was not modeled because (i) FE analysis showed that the toe region was not as highly loaded as the heel notch region; (ii) thermosonic inspection from seeded fault test No. 2 showed that the crack initiated at the heel notch; (iii) propagation of the heel notch crack would cause catastrophic failure of the gear (as opposed to localized tooth breakage with the toe notch), and (iv) a crack propagating from the toe notch would not be expected to influence the heel notch propagation path.

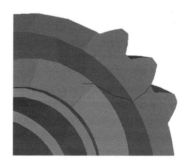

Figure 19.22   Computed crack surface trajectory

Figure 19.23   Observed crack surface trajectory

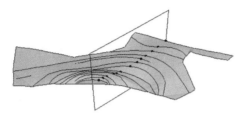

Figure 19.24   Crack front and crack length metric

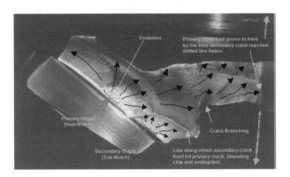

Figure 19.25   Post mortem analysis

The crack propagation model was applied to determine the cycles required to propagate the crack from the a_i inches discussed previously to the critical crack length. During a crack growth simulation, at each step of propagation, the direction and extent of crack growth at each point along the crack front had to be determined.

The best crack growth data for 9130 was used [12]. The data consists of Paris parameters for three different R ratios in both laboratory air and 250 °F (121 °C) oil subjected predominantly to plane stress loading. The R ratio in the seeded fault tests was 0.0. Both the analysis and the model considered the material properties of the AISI 9310 steel, case hardened to a case hardness of Rc 60–64 and core hardness of Rc 30–45 to have intrinsic Paris parameters of: $n = 3.2$; $C^* = 2.738e - 19$ (in/cycle)/(psi*in$^{0.5}$)^n.

When these values are applied to the Paris law which states that cumulative load cycles (N) to reach a certain crack size (a) can be found be using:

$$\frac{da}{dN} = C^*(\Delta K_{eff})^n \qquad (19.14)$$

Known sources of error were cumulatively applied to this deterministic model to create a probabilistic model where input parameters are represented as statistical variants. In all cases, Normal distributions were assumed. Ultimately, the plot shown in Figure 19.26 was generated.

This analysis enabled determination of the transition point between initiation and propagation. The calculated da/dN rate exceeded subzone propagation rate at approximate 0.00075 in. This distance was divided into six subzones and the deterministic calculation of crack initiation life was found to be approximately 11 h at 2000 ft-lbs (recall test No. 1 was believed to have initiated a crack at approximately 15 h).

19.5.6 Gear Vibration Diagnostic Information Fusion

This section describes the method employed to 'calibrate' the physics-based prognosis model, given diagnostic evidence. Certain types of vibration feature have been found to detect small cracks at early stages of crack propagation when supplied with speed, torque and time-domain vibration data. Such features provide valuable information

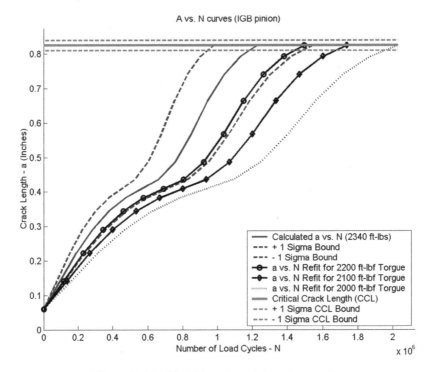

Figure 19.26 Model-based crack length vs cycles

that, when used properly, can serve to calibrate the model-based aspects of the prognostic module and reduce uncertainty levels. This means that tighter confidence bounds can be calculated for remaining component life.

Test results shown in Reference [2] have indicated that residual kurtosis and Residual Pk2Pk (peak-to-peak) features provided the earliest detection of a gear tooth cracking and are defined below.

- *Residual kurtosis* The residual signal is defined as the signal minus the gear mesh frequency (and harmonics) and one shaft modulated side lobe. The kurtosis of the residual time series is then computed [11].

$$kurtosis = \frac{\Sigma(x - \mu)^4}{N\sigma^4} \tag{19.15}$$

- *Residual Peak-to-Peak (Pk2Pk)* The signal peak-to-peak is computed using the residual signal normalized and rectified.

Making use of these crack detection features in the prognostic model calibration requires that the following functional mapping be created:

$$[a_{norm}(F_i), \sigma(F_i)] = f[mag(F_i), Torque] \tag{19.16}$$

where:

$$a_{norm} = \frac{a}{a_{critical}}$$

and: $a =$ crack length, and $a_{critical} =$ critical crack length (estimated to be about 0.5 inches)

In other words, applying a fusion method such as weighted averaging to the magnitude of the vibration features corrected for loading effects can be an alternate means of estimating current crack length. At the very least, an a_{norm} beyond a threshold value can be used to assess confidently that a crack has initiated to an engineering length.

The best estimates of the two vibration features vs. normalized crack length from the two tests are provided in Figure 19.27. Note that residual peak-to-peak (Pk2Pk) was multiplied by 5 for clarity and that both features 'drop off' after an a_{norm} of about 0.3 (a_{actual} of 0.15 in. or 0.381 cm), most likely due to load sharing effects.

Using the key assumption that the vibratory feature response, as the failure progresses from the two tests, is representative of the response for another test with identical conditions, the potential exists to reduce the confidence bounds on the failure prognosis.

The total probability of failure at any given time is the combination of two independent events: the initiation of a crack and the propagation of that crack to failure. For independent events, the total probability is:

$$P_{total} = P(i)^* P(p) \tag{19.17}$$

where: $P(i) =$ Probability of crack initiation (to 0.00075 in); $P(p) =$ Probability of propagation to failure (to 0.82 in).

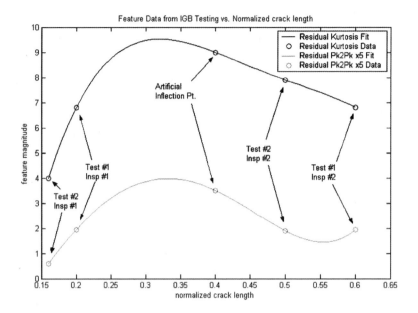

Figure 19.27   Normalized crack length vs feature magnitude

As already stated, $P(p)$ is a function of the model estimated crack length and when fused feature crack length estimates are available, the model estimated crack length distribution may theoretically be replaced with the feature-based crack length distribution. However, for a number of reasons, the initiation probability was simply set to 1 when residual kurtosis exceeded 4 in this study. Hence, a fundamental link was formed between state awareness and material capacity.

19.5.7 Gear Prognosis Module Results

A consolidated results table is provided in Figure 19.28 showing the time/cycle predictions of the individual elements of the prognostic model with respect to the estimated time/cycles to crack initiation and propagation from the two tests.

The second to last column in the table is the difference between predicted and actual in terms of standard deviations. The last column is the percentile the actual point falls

Crack Initiation and and Propagation to 0.75 mils								
Test #	Torque	Predicted Initiation Time (hrs)	Standard Deviation	Predicted Initiation Time (Cycles)	Estimated Time to Initiation (hrs)	Estimated Time to Initiation (Cycles)	Difference (# Stdevs)	Percentile
1	2000	11.09	1.64	2738786.40	15.00	3704400	2.38	98.00
2	2340	2.38	0.36	587764.80	2.50	617400	0.33	63.00

Crack propagation to Critical Crack Length (0.825 inches)								
Test #	Torque	Predicted Propagation Time (hrs)	Standard Deviation	Predicted Initiation Time (Cycles)	Observed Time to Failure (hrs)	Observed Time to Failure (Cycles)	Difference (# Stdevs)	Percentile
1	2340	4.85	1.13	1197190.00	6.00	1481760	1.02	82.00
2	2340	4.85	1.13	1197190.00	3.00	740880	-1.64	4.20

Figure 19.28   Predictions vs ground truth

on the predicted distribution. In other words, this number is the probability of either crack initiation or propagation to failure the prognostic model would have predicted at the actual time/cycles. Clearly the prognostic model distributions contain the ground truth observations while still retaining relative tight confidence bounds.

The prognostic modeling approach illustrated herein is an example of how multiple sources of data, models and knowledge are used to predict more accurately the remaining useful life of a critical component. Specifically, state awareness indicators in the form of extracted vibration features were fused with a stochastic, physics-based model to produce a prognostic module for accurately predicting the health of a transmission gearbox. Using system level observable features to orient and update the models responsible for tracking the component's material condition, provides a sound approach for improved awareness of the current health state. The real-time calibrated model is then also available to predict future performance over a range of potential loads and environmental conditions. Deployment of calibrated prognostic tools within a tactical decision-aiding environment will provide intelligent asset management and lead to improved readiness and mission assurance.

The following conclusions could be reached through this analysis of the gear prognosis module:

(1) Prognostic modeling efforts can benefit from a larger population of seeded-fault, run-to-failure tests with highly controlled observation of failure progression.
(2) High fidelity prognostic models for gears that are capable of predicting remaining life as a function of speed and load are achievable.
(3) Vibration features can be used effectively to calibrate physics-of-failure based prognostic models and reduce their intrinsic uncertainties.
(4) Uncertainty in a variety of factors, ranging from gear geometry/contact, gear loads to material properties, limit the ability of prognostic systems to generate reliable information. Within these factors, material properties is a core group that has a rather dominant impact on life prediction.
(5) Conventional design practices have relied on simpler statistical models for material properties for a given population of gears. However, these statistical models may not provide the information necessary for accurate gear prognostics. For example, one would need statistical variation of fatigue and fracture properties within the case-hardened layer.
(6) The statistical variation of compressive stresses within the case-hardened layer would be very valuable information for more accurate gear prognostics. Though such statistical databases will represent properties across a population, they are necessary in order to be able to address variations in the parameters that directly affect life calculations.

19.6   BEARING AND GEAR PROGNOSIS MODULE DISCUSSION

Real-time algorithms for predicting and detecting bearing and gear failures are currently being developed in parallel with emerging sensor technologies, including in-line oil debris/ condition monitors, and vibration analysis MEMS. These advanced prognostic/diagnostic algorithms utilize intelligent data fusion architectures optimally to combine sensor data with probabilistic component models to achieve the best decisions on the overall health of oil-wetted components. By utilizing a combination of health monitoring data and model-based techniques, a comprehensive component prognostic capability can be achieved throughout a component's life, using model-based estimates when no diagnostic indicators are present and monitored features such as oil debris and vibration at later stages when failure indications are detectable.

19.7 UTILIZATION OF PROGNOSIS INFORMATION IN NAVY ICAS SYSTEM

With the developments of prognostic technologies such as those described in the previous sections, the benefit from such predictions will only be realized if an integrated asset management system is in place to utilize the information correctly for planning and performing maintenance. Since the early 1990s, the United States Navy has been augmenting ship control systems by installing the Integrated Condition Assessment System (ICAS) to enable automated monitoring, trending, and condition based maintenance (CBM). With this system, the Navy is capable of utilizing such information by linking it with the proper maintenance tasking software. The Navy, capitalizing on commercial off-the-shelf (COTS) software, has successfully deployed ICAS on 95 surface ships.

ICAS is the Navy's CBM Program of Record and is the shipboard system for machinery monitoring and analysis, supporting various levels of condition-based maintenance, including prognosis. For ICAS to realize its full potential, enhancements continue to be deployed in the fleet. Examples of such enhancements would include a seamless integration with the shipboard control system and computer maintenance management systems (CMMS) such as an enterprise resource planning (ERP) system.

The Navy foresees a maintenance and operation vision in 2010 that fully integrates ICAS with the control and logistical support systems, as depicted in Figure 19.29. ICAS will have the ability to predict a fault on a particular machine and, at the appropriate time, alert the control system to secure the unit and bring an alternative on line. At the same time, ICAS will alert the integrated logistic system of the required maintenance to be accomplished in order to prevent the impending casualty. This will allow the ILS to ensure that necessary parts and specialized labor are available at the optimal time, thus maximizing readiness while reducing maintenance costs.

19.7.1 ICAS Installations Today

A typical US Navy ICAS installation consists of four to five workstations, one in each major machinery compartment as depicted in Figure 19.30, connected by an active local area network (LAN).

Each workstation accommodates a unique configuration data set (CDS), which contains the engineering information. ICAS converts uploaded data into useful information.

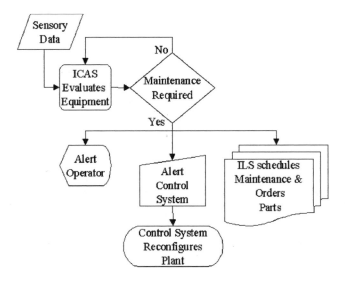

Figure 19.29 Integration of prognosis information into future ICAS systems

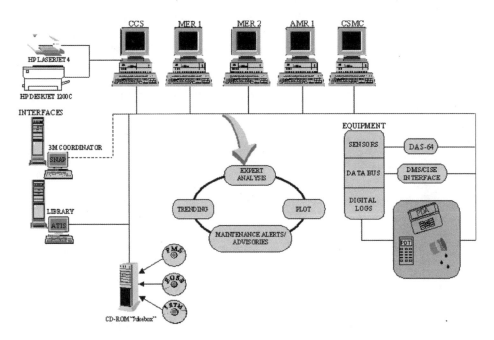

Figure 19.30 Typical ICAS installation

Data is trended, evaluated, and fused to allow for maintenance to be accomplished based on evidence of need. Typically ICAS will interface to an existing machinery control data bus to receive pertinent information without duplicating sensor or processing hardware. For additional data points, ship's force utilizes a portable data collector to upload data, via a serial interface, to the workstations.

An ICAS installation on a CG-47 class hull monitors the following machines/systems: Main propulsion; reduction gear; line shaft bearings; controllable pitch propeller; ship service gas turbine generator; fuel oil service; main propulsion lube oil; lube oil fill, transfer, and purification; air conditioning; refrigeration; distilling plant; auxiliary boiler; firemain; seawater pumps; fuel oil fill and transfer; high pressure compressed air, and low pressure compressed air.

ICAS also contains links to digital Navy logistic products such as the engineering operational sequencing system (EOSS), planned maintenance system (PMS), and integrated electronic technical manuals (IETMs). These link directly to the appropriate section or card as well as enabling browsing of the entire books.

Data automatically collected on these ships has annually saved thousands of man-hours through the automation of performance monitoring, as well as saving time through the software's automated diagnostic features.

An effort has been made to gather this data in a common database, maintenance engineering library server (MELS), so that statistical analysis can be accomplished to gain a better knowledge of equipment operation in a marine environment and to improve maintenance savings. As more data is gathered, the failure rates and causes of the failures are better understood and thereby more predictable. This knowledge is used to effect maintenance periodicities, design changes, and operational practices.

Figure 19.31 is a graphical plot of a gas turbine generator start on a CG-47 Class hull; the data was retrieved from the MELS database. The plot can be used to determine if the specific start was successful and further used in determining the overall start reliability of the engine.

Technicians can use this type of information, processed data, in prescreening systems for review by maintenance visit assessment teams. These teams visit the ship

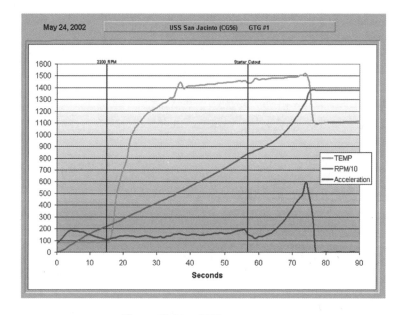

Figure 19.31   GTG start screen

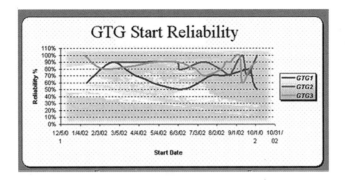

Figure 19.32   GTG start reliability

prior to deployment to identify issues and perform needed maintenance. Figure 19.32 represents a trend of the three gas turbine generators (GTG) start reliabilities over a 10-month period. By reviewing this trend, the technician gains immediate knowledge of the operational capability of all three units. Trending this type of processed data (start reliability in percentage format vice parametric) has allowed maintainers not only to identify maintenance, but also to capture a system's operational capability.

19.7.2   ICAS Installations for Tomorrow

The ICAS of 2010 will be a continuous evolution of the 'ICAS of today' with new technologies and best business practices implemented. It will leverage off commercially available electronic business applications, wireless networks and devices, while integrating with specific military control systems and integrated logistic products in a secure environment. The following paragraphs detail the evolution of the hardware and software needed to achieve the 'ICAS of Tomorrow' (Figure 19.33).

19.7.3   Open Systems Software Architecture

An open systems architecture will be achieved on different levels at varying resolution. Presently, ICAS allows for open exchange of information at certain levels, data processing and presentation. ICAS will also need to provide for open exchanges at the diagnostic and prognostic levels to meet the challenges of distributed systems. Figure 19.34 depicts these levels overlaid on the open systems architecture for condition based maintenance (OSA-CBM^{TM}). The OSA-CBM model was developed under a Dual Use Science and Technology (DUST) program between the Office of Naval Research (ONR) and industrial partners.

Open system architecture allows many benefits in the system's life cycle, from system development through deployment. The Navy can leverage off of other installed open systems to share data and/or information without costly hardware and software development, testing, and installation. This architecture reduces in-service costs by replacing obsolete hardware and software with those of a 'plug and play' philosophy. These

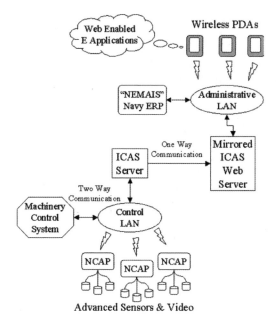

Figure 19.33 'ICAS of Tomorrow'

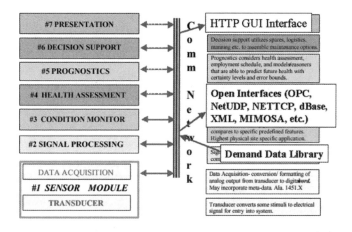

Figure 19.34 OSA-CBM hierarchy

savings are realized when the new piece of hardware or software is seamlessly integrated with the legacy system.

19.7.4 Integration with the Navy's Enterprise Resource Planning System

ICAS should be fully integrated with the Navy's Enterprise Maintenance Automated Information Service (NEMAIS), which is focused on improving maintenance and modernization processes aboard ship and in shore based industrial activities. ICAS will provide maintenance information and recommendations to NEMAIS so the work order can be scheduled, parts can be ordered, technical documents can be provided, and resources assigned. The information from ICAS will allow for the scheduling of work on the condition of the equipment, thus helping to streamline the process and reduce maintenance costs.

With ICAS receiving information from the logistics management software, it can further provide information to the operators on what maintenance is planned and inform the control system to shift equipment to allow for the planned maintenance. Hence, through the integration of prognosis information into the Navy's ICAS system, the Navy is expecting to drastically improve its Asset Management Solution for all of its equipment aboard the ship.

REFERENCES

[1] Lundberg and Palmgen (1947) "Dynamic Capacity of Rolling Bearings", *Acta Polytechnica Mechanica Engineering Series 1*, Royal Swedish Academy of Engineering Sciences, No 3, 7.

[2] Ioannides and Harris (1985) 'A new fatigue life model for rolling bearings', *Journal of Tribology*, **107**, 367–378.

[3] Yu and Harris (2001) 'A new stress-based fatigue life model for ball bearings', *Tribology Transactions*, **44**, 11–18.

[4] Sines and Ohgi (1981) 'Fatigue criteria under combined stresses or strains', *ASME Journal of Engineering Materials and Technology*, **103**, 82–90.

[5] Kotzalas and Harris (2001) "Fatigue Failure Progression in Ball Bearings", *Transactions of the ASME*, Vol. 123, pp. 238–242.

[6] Lewicki, D.G. (2001) *Gear Crack Propagation Path Studies – Guidelines for UltraSafe Design*, NASA/TM–2001-211073, July.

[7] Toth, D.K., Saba, C.S. and Klenke, C.J. (2001) *Minisimulator for Evaluating High-Temperature Candidate Lubricants Part I-Method Development*, University Dayton Research Institute-Aero Propulsion Directorate USAF, Dayton, Ohio.

[8] Harris, T. (2001), *Rolling Bearing Analysis*, Fourth Edition, John Wiley & Sons, Inc., New York.

[9] Wensig, J.A. (1998) *On the Dynamics of Ball Bearings*, PhD Thesis, University of Twente, The Netherlands.

[10] Socie, D. (1979) 'A procedure for estimating the total fatigue life of notched and cracked members', *Engineering Fracture Mechanics*, **11**, 851–859.

[11] Roemer, M. and Kacprzynski, G. (2001) 'Development of Diagnostic and Prognostic Technologies for Aerospace Health Management Applications' *IEEE Aerospace Conference*, Big Sky, Montana.

[12] Kacprzynski, G. Roemer, M.J., Sarlashkar, A. and Palladino, A. (2003) 'Calibration of Failure Mechanism-Based Prognosis with Vibratory State Awareness Applied to the H-60 Gearbox', *Proceedings of the 2003 IEEE Aerospace Conference*, Big Sky, Montana.

20

Application of Simplified Statistical Models in Hydro Generating Unit Health Monitoring

Geraldo C. Brito Jr

Itaipu Binacional, Brazil

20.1 INTRODUCTION

20.1.1 The Itaipu Power Plant

Itaipu Hydroelectric Power Plant is a binational project developed by Brazil and Paraguay in the Paraná River. The plant has 20 generating units of 700 MW each, and their technical characteristics are available in References [1] and [2]. Eighteen units entered into operation in the period of 1984 to 1991; the two remaining units are presently in assembly and should start operation in 2005. Itaipu produces more than 90 billion kWh annually, supplying 95 % of the electric energy consumed in Paraguay and more than 20 % of the Brazilian electrical system, which makes it a plant of high strategic importance for the Brazilian and Paraguayan electric grids.

20.1.2 Structural Health Monitoring in Itaipu

The Itaipu technical team was concerned with structural health monitoring since the very beginning of the power plant. The monitoring of the civil structure began

---

Damage Prognosis – For Aerospace, Civil and Mechanical Systems Edited by D.J. Inman, C.R. Farrar, V. Lopes Junior and V. Steffen Junior © 2005 John Wiley & Sons, Ltd

simultaneously with the construction. The monitoring of the equipment of the plant started on 1991, with the monitoring of partial discharges in the SF_6 gas insulated station, using the enclosures' vibrations as descriptors [3], and with the monitoring of temperatures and vibrations of the generating units [4]. The following sections describe the most important aspects of the generating-unit monitoring.

20.1.2.1   Generating-Unit Temperature Monitoring

Certain types of damage are well known, or at least one can easily predict their occurrence, in machines like the Itaipu generating units. To eliminate or to minimize the possibility of occurrence of these types of damage, the manufacturers took special care and preventive action in the design. Despite that, detected some of these kinds of damage were detected damages in the Itaipu generating units.

Examples of the above mentioned damage are the clogging of the stator winding cooling water conductors [1] and the deformation of the guide bearings bracket [5]. This damage could have had serious consequences to the availability of the generating units, but in both cases temperature monitoring instilled confidence in continuing operation after the damage detection.

Due to these antecedents, there were no doubts regarding to the necessity of monitoring the temperatures of the generating units. For each generating unit around 400 are monitored temperatures (windings, stator, bearings and cooling systems). Also monitored are electrical (power, voltages, currents) and mechanical (water flows, head, positions) parameters, in order to characterize the environmental and generating-unit operating conditions.

20.1.2.2   Generating Units Vibration Monitoring

The commissioning of the first four units, all 50-Hz units, showed that apparently the Itaipu generating units should all have a very nice vibratory behavior. The initial shaft vibrations of less than 200 μm peak-to-peak dropped to one third of this value after one plane balancing of the generator rotor (the balancing masses varied around 200 kg).

The first serious problem with vibration appeared during the commissioning of the first two 60-Hz units, one each from the two manufacturers involved. One generator was producing a high audible noise. Tests demonstrated that this noise was coming from intense stator core vibrations (18 mm/s RMS) at 720 Hz. Further investigations in the quiet generator, produced by the second manufacturer, also showed very high vibrations in the stator core (15 mm/s RMS), but now at twice the line frequency (120 Hz). These vibrations could lead to serious damage to the stator core lamination and to the winding insulation.

The stator core is a large steel laminated ring with 16 meters of internal diameter and 3 meters high. Vibrations tests, including some elementary modal tests, indicated that the stator core had a 24 nodes natural frequency near to 120 Hz. Some harmonics of the magnetic forces produced by the interaction of rotor and stator were exciting this natural frequency. The manufacturer solved the problem by modifying the stator winding distribution in order to reduce these magnetic forces harmonics. Further information about this subject is available from References [6], [7] and [8].

With this retrospective, the maintenance team was convinced of the importance of vibration monitoring in the most important components of the generating units: shafts, bearings, stator core and other static parts. Vibration signals processing produces a considerable amount of information. For each signal, descriptors like the overall value, root mean square value, harmonics of the rotating speed or other components of most important frequencies are monitored.

20.2 INFLUENCES OF THE ENVIRONMENT AND OPERATING CONDITIONS IN THE BEHAVIOR OF THE GENERATING UNITS

The following sections describe briefly the influences of the variations of environment conditions and of operating state in the behavior of the generating units and consequently in the health monitoring process. Further information is available in references [5], [9], [10], [11], [12] and [13].

20.2.1 Influence of the Lake Water Temperature

The main losses of the generating units are the losses by Joule effect in the windings and stator core (Foucault currents), ventilation losses, friction losses in the bearings and the excitation power. The losses in the Itaipu generating units exceed 5 MW. Most of this power it transformed in to heat and the cooling system must remove this heat from the generating unit. The bearings and stator core heat exchangers use a small part of the turbine water flow to do that. The water flow is constant in all cooling systems, with the exception of the stator winding cooling system, where there is an automatic control valve.

Contrary to what occurs on most of the Brazilian rivers, the water of the Paraná River near to Itaipu experiences expressive seasonal variation of temperature. Figure 20.1 shows variations of both Paraná River water temperature (dashed crossed curve) and of the dam-lake water temperature (continuous curve). The transient occurred in the period of 1980 to 1982, and was due to the construction of the concrete dam where the temperature sensor is embedded. The transitory behavior of the period 1982 to 1990 is the reflection of the dam lake formation.

Figure 20.1 also shows that the lake water temperature measured near to the turbine water intake varies from 17 to 28 °C. The cooling water diverted from the turbine varies across a similar temperature range. As will be seen in the following sections, the lake water temperature has significant influence in the temperatures and vibrations of the generating units.

20.2.1.1 Influence of the Lake Water Temperature in the Temperatures of the Generating Units Components

The mechanism of influence of the cooling water temperature over the temperature of the components of the generating unit is obvious. As the water flow is constant, more heat is removed from the generating unit in the winter season, when the lake water

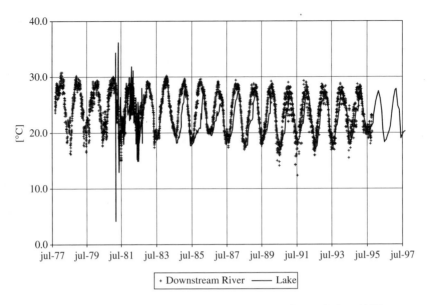

Figure 20.1   River and lake water temperatures from 1977 to 1997

temperature is lower. This mechanism usually produces a reduction in the temperature of the components in the winter. The exceptions are temperatures of some guide bearing pads, which can increase in the winter due to excessive reduction on the bearing clearances, which increases the bearing pad losses.

20.2.1.2   Influence of the Lake Water Temperature on the Vibrations and Other Dynamic Signals of the Generating-Unit Components

The influence mechanism is different for each considered vibration. For instance, when considering shaft vibration, the influence of the cooling-water temperature acts mainly through the thermal expansions or contractions of the bearing brackets, causing variations in the guide-bearing clearances, as can be seen in Figure 20.2. Additionally, the bearing oil temperature follows the cooling-water temperature variation. As a bearing may contain tens of thousands of liters of oil, the viscosity of the oil in the pads clearances will vary inversely with the cooling water temperature. These two factors combined cause noticeable variation in the oil film stiffness.

Due to that, in the winter the guide-bearing clearances are lower and the oil viscosity is higher, which increases the oil film stiffness. Thus, the shaft vibrations are lower in the winter, as can be seen in the orbits comparison in Figure 20.3. As a conclusion, the lake water temperature drops in the winter and the overall shaft vibrations follow it. On the other hand, the amplitudes of the harmonic components of shaft vibrations could increase, due to intensification of the nonlinear behavior of the oil film stiffness, especially in the directions of the bearing with low clearances.

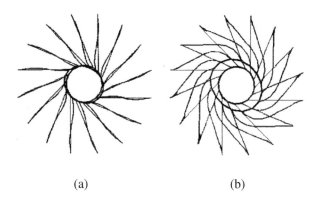

(a)   (b)

Figure 20.2 Finite-element simulation of the upper bearing bracket deformation, using temperatures measured at generating unit U06: (a) during winter, and (b) during summer

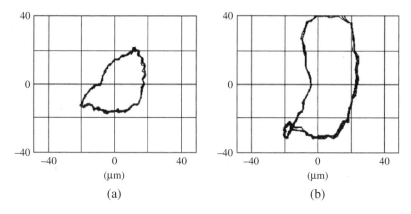

(a)   (b)

Figure 20.3 Shaft orbit in the upper guide bearing of generating unit U06 (a) during winter (720 MW, 140 Mvar, 118.6 m and 640 m^3/s), and (b) during summer (670 MW, 110 Mvar, 112.1 m and 640 m^3/s)

20.2.2 Influence of the Active Power

20.2.2.1 Influence of Generating-Unit Power on the Temperature of its Components

The influence mechanism of the active power unit on the temperatures of the generating unit is also clear. For instance, when the active power increases, the losses in the stator winding by Joule effect will also increase. With this, the temperatures of several other components (winding temperatures, winding cooling water and cooling air temperatures) will increase proportionally.

Less obvious is the influence of power in the stator-core temperatures. One could suppose that there is no influence between the stator winding and stator core temperatures. As the cooling water removes heat directly from where it is generated (winding bars), and as the stator winding electrical insulation has good thermal insulation, one could suppose that there is no heat transfer between core and winding. However, measurements showed that this is not completely true, especially in summer, when there is a significant temperature difference between stator winding and core.

On the other hand, there is an increase in the armature reaction with power, increasing the losses in the stator by Joule losses of the Foucault currents. Consequently the power also influences the temperatures of the stator core, of the stator cooling system (air and water), and of the rotor.

In the thrust bearing, the influence mechanism comes from the increase in turbine-head cover pressure with increase of power or with increase in the water flow in the turbine. The hydraulic load increases and with that the losses in the thrust bearing and combined guide bearing. Turbine and upper guide-bearings temperatures are usually subject to much less influence from power.

20.2.2.2 Influence of the Generating-Unit Power in the Unit Vibration

The influence of the generator active power (or the influence of the water flow in the turbine) in the vibratory behavior of generating units equipped with Francis turbines is well known. The Karmann vortices have huge influences on some shaft vibration descriptors, like overall value of a vibration or in the low-frequency component value in the frequency range from a fifth to a quarter of the rotating frequency. This influence is especially significant on partial loads, from 30 % to 70 % of the nominal load. Other vortices can also affect vibration descriptors at high loads, but neither phenomenon should have influence on the vibration at the rotating speed.

Figure 20.4 shows expressive influences of the active power in the first harmonic of the shaft vibrations at combined and turbine guide bearing planes. There is a similar

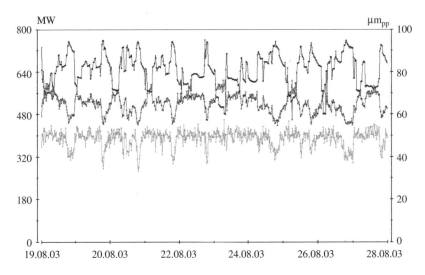

Figure 20.4 Influence of active power (upper curve) in the first harmonic of shaft vibrations of combined bearing (middle curve) and turbine guide bearing (lower curve) of U04

Table 20.1 Correlation between active power with shaft and bearing vibrations

Descriptor	Place	Correlation factor with active power	
		U04	U14
Shaft vibration	Upper bearing plane	+0.504	+0.751
(first harmonic)	Middle bearing plane	−0.915	−0.869
	Turbine bearing plane	−0.645	+0.855
Bearing vibration	Upper bearing housing	+0.743	+0.188
(RMS value)	Middle bearing housing	+0.895	+0.934
	Turbine bearing housing	+0.904	+0.941

influence in the root mean square value of the bearing-bracket vibrations. To evaluate how much stronger are these influences, Table 20.1 shows the correlation factors between active power and the descriptors of the generating unit vibrations. This table shows that the intensity of the influence can vary significantly from different places of a machine or for the same place in different machines. For instance, it is curious that the correlation factor between active power and first harmonic of shaft vibration at the turbine bearing plane is positive at unit U14 (+0.855) and negative for unit U04 (−0.645).

So far it is not important to detect if there is damage in units U04 and U14 or to diagnose it. The important thing is to stress that there are unexpected very good correlations between active power and several descriptors of the vibration signals. Anyway, a possible explanation for the opposite behavior of the shaft vibration at turbine bearing planes of units U04 and U14 is the existence of hydraulic imbalance. Hydraulic imbalance produces a vibration at the rotating speed that increases with the active power [14], which would explain the behavior of the vibrations. If the hydraulic unbalance were in phase with a residual mechanical imbalance of the turbine, the behavior would the similar to unit U14. On the other hand, if the hydraulic and mechanical imbalances were in phase opposition, the behavior would be like that of unit U04.

20.2.3 Other Influencing Parameters

There are other influencing parameters, like the turbine water head. Variations in this parameter will change the relationship between generator active power and turbine water flow. This strongly influences the dynamic behavior of the Francis turbines by changing the relationship between active power and some of the descriptors of vibrations and pressure fluctuations.

Another important influencing parameter is the reactive power. The losses by Joule effect in the rotor winding will vary with the square of the excitation current, a parameter that has significant correlation with the generator voltage and with the reactive power. The reactive power can vary over a wide range, according to the generator capability curve, causing significant variations in the stator and rotor temperatures, as well as in some descriptors of the stator core vibration and of the generator air gap (rotor magnetic stiffness).

There are other less important influencing parameters, like the temperature of the stator cold air, the temperature of the stator cooling-outlet water, bearing oil temperature, etc. However, all these parameters are secondary and vary with the variation of other influencing parameters. The four most important influencing parameters in the descriptors values are the cooling-water temperature, the water head, the active power and the reactive power. The first two are environmental influencing parameters and the last two are operational influencing parameters. Nevertheless, it is important to note that experience shows that errors in estimating the descriptors usually decrease when the number of influencing parameters is increased.

20.3   STATISTICAL MODELS FOR STRUCTURAL HEALTH MONITORING

20.3.1   Statistical Models for Damage Detection

The previous section showed how significant could be the influences of the environmental and operating conditions on the generating-unit behavior. A descriptor used for a certain type of damage detection can vary many times its initial value and the monitored machine could still be in a healthy condition. Consequently, the application of an elementary limit-violation checking for a descriptor could lead to a false positive or false negative damage indication. Therefore, it could be very difficult to answer even the apparently easiest question in structural health monitoring: is there an incipient damage in the generating unit? Damage detection implies an adequate comparison of the measured value of a descriptor with its expected value for a healthy machine, under the same operating and environmental conditions.

This section will describe an easy solution for this problem, the application of regression analysis in temporal series to determine the expected value of a descriptor. This procedure requires nothing but basic statistics and can be applied even using ordinary electronic spreadsheet. Nevertheless, it can produce very satisfactory results.

20.3.1.1   Using Regression Analysis for Damage Detection in Temperature Monitoring

Figure 20.5 shows the variation of some temperatures of the generating unit U04, for the period of April 1992 to July 2002. The maintenance team measures all temperatures monthly, with the load of the generating unit (MW – upper dashed curve) stable during a previous period of at least 5 hours. The continuous curve with circles represents the cold-air temperature of the stator core cooling system (T_{CA}), also varying seasonally over the range of 25 to 40 °C, due to the influence of the lake water temperature.

Below the active power, the continuous crossed curve represents the average of the measured stator core temperatures (T_{meas}). Near to this curve is there another continuous line representing the average temperature of the stator core (T_{calc}) calculated using the regression analysis function of the MS-Excel®.

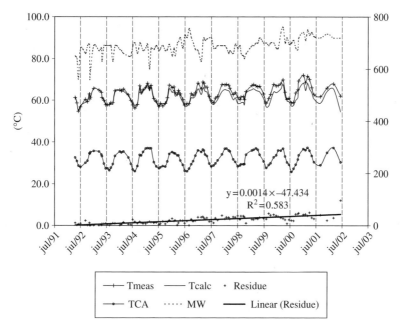

Figure 20.5 Regression analysis application in the U04 stator core temperatures (learning period: 04/92 to 06/94)

Table 20.2 Coefficients of regression analysis application in the stator core temperatures (learning period: around 24 months)

Unit	kV m_9	I_{exc} m_8	I_{phase} m_7	MVA m_6	Mvar m_5	MW m_4	T_{HW} m_3	T_{CA} m_2	T_{CW} m_1	b
U01	0.43	0.36	0.11	−0.02	0.01	0.02	1.26	0.06	−0.32	6.8
U02	−10.43	−0.28	−9.31	0.28	0.02	0.04	1.36	−0.16	−0.14	205.5
U03	−2.45	6.13	−2.87	0.05	0.00	0.05	0.88	0.08	0.07	51.8
U04	0.09	4.08	0.83	0.02	−0.02	−0.04	1.39	0.33	−0.87	7.9
U14	10.06	5.34	8.20	−0.40	0.00	0.14	−0.82	0.94	0.84	−157.9
U15	5.01	−0.32	2.39	−0.01	0.01	−0.04	0.13	1.36	−0.55	−68.4
U16	4.47	1.86	2.67	−0.07	0.00	0.00	−0.47	1.30	0.12	−61.2
U17	0.81	4.19	−0.24	0.05	0.00	−0.03	0.45	0.81	−0.29	2.5

kV, Generator voltage; I_{exc}, excitation current; I_{phase}, phase current; MVA, apparent power; Mvar, reactive power; MW, active power; T_{HW}, stator hot water temperature; T_{CA}, stator cold air temperature; T_{CW}, stator cold water temperature; θ_{CORE}, stator core average temperature.

$$\theta_{CORE} = m_9 \cdot kV + m_8 \cdot I_{exc} + m_7 \cdot I_{phase} + m_6 \cdot MVA + m_5 \cdot Mvar + m_4 \cdot MW + \ldots \ldots$$
$$+ m_3 \cdot T_{HW} + m_2 \cdot T_{CA} + m_1 \cdot T_{CW} + b$$

To obtain the regression coefficients listed in the fourth line of Table 20.2, the temperatures measured during the first 26 months of monitoring were used. One can see that in the learning period (04/92 to 06/94) the curves of both measured (T_{meas} – curve with crosses) and calculated (T_{calc} – continuous curve) temperatures are practically coincident.

After that, a small but increasing discrepancy begins, indicating either that the generating unit is very slowly producing more heat or the cooling system is losing efficiency, or both damages are taking place simultaneously. Is important to remark that the increasingly discrepancy observed in unit U04 does not exist or is almost imperceptible in the case of unit U15 (Figure 20.6).

Table 20.3 shows the errors obtained with the application of the regression analysis in the stator core temperatures of eight generating units. The root mean square error

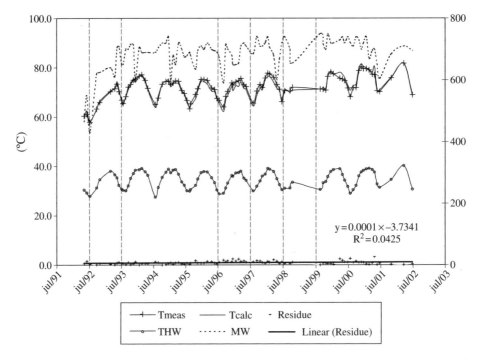

Figure 20.6 Regression analysis application in the U15 stator core temperatures (learning period: 05/92 to 12/94)

Table 20.3 Errors in the estimation of stator core temperatures using regression analysis, during learning period

Unit	RMS error (%)	Average error (%)	Maximum error (%)	Minimum error (%)
U01	0.94	0.01	3.05	−1.53
U02	1.30	0.02	3.59	−1.84
U03	0.81	0.01	2.91	−1.46
U04	0.84	0.01	1.27	−2.35
U14	0.89	0.01	2.35	−1.90
U15	1.06	0.01	2.52	−1.99
U16	0.76	0.01	2.04	−1.49
U17	0.45	0.00	1.16	−0.80
Average	0.88	0.01	2.36	−1.67

varies from 0.45 % to 1.30 %. Table 20.2 shows the coefficients obtained for different generating units. It is interesting to note that these coefficients vary over a wide range. The temperatures estimated using these coefficients for the same operating condition (kV, Iexc, Iphase, MVA, Mvar, MW, THW, TCA and TCW) for the different generating units varied up to 15 %. These results are similar to those observed in the measured temperatures of the real units.

This indicates that despite having identical design and being manufactured and assembled in the same way, the behavior of the generating units could differ significantly. The question is if thermal mathematical models could take into account the parameters responsible for such differences in the final behavior of the generating units.

On the other hand, the results with the application of regression analysis on damage detection in the stator core temperatures are very satisfactory for maintenance purposes. This is especially true when considering the poor resolution of the instrument used to measure temperatures (1 deg. C) and the reduced amount of data (only one measurement per month). When using more samples for the learning period, with better resolution and accuracy, better results would be expected. It is important to note that despite the mentioned limitations, the application of regression analysis in the temperatures of other components of the generating units (stator winding, rotor winding, bearings, heat exchangers, etc.) also produced similar good results.

20.3.1.2 Using Regression Analysis for Damage Detection in Vibration Signals

The monitoring team has still not extensively applied the regression analysis for damage detection in vibration signals. However, the high correlation factors between the active power and some of the descriptors of the vibrations (shaft, bearing and of other components of the generating units) are a very good indication of success in the case of application.

20.3.2 Statistical Models in Damage Prognosis

It seems premature to discuss the application of simplified statistical models to damage prognosis. Nevertheless, one can do an exercise on its application in the case of very simple damage, as in the case of the stator core temperature of generating unit U04 (see Figure 20.5).

The fault or the damage in the stator might be deterioration in the varnishing of the stator core lamination, which increases the Foucault currents and consequently the losses. It might also be a reduction in cooling system efficiency, by clogging of piping or other reason. Incidentally, both kinds of damage could be present simultaneously. Anyway, in all these cases the most probable failure (i.e., the reason for the interruption in the generating unit operation) would be excessive temperature in the stator core. Eventually, the maintenance team will decide to stop the generating unit in order to repair the cooling system or, in the worst case, to restack the stator core or to replace it.

In this process, 'near-to-real-time' means a period of several weeks or even some months. Therefore, a simplified statistical model might be used for 'the prediction in near-real-time of the remaining useful life of' the stator core or to predict the remaining time before failure. Considering the maximum allowed temperature for the stator core to be 100 °C [1], the present temperature value around 70 °C and a rate temperature increasing of 0.6 deg. C per year, the estimated remaining time for failure is of the order of some decades. The continuous recalculation of this estimation, in order to detect acceleration in the deterioration process, can give the adequate confidence to the health condition monitoring.

20.4 CONCLUDING REMARKS

Analysis based on experimental data showed that large hydrogenerating units could have a complex dynamic behavior, from both thermal and rotor dynamic points of view. Two apparently identical generating units might have completely different behaviors, even under similar environmental and operating conditions. Only detailed and complex thermal and rotor dynamic mathematical models could reproduce adequately the different behaviors of these machines, giving enough accuracy in the determination of the expected value of a descriptor monitored for damage detection purposes. The difficulties of using mathematical models resulted in using the models based on neural networks and fuzzy logic. These models also have their difficulties in application and problems like the lack of physical correlations proportioned by the theoretical mathematical models.

The application of regression analysis on the temperatures of the Itaipu generating units showed that this technique could provide very good results for damage detection purposes. Simulations with this type of analysis on vibration data showed significant correlation factors between some vibrations descriptors and the environmental and operating influencing parameters. This indicates good possibilities of success on the application of these simplified statistical models also for vibrations descriptors. The using of regression analysis for estimation purposes is of course not new. It was, and still is, widely applied for several estimation purposes, including weather forecasting for a long part of last century. The aim of this paper is to remark that the tests in the monitoring of Itaipu generating units showed that these simplified statistical mathematical models can still be a good third option for damage detection purposes, especially when 'near-to-real-time' decisions are not necessary and when it is possible to apply steady state evaluations. They can at least be a good complementary tool for the monitoring engineer.

20.5 TERMINOLOGY

This paper employs, as much as possible, the terminology defined in references [15], [16] and [17]. As there are some differences in the terminology of these references, the following remarks are important to avoid unnecessary confusion.

Reference [16] defines **fault** ('*of a component of a machine, in a machine*') as the '*condition of a component that occurs when one of its own components or their assembly is degraded or exhibits abnormal behaviour, which may lead to the failure of the machine*'.

References [15] and [17] define: '*damage* in engineering systems is defined as intentional or unintentional changes to the material and/or geometric properties of these systems, including changes to the boundary conditions and system connectivity, which adversely affect the current or future performance of that system'.

Reference [16] defines *failure* ('*of a machine*') as the '*state of a machine that occurs when one or more of the principal functions of the machine are no longer available and which generally happens when one or more of its components are in a fault condition*'.

Reference [17] defines '*damage prognosis*' as '*prediction in near-real-time of the remaining useful life of an engineered system given the measurement and assessment of its current damaged (or aged) state and accompanying predicted performance in anticipated future loading environments*'.

In this paper, the terms *fault* and *damage* are indistinctly used. Additionally, *damage prognosis* means the prediction (in the same above-described condition) of the remaining time for *failure*.

REFERENCES

[1] Itaipu Binacional (1994) 'Itaipu Hydroelectric Project – Engineering Features', Chapter 12, *Powerhouse Generation Equipment*, Gráfica Editora Pallotti, Porto Alegre, Brazil.

[2] Itaipu Binacional, http://www.itaipu.gov.br/

[3] P.H. Teixeira, M.F. Latini and G.C. Brito (1995) 'Partial Discharges Detection in Gas Insulated Stations: Consolidating the Mechanical Method' (in Portuguese), Thirteenth National Seminary of Production and Transmission of Electrical Energy (see www.xviisnptee.com.br), biennial seminary organized by CIGRÉ (see www.cigre.org) in Brazil, Camboriú.

[4] P.H. Teixeira, M.F. Latini and G.C. Brito (1993) 'The Simplified Monitoring: Low Costs, High Benefits' (in Portuguese), Twelfth National Seminary of Production and Transmission of Electrical Energy (see www.xviisnptee.com.br), biennial seminary organized by CIGRÉ (see www.cigre.org) in Brazil, Recife.

[5] G.C. Brito, O. Bacchereti, J.W. Boveda V and F. Chiesa (1995) 'Phenomenon of Ovality/Deformation of the Itaipu Generating Units Guide Bearings' (in Portuguese), Thirteenth – Camboriú – 1995.

[6] G.C. Brito and C.C. Cuevas (1988) 'Problems with Noise and Vibrations of Magnetic Origin in the Itaipu Power Plant 60 Hz stators' (in Portuguese), *III Brazilian Congress of Maintenance*, ABRAMAN, Salvador.

[7] M. Dias, E. Marconi, M. Uemori, C.A.D. Costa, E.I.B. Ang and P.R.D. Oda (2001) 'Vibration in the Stator Core of the Generator of GU-05 of Engineer Sérgio Motta Power Plant' (in Portuguese), Seventeenth National Seminary of Production and Transmission of Electrical Energy (see www.xviisnptee.com.br), biennial seminary organized by CIGRÉ (see www.cigre.org) in Brazil, Campinas.

[8] R. Cardinali, R. Pederiva, M.D. Silva, G.C. Brito, E.C.C. Maimone and A.B. Feliú G (1989) 'Vibrations of Magnetic Origin in Stators of Large Hydrogenerators' (in Portuguese), Tenth National Seminary of Production and Transmission of Electrical Energy (see www.xviisnptee.com.br), biennial seminary organized by CIGRÉ (see www.cigre.org) in Brazil, Curitiba.

[9] G.C. Brito, H.I. Weber and A.G.A. Fuerst (1996) 'Dynamics of Large Hydrogenerators', *Proceeding of the XVIII IAHR Symposium on Hydraulic Machinery and Cavitation*, Valencia E. Cabrera *et al.* (Eds), Kluwer Academic Publishers, Dardrecht, The Netherlands.

[10] G.C. Brito and A.B. Feliú G (1992) 'Technical Aspects in Low Speed Hydrogenerators Vibration Monitoring', *Euromaintenance 92 – European Federation of National Maintenance Societies*, Lisbon.

[11] G.C. Brito and A.B. Feliú G (1991) 'Subsidies for Continuous Vibration Monitoring in Hydrogenerators' (in Portuguese), Eleventh National Seminary of Production and Transmission of Electrical Energy (see www.xviisnptee.com.br), biennial seminary organized by CIGRÉ (see www.cigre.org) in Brazil, Rio de Janeiro.

[12] G.C. Brito (1996) *Dynamic Behavior of Large Hydrogenerators Guide Bearings* (in Portuguese), MSc Dissertation, Faculty of Mechanical Engineering, Campinas State University (UNICAMP).

[13] H.I. Weber, A.G.A. Fuerst and G.C. Brito (1996) 'Simplified Calculation of Dynamic Coefficients for Hydrodynamic Bearings in Large Hydrogenerators', *IMECHE Conference Transactions – Sixth International Conference on Vibrations in Rotating Machinery*, Mechanical Engineering Publication Ltd, Bury St Edmunds, Suffolk, UK.

[14] L.A. Vladislavlev (1979) *Vibration of Hydro Units in Hydroelectric Power Plants* – Amerind Publishing Co. PVT, Ltd, New Delhi.

[15] C.R. Farrar and S.W. Doebling (1999) 'Structural Health Monitoring at Los Alamos National Laboratory', *Institute of Electrical Engineers Colloquium on Condition Monitoring: Machinery, External Structures and Health*, Birmingham, UK, April 22–23, LA-UR-1582.

[16] International Organization for Standardization (2001) 'Mechanical vibration – Condition monitoring and diagnostics of machines – General guidelines on data interpretation and diagnostic techniques', *ISO/TC 108/SC 5, Draft International Standard ISO/DIS 13379*.

[17] Los Alamos National Laboratory, *Structural Health Monitoring – SHM Introduction* http://www.lanl.gov/projects/damage_id/

Index

Index compiled by Paul Nash